The Quantum Quark

The world you can feel and touch is built of atoms, the smallest identifiable chunks of matter. Yet the heart of each atom is itself a whole new world, a world populated by quarks: indivisible, vanishingly small, the ultimate building blocks of our Universe. This inner world where quarks reign is subject to new and unfamiliar rules, the rules of the quantum world. Colossal particle accelerators enable physicists to bring this inner world into focus, and have helped them shape a theory respectful of quantum rules that explains how quarks feel one another's presence. *The Quantum Quark* is the story of that theory: quantum chromodynamics.

Bypassing the need for a background in mathematics or physics, this book offers the reader an intimate look at quantum chromodynamics, its genesis, its successes, and its role in reshaping our understanding of the Universe. This is an accessible and inspiring account of how scientists have come to understand one of nature's four fundamental forces. It will appeal both to those with a general interest in cutting edge science and to students of particle physics who wish to broaden their perspective.

ANDREW WATSON is a freelance science journalist and writer. After his Ph.D. in theoretical physics, from the University of Manchester, he went on to do physics research and lecturing. He now combines working part-time as a scientist at the Institute of Food Research with writing. He has written for a range of publications from popular magazines and newspapers to scientific journals, and is a regular contributor to *Science* magazine.

The Quantum Quark

ANDREW WATSON

PUBLISHED BY THE PRESS SYNDICATE OF THE UNIVERSITY OF CAMBRIDGE
The Pitt Building, Trumpington Street, Cambridge, United Kingdom

CAMBRIDGE UNIVERSITY PRESS
The Edinburgh Building, Cambridge, CB2 2RU, UK
40 West 20th Street, New York, NY 10011-4211, USA
477 Williamstown Road, Port Melbourne, VIC 3207, Australia
Ruiz de Alarcón 13, 28014 Madrid, Spain
Dock House, The Waterfront, Cape Town 8001, South Africa

http://www.cambridge.org

First published 2004

Printed in the United Kingdom at the University Press, Cambridge

Typeface Trump Mediaeval 9.5/13.5 pt   *System* LaTeX 2_ε [TB]

A catalog record for this book is available from the British Library

Library of Congress Cataloging in Publication data

Watson, Andrew, 1958–
The quantum quark/Andrew Watson.
   p.   cm.
Includes bibliographical references and index.
ISBN 0 521 82907 0 (hardback)
1. Quantum chromodynamics.   I. Title.
QC793.3.Q35W38   2004
539.7′548 – dc22   2004040660

ISBN 0 521 82907 0 hardback

To my daughters

Contents

Preface

Feel the weight of this book and you will experience gravity. See the words on this page and you will know light: the forces of gravity and electromagnetism are pretty familiar to us all. But nature has two less familiar forces at her disposal. To see how these other two forces work means viewing the world at a very high magnification, on a scale so fine that everyday matter is revealed as composed of more elementary components. The two less familiar forces at play here are called the strong force and the weak force. The strong force binds together the ultimate constituents of matter, and powers our Sun, whereas the weak force is manifest in radioactivity. Physicists have developed a theory of how the strong force works at its most elemental level, and this book is about that theory. The theory is called quantum chromodynamics, or QCD for short. It is one of the greatest achievements of the human intellect.

The branch of physics that embraces the study of this strong force is called particle physics, or equivalently high-energy physics. High-energy physicists also explore the weak force, but we will cheat a little and leave that to one side, along with attempts to embrace gravity and to create so-called theories of everything. Like a history book that covers just the history of England rather than the whole of Europe, *The Quantum Quark* focuses on quantum chromodynamics, the essentials of how it works and how it came into being. At the modest cost of an incomplete story, just as a history of England is incomplete, *The Quantum Quark* explores QCD in more detail than all other books on particle physics intended for a wider readership.

The Quantum Quark is intended to be broadly accessible: no background in physics or mathematics is assumed, and symbolism is at a bare minimum. Like a Russian classic, there are many characters with unfamiliar names, so *The Quantum Quark* includes a comprehensive glossary to smooth the journey. There's also a chronology to help picture the way the story of QCD has unfolded over time. Quantum chromodynamics is a vast subject and a huge number of people have contributioned over the years – too many to all be mentioned here. I hope specialists will forgive the numerous omissions of ideas, people, and experiments, and also the decision to leave out references on the grounds that this is a pleasant stroll through great ideas, and not a text book.

Acknowledgements

Many people, too many to list in full, are due thanks for the help they have provided. However, I should like to single out a few for special mention. I am grateful to Tracey and Mark Williamson, and to Graham Keeley, for providing accommodation during the early days of this book. Thanks are also due to Arturo Sangalli and Rob Penfold for their support and encouragement throughout. I am indebted to Bryan Webber and the Cavendish Laboratory, University of Cambridge, for access to their facilities. I should also like to acknowledge the input of several anonymous expert reviewers, who identified shortcomings in earlier drafts. I also offer special thanks to Elke-Caroline Aschenauer, Christine Davies, Gordon Kane, Donald Perkins, Alexander Polyakov, and Mikhail Shifman who gave generously of their time to look over all or part of the manuscript, and who offered numerous suggestions for improvements. Of course any shortcomings or inaccuracies are mine alone. Finally, I am particularly grateful to those scientists who supplied images from their personal collections.

I Introduction

OF FORCES AND THEORIES

There's an unusual "L" shaped building in Livingston, Louisiana. Laser light flies back and forth along both of the 4 km long arms, reflected between mirrors at the distant ends of the arms, and at the elbow where the arms join. When, after many to and fro journeys, the beams are recombined, the result measures amazingly small changes in the lengths of the arms.

Why should the arms change length? Is the state of Louisiana prone to stretching and shrinking? Scientists believe it is. That is, if Albert Einstein's theory of gravitation is correct, then ripples will pass through the state of Louisiana, just as they will pass through Washington state, where at Hanford the twin of the Louisiana instrument will be ready and waiting. Ripples will pass through the whole of planet Earth. In fact according to Einstein's theory, called general relativity, the ripples pass through the very fabric of space and time itself. Louisiana and Washington are just convenient places to look for them.

The ripples come from events so violent they cause the universe to shudder, events such as the collapse of stars or even the creation of the universe itself. General relativity predicts that these will give off gravitational waves, the ripples in space and time that scientists will be waiting to see in Louisiana and Washington, and also at other detectors such as those in Italy and Australia.

It's general relativity that led to the prediction of black holes – collapsed stars whose intense gravitational pull is enough to prevent even light escaping. Astronomers, unable to image black holes directly, have nevertheless discovered many, such as the "nearby" Cygnus X-1 system. They know there's a giant black hole lurking at the center of our own galaxy. Astronomers' stargazing also backs another of general relativity's predictions, gravitational lensing. This says that massive galaxies, for example, should form multiple images of suitably aligned and more distant objects as viewed from Earth. Lensing, first seen in 1979, is a variation of Einstein's original idea that the Sun should bend starlight. It was this prediction, confirmed in 1919, which put Einstein and general relativity on the international map.

The starting point of Einstein's theory is the lesson of the elevator. When an elevator starts to rise you feel heavier, and when it begins to descend you feel lighter. According to Einstein's idea, the effects of gravitation are indistinguishable from acceleration. He blended this principle with the newly invented mathematics of curved surfaces to give a theory of gravitation in which gravitational force is the result of a distortion of space and time due to the presence of mass. His general relativity superseded Isaac Newton's original seventeenth-century law of gravitation. Newton's version emerges as a special case applicable to such things as falling apples and planetary motion.

Pre-eminent on cosmic scales, gravitation turns out to have a nominal role at the smallest scales. Though an all-pervading attractive force between objects having mass – including the smallest fundamental particles of which the material universe is composed – the gravitational force at short ranges is swamped by nature's other three forces. Not only that, but so far nobody has been able to piece together a theory of how gravity works at these tiniest ranges. So it does little harm to drop gravity from the story of quantum chromodynamics.

Electricity everywhere

The story goes that William Gladstone, at the time Britain's Chancellor of the Exchequer, asked Michael Faraday, inventor of the electric generator and motor, what would be the practical use of electricity. Faraday had at least some inkling of the social consequences of his discoveries: "One day, Sir, you may tax it," he replied.

Mastery of electricity and magnetism, collectively electromagnetism, has come a long way since then. For example, popular soap operas are now available in almost every corner of the globe: versions of Faraday's dynamo provide the mains electricity, electromagnetic waves called radio waves carry the signal, and electromagnets control the picture-forming beam in television sets.

After gravitation, electromagnetism was the second of nature's forces to begin yielding up its secrets to scientists. Electric charge is the source of electrical influence, and magnetic effects can always eventually be traced to moving electric charges. James Clerk Maxwell achieved the complete unification of electricity and magnetism, hinted at in Faraday's discoveries, in his theory of electromagnetism. This in turn led to the idea that electromagnetic waves are oscillating electric and magnetic waves propagating together at a fixed speed. There's a whole spectrum of these waves, ranging from radio waves at the low-frequency end, to visible light through to X-rays at the high-frequency end. Light, then, is an electric–magnetic wave.

Motors, generators, and radio waves are the stuff of what physicists call "classical electromagnetism." This is to differentiate it from a new electromagnetism that began to unfold in the early years of the twentieth century with the realization that light, amongst other things, is not a simple smooth quantity as had been supposed. In fact, according to this view, light behaves at times as though it's made of particles, and at the finest level – at the scale of individual atoms – the interaction between charged objects occurs via the transfer of these individual electromagnetic carriers, called photons and denoted by γ. In essence, tiny charged particles interact via the transfer of "particles of light," the photons.

Photons blaze around us all the time. Using notation in which a hundred is written 10^2, and a thousand as 10^3, about 10^{20} photons burst out of a household light bulb every second. In everyday experience photons are so plentiful and the energy each carries so minuscule that they are treated *en masse* as "light." But on the scale of individual atoms, both the interaction of light with matter and the mutual interaction of charged matter particles via electromagnetic forces are dominated by this particle nature of light.

Creating the currently accepted theory, now called quantum electrodynamics or QED, describing the micro-world rules of engagement for photons with matter particles was a Herculean task begun in 1927 and completed around 1950. The story of this achievement is not really the story of this book, which is the story of the strong force theory called quantum chromodynamics, or QCD. But QED, which is simpler, and which predates QCD by several decades, provides an essential template. In this sense, understanding the fundamental interactions of light with matter is the route to understanding QCD.

Subatomic searching

Chicago, Geneva, Stanford, Hamburg, and Tokyo – a roll call of the cities that are home to some of the world's great particle accelerators. These, and other accelerators at several other spots around the globe, are at the cutting edge of experiments that scientists hope will unravel the secrets of nature's other two forces. The mission is huge, both in terms of the job to do and the resources required to do it. Apart from the major experimental sites, there are many other institutions and universities – more than 6000 in all – across the globe providing backup for those experiments, analyzing the results and developing theoretical ideas. All told, thousands of men and women are devoting their hearts and souls, and billions of tax dollars, to a voyage of discovery.

So what are they looking for? Atoms, the smallest chunks of matter recognizable to a chemist as identifiable substances, are mostly empty space. Naively speaking, atoms are fuzzy balls of negatively charged particles called electrons, and a hard, positively charged core called the nucleus. An atom is about 10^{-10} m in diameter, where now the shorthand is 10^{-2} for a hundredth and 10^{-3} for a thousandth. In other words, it takes about ten billion atomic diameters to span one meter. A nucleus is about a hundred thousandth the diameter of an atom, a mere 10^{-15} m across, yet it comprises about 99.975% of the atom's mass.

Ernest Rutherford deduced the existence of the atomic nucleus in 1911, and within a few years surmised that it contains particles called protons. In 1932, James Chadwick discovered the other component of the atomic nucleus, the neutron. A new layer of matter lay exposed.

Protons have a tiny mass, about 1.673×10^{-27} kg, and a positive electrical charge equal in magnitude to the negative charge of an electron. The proton is represented by the symbol p. Atoms of the simplest element, hydrogen, comprise a single electron and a nucleus containing just one proton. Neutrons have a similar mass to that of the proton, about 1.675×10^{-27} kg, and the fact that it is so close yet not identical to the mass of the proton is no coincidence. Neutrons, represented by the symbol n, carry no electrical charge. The nucleus of a helium atom, the next simplest element after hydrogen and the gas in floaty party balloons, contains two protons and two neutrons: nuclei contain roughly equal numbers of protons and neutrons, with the number of electrons in an atom equaling the number of protons in its nucleus.

Protons and neutrons are major players in the story of QCD; they will appear often. And we have already encountered in passing the electron, symbol e, which has a mass of just 9.11×10^{-31} kg, or about 1/2000 that of the proton and neutron. Its electrical charge is negative and very small, about -1.602×10^{-19} C. Something like 10^{19} times this amount of charge flows along a household electrical cable every second. Electrons are accepted as truly fundamental particles, indivisible and having no measurable size, so far as anybody knows.

The same is not true of protons and neutrons, and that is the point of it all. There are two other forces that determine what goes on inside neutrons and protons, and how those protons and neutrons stick together. Nuclear physicists are more interested in the sticking together of protons and neutrons to give nuclei. Particle physicists delve one level deeper, into the structure of the protons and neutrons themselves, and the fundamental properties of the forces at play in this subnuclear world. This is what the researchers using those huge accelerators are doing.

Weak revelations

The year is 1983, the place Geneva, Switzerland. Scientists at CERN, the European Laboratory for Particle Physics, make one of the scientific discoveries of the century.

Beneath the ground is a circular tunnel almost 7 km in circumference. It houses an evacuated pipe surrounded by vast magnets. The magnets are there to guide charged particles traveling along the pipe, making sure they follow the line of the pipe, never once touching the walls and always staying tightly grouped. Huge radio-frequency generators pulse energy into the particles, accelerating them each time those particles pass by until they are racing round the loop at incredible speeds. There are two types of particles in the one machine, protons and the proton's opposite number, the antiproton, denoted $\bar{p}$. Antiprotons will feature in detail later: for the moment it's sufficient to know that they are the same mass as protons but have opposite electrical charge. In the pipe, under the influence of magnetic fields and accelerated by radio-frequency pulses, they move in the opposite direction to the protons, round and round the ring.

Great care is taken to ensure that most of the time the two counter-rotating beams do not see one another. But at certain points the beams are magnetically manipulated and forced to intersect. Protons and antiprotons collide. They smash into one another, giving a spray of radiation into huge detectors placed around the intersection regions. The detectors, each as big as a house and weighing rather more, contain several radiation-detecting layers to monitor the passage of particles emanating from the collision point.

The radiation is made up of many particles. Arrays of powerful computers analyze the detector signals, reconstructing the paths of particles through the detector, and calculating the particle masses and charges. And there it is, the first W particle, followed shortly after by the first Z particle ever seen by experimenters.

The discovery of the two W particles, the positively charged W^+, the negatively charged W^-, and the charge-free Z particle, netted Italian physicist Carlo Rubbia and Dutchman Simon van der Meer the 1984 Nobel prize in physics.

In fact, theorists Sheldon Glashow, Abdus Salam, and Steven Weinberg had previously devised a theory based on these W and Z particles, a feat that had already earned them their Nobel prize. Their theory came with a built-in acid test: the mass of the W must be related to that of the Z in a prescribed way. In addition, both masses are linked to an experimental quantity that was first measured in 1973. So theorists knew what the W and Z masses should be even before Rubbia had built his detector. That's why the discovery of the W and Z particles with the expected masses was so important and so thrilling. Particle accelerators now mass-produce these particles, allowing experimenters to carefully measure their properties. And all's well, but for one intriguing, unresolved question.

Central to the Glashow–Weinberg–Salam theory, or simply the electroweak theory as practitioners call it, is the explanation of how it is that the W and Z particles have mass at all. The answer, according to theorists, involves the introduction of another particle called the Higgs particle, named after British theorist Peter Higgs. Yet nobody knows precisely where to find this elusive creature, or even exactly what it will look like. Certainly nobody seriously thinks they have seen one, though armies of people are still looking.

It turns out that in QCD the Higgs particle has essentially no role. It really belongs to a different story, the story of the weak force, in which the W and Z particles are leading players and the electroweak theory the accepted description. The weak force is manifest, for instance, in a form of radioactivity called beta-decay, in which an atomic nucleus spontaneously transmutes with the emission of an electron – the beta particle as it is called in radioactive decay – and another particle called an antineutrino. The W and Z particles are the transmitters of this weak force, where they play a role similar to that played by the photon in electromagnetism.

Apart from its role in radioactivity and particle accelerators, the weak force would seem to be something the universe could live without. If you were marooned on a desert island and had to dump one of the four forces, the weak force would be it. Why does nature go to the trouble of having one whole force that does little more that break up a few atomic nuclei?

This is not a question for mere mortals perhaps. But the antineutrino given off in radioactive beta-decay, and its partner particle, the neutrino, may provide a clue. The neutrino, usually represented v, and the antineutrino, $\bar{v}$, are very special particles. In some ways they are like electrons, sharing with the electron membership of the exclusive club of truly fundamental particles. But they have no electric charge and only the tiniest mass. So out of the choice of strong, weak, and electromagnetic forces, it experiences only the weak force. Strangest of all, perhaps, is that neutrinos are overwhelmingly left-handed and antineutrinos are similarly right-handed.

Which means that of all nature's forces, the weak force is the only one that knows the difference between right and left. So maybe the weak force is more important than it at first appears, as it gives nature a way of distinguishing left from right, and has at its disposal the mysterious neutrino. The question then is why does nature need these?

Introducing the strong force

In contrast, nature's need for what physicists call the strong force is plain to see. A typical atomic nucleus contains many protons and neutrons, the number of each depending on the nucleus in question. The protons have a positive electrical charge yet live in close proximity to one another. The neutrons, though they have no charge, also live in the same tiny enclave. What holds them all together? Whatever does the job is strong enough to bind protons together in the face of a powerful electromagnetic repeling force that would otherwise hurl the protons apart, causing the nucleus to disintegrate in a fraction of a second.

It's the strong force that does the job. This is the force that also governs the nuclear burning process, called nuclear fusion, which powers the Sun and other stars. And it's the strong force at work inside a nuclear reactor, where in the nuclear fission process nuclei break up into smaller nuclei with the release of energy. One type of radioactivity, called alpha-decay, in which the nucleus breaks off a

"nuclear fragment" called an alpha particle that is made of two protons and two neutrons, is also controlled by the strong force.

Most important of all, though, is the fact that without the strong
force the nucleus would not stay together. No atomic nucleus would
mean no atoms, and if there were no atoms then we would not exist,
though we would not be here to worry about it.

In recent years, physicists have developed and refined what they
believe to be the true theory of the strong force. That theory is QCD,
the theory of the strong force and the theory that is the subject of this
book.

Protons and neutrons themselves are composed of more fundamental particles: QCD is the script for these more elemental actors.
The stage of the strong force is no larger than the nucleus in which the
players live. The subatomic world is a world where what physicists call
"classical rules," applicable to objects large enough to be held in the
palm of the hand, do not apply. To describe and understand this inner
world, physicists use a different set of rules, called quantum mechanics.
They also have come to learn that patterns and symmetries are central
to their efforts. For their part, experimentalists are looking at the smallest objects ever studied, and for this they need the largest microscopes.
Those "microscopes" are particle accelerators.

2 Symmetry

Physicists like symmetry, and with good reason. Without it, they would be hard-pressed to make much sense at all of what goes on in the subatomic world. So far as theories of fundamental forces are concerned, symmetry enters at the very start as one of the pillars on which these theories are built. Without explicitly using symmetry, physicists would probably have made little progress in understanding the weak and strong forces.

Mathematicians like symmetry too. They discovered it long before particle physicists realized they couldn't live without it. The mathematics of symmetry, called group theory, has some of its roots in the attempts of nineteenth-century Norwegian mathematician Niels Abel to understand equations containing all powers of a variable up to and including five, so-called quintic equations. Abel's proof that there is no simple way of writing down solutions for quintic equations came to the attention of one of his contemporaries, Frenchman Evariste Galois, one of the most tragic figures in mathematics. Where Abel's life was a wretched saga of poverty and obscurity – he was finally offered his first university post two days after he died – Galois' was a more colorful, almost romantic affair. Political intrigues saw Galois in and out of jail until he finally met his end in a duel. However, the night before the duel he wrote a long letter to a trusted friend in which, building on Abel's result, he laid the foundations of modern group theory: it was Galois who coined the term "group."

A related but distinct thread, more in tune with the physicist's world, came from French scientist Auguste Bravais' work of the classification of crystal symmetry. These and other threads were woven together by French mathematician Camille Jordan, whose work did much to bring group theory to the attention of the mathematics community towards the end of the nineteenth century. Group theory's sweeping entry into modern physics owes much to the Hungarian-turned-American physicist Eugene Wigner, who in 1926 introduced group theory to the study of the inner workings of the atom.

Group theory is a vast subject: for example the classification of so-called "finite simple groups," completed in the 1980s, has consumed the efforts of numerous mathematicians over many years and runs to around ten thousand pages in total. Group theory is also a very beautiful and active subject, with tentacles that reach into many other branches of mathematics. It is the fragments of group theory used by particle physicists that's the subject of this chapter.

SCIENCE OF SAMENESS

Symmetry to mathematicians and physicists is pretty much the same as symmetry for everybody else. The intuitive idea of symmetry is that of a "sameness" between two or more parts of some whole that are related in some way. The human body has an approximate symmetry about a line down the middle, the left-hand side of the body being (roughly) a mirror reflection of the right-hand side. An old favorite that reflects the interests of some of the earliest workers in group theory is the snowflake, an exotic ice crystal which, under a microscope, is seen to have six identical portions arranged about a central axis. Turn the snowflake through 1/6 of a complete rotation, in other words through 60°, and it's indistinguishable from the starting position. Finally, an example closer to the way symmetry enters the story of QCD is the circle. Rotate a circle by any number of degrees, or any fraction of a degree, about its center and it will still look the same.

This innocent observation, that changes can leave something looking as though nothing has happened, contains the germ of a hugely important idea. A symmetry means there is some operation that can be performed on the system which leaves it indistinguishable from the starting state. Correspondingly, there is something about the system that is unchanged under this operation. Something that remains the same is described as invariant with respect to that operation: a symmetry implies a corresponding invariance, something that is unchanging. In real physical situations, invariant quantities mean that something is conserved, such as the conserved energy and momentum of everyday mechanics. Conserved attributes are central to the way physicists understand and label the world. The all-important conserved quantities, and the rules for their use, called conservation laws, without which physics would be rendered almost powerless, are therefore rooted in symmetry. Symmetry with respect to changes in time corresponds to energy conservation, and symmetry under changes in position along a line corresponds to momentum conservation. And symmetry under rotations maps to the conservation of angular momentum, the special type of momentum that takes account of the rotational rate and distribution of mass in a rotating object. Symmetry in QCD is related to exotic new conservation laws of the strong force: group theory is one of the two main ingredients of QCD.

Tricks with squares

Turn a square through 90° and nobody can tell that you have done anything. You can do any number of such turns and still nobody can spot the difference. A square has fourfold symmetry of rotation about an axis through its center, corresponding to the set of rotations by 90°, 180°, 270°, and a "do nothing" option, which corresponds to 0°, or no rotation at all.

Firstly the "do nothing" operation. Mathematicians call this do nothing option the identity: in the present case it's a rotation of 0°, though in other contexts it might be multiplication by one. Next comes the observation that one rotation followed by another rotation in the set is just another rotation, a property called closure. In other words, rotation of a paper square laying flat on a table, followed by a second rotation leaves the paper square rotated but still flat on the table and not sticking out from it. Thirdly comes a "technical detail" that says for a sequence of three rotation operations one, two, and three, if rotations one and two are completed and the resultant followed by rotation three, then the final result is the same as rotation one performed on the resultant of rotation two with three. Though overall the order is still one, two, and three, the groupings are different. This property is called associativity: multiplying ordinary numbers together follows this associativity rule, but mathematicians know plenty of instances where the rule doesn't hold. Fourthly, there is an operation that undoes what has been done, a further rotation so that the end result is a rotation of 0°. This is an example of an inverse: each rotation has an inverse rotation, one of the collection of rotations, such that the combination gives the identity.

Because the rotations of the square satisfy these four simple rules, mathematicians say they form a group. In fact any set of elements having these four properties – an identity element, closure, associativity, and an inverse for each element – is a group. This is the definition of a group. It doesn't seem much, but mathematicians have a knack for getting a lot of mileage out of a few simple rules. The rules *are* as simple as they look, yet they are enough to capture the essence of symmetry. And the rules are not automatic in the sense that there are cases known to mathematicians that violate one or more of the rules.

Though the four rotations of a square in the plane form a group, they do not exhaust the list of possibilities for transforming the square. There are four other operations to include. Two of these are the rotations through 180° about each of the two diagonal lines across the square. The remaining two are rotations about the two lines joining the mid-points of opposite sides. The full symmetry group of a square thus contains eight elements. The rotations in the plane about the axis through the center form a group on their own, a "group within a group" that mathematicians call a subgroup of the larger group.

Additionally, both are examples of a discrete group, since each of the symmetry operations is a jump from one configuration to the next. This is in contrast to the example of a circle, where any of infinitely many smooth rotations leaves the circle indistinguishable from the original. The corresponding group in this case is called a continuous group; the set of rules that defines a group works here too. The infinitely many elements, one for each possible smooth rotation between 0° and 360°, are described in terms of a smoothly varying angle that can take all values between 0° and 360°. The difference between discrete and continuous is the difference between the positions of the 3, 6, 9, and 12 on a clock face, and the position of the point of the smoothly turning second hand as it sweeps continuously round.

Both discrete and continuous groups are useful in physics. A good place to find discrete groups at work, for instance, is in the study of crystal structure. In crystals, the ordered rows of atoms are symmetric under rotations through specific angles, for example 90°, or step-like translations along the atomic rows, discrete symmetries that are encoded using discrete symmetry groups. Usually, however, in physics continuous groups are the more important. It's easy to see why, as many real physical systems involve continuous rotations or translations, or continuous changes of some other quantity related to a continuous symmetry. And of all continuous groups the most important in particle physics are called Lie groups, after another Norwegian mathematician, Sophus Lie. Lie groups are essentially continuous groups for which quantities vary not only continuously but also smoothly. This probably embraces all continuous groups that a typical physicist could think of. All of the groups that feature in the building of theories of fundamental forces are Lie groups.

The group of rotations of the circle about its center is an example of a Lie group. In fact the group theory of rotations is the template for much of the group theory used in particle physics, both in QCD and in symmetry schemes for explaining particles such as protons and neutrons in terms of their most fundamental components.

Squares to circles

The tip of the second hand sweeping round a clock face maintains a constant distance from the center of the face. The clock face itself is a portion of a flat plane; it's easy to imagine the clock mechanism mounted in a flat tabletop so that the end of the second hand still marks out a circle on the tabletop. The tabletop itself provides a backdrop for a whole class of transformations, namely rotations and translations, which in principle would allow an object at any point to be moved to any other. Compared to this, the smooth circular motion of the end of the clock's second hand is very limited. In fact the motion of the end of the second hand is an example of a restricted class, called orthogonal transformations, of all possible transformations. "Orthogonal" means "at right angles" – an orthogonal transformation moves an object or point in such a way that the distance between it and some fixed spot remains constant. Since a simple orthogonal transformation applied to a point in the plane just takes that point to some other location on a circle, orthogonal transformations form a continuous group, to be identified with the continuous rotational symmetry group of the circle.

Orthogonal transformations in a two-dimensional plane form a continuous group called $O(2)$. The "O" is for orthogonal, which translates in some sense to transformations of "size one." The "2" is the number of dimensions – a two-dimensional plane – in which the rotation is performed. Measured against the tabletop, such a transformation takes a point from its starting position to a new resting place, which is perhaps less distance along the table and more distance across it, but with both initial and final positions the same distance from some fixed point. In other words, the transformation takes two position co-ordinates (along and across measures) to two new ones. Mathematically, each of the new co-ordinates depends on both of the two original ones, so that to get from the original to the final values means using four numbers. These numbers can be written as a box of numbers, having two rows and two columns, which expresses the way the original co-ordinates mix to give the final co-ordinates. The box of numbers is called a matrix; in general, in mathematics a matrix can have any number of rows and any number of columns. The "2" of O(2), originating as the number of dimensions in which the rotation was performed, is therefore also the size of the smallest matrix, a two-by-two, able to do the job.

In fact, although there are four numbers in this matrix, they are all related, so that all four can be expressed in terms of just one. This coincides with the fact that for a rotation in the plane, there is really only one parameter required, the angle of rotation about the origin. For the most general transformation of one point of the tabletop to any other point, such that the requirements of group theory are satisfied, the two-by-two matrix has four independent numbers corresponding to four independent parameters.

Setting the tabletop at some obscure angle relative to the room it sits in introduces a new twist. The problem is really just the same, describing the transformation of a point on a tabletop to a new position that's a fixed distance from some special selected point. But the new challenge is to do this by tracing the point's location in terms of the up, down, and across measurements relative to the room itself. Doing this will need a three-by-three matrix to track up, down, and across co-ordinates transforming into new up, down, and across co-ordinates. Though there are nine entries in the matrix, there's still only one parameter that actually changes as the point moves.

This looks like a more complicated way of doing exactly the same thing, which it is. The two-by-two and the three-by-three matrices are both associated with the same group, O(2). They are just two different ways of actually implementing an O(2) transformation, for which reason they are called representations of the group O(2). That is, what mathematicians have done is to realize that there is an abstract mathematical entity called the group O(2), then introduce the idea that to make it do something, for example move a point about, means writing down a particular representation of that abstract entity. So using the co-ordinates of the plane, a two-by-two representation is adequate. Achieving the same transformation referred to the room co-ordinates requires a three-by-three representation of the same group, O(2). The subject of a group representation's attention may be points on a table, or elementary particles that are governed by a symmetry relationship. There are infinitely many possible group representations corresponding to a particular group, though many will not be independent.

There's one property of a clock's second hand, common in many real physical systems, which the group $O(2)$ fails to capture. The second hand never jumps out through the glass cover of the clock face to go straight from, say, three to nine. Merely requiring that the point of the second hand is a constant distance from a central fixed point does nothing to inhibit this possibility. Yet in the context of physical systems, allowing this type of a jump is undesirable, since it implies instantaneous leaps from place to place. So physicists prefer to eliminate all these jump options. The result is a subgroup of $O(2)$, in other words a group within a group, called $SO(2)$. The "S" represents special. In terms of the matrix that represents the group moving one point to another, the S translates into an additional overall condition on the plus and minus signs of the numbers appearing in the matrix. Again, all the groups featuring in the building of fundamental theories are of this "special" jump-free variety.

An obvious extension of these ideas is to go from points moving on a tabletop to points moving around in a room. Insisting in a similar way that a point always maintains a constant distance from some fixed origin leads to the group $SO(3)$, of which $SO(2)$ is in fact a subgroup. The smallest matrix that can be used to represent this group is a three-by-three matrix, as denoted by the "3" in $SO(3)$; larger matrix representations are possible, just as they were for $SO(2)$. In addition, $SO(3)$ has three possible independent rotation angles, one each about the three directions up, along, and across the room. So although the three-by-three matrix has nine possible entries in it, they can all be expressed in terms of just three changing angles.

In addition to being one of the most important groups in physics in its own right, $SO(3)$ provides an introduction to two of the principal groups appearing in particle physics, including the group underlying QCD.

The reason $SO(3)$ is so important is that it expresses the symmetry of real rotational systems, for example spinning tops, gyroscopes, and pirouetting ballerinas. All three are about the conservation of angular momentum. It's this that keeps tops and gyroscopes pointing in a fixed

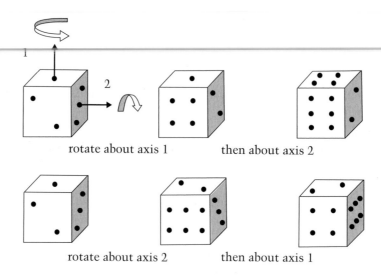

rotate about axis 1   then about axis 2

rotate about axis 2   then about axis 1

FIGURE 2.1 Rotations of a die about two different axes give a different result depending on the order of rotation. In this example, the results of two rotations are related by a further rotation about an axis diagonal through the corner of the die "protruding" from the page.

direction, and allows the dancer to alter the speed of a pirouette by arm movements that change his or her mass distribution. Mathematically, the conservation of angular momentum corresponds to invariance of the equations of the system under $SO(3)$ transformations.

Rotations limited to the plane combine in a simple way: any two rotations give the same end result, no matter which one is performed first. In other words, the ordering is not important. Groups that have this property are described as Abelian, after Niels Abel: the group $SO(2)$ is an example of an Abelian group. Another example is the symmetry group of electromagnetism.

For rotations about three independent angles, things are more complicated, and the order in which the rotations are performed does matter: changing the order gives a different outcome. One way to see this order-dependence of rotations is to turn a dice through 90° about two different axes, Figure 2.1. A rotation about axis one followed by a rotation about axis two gives an end result different from a rotation about axis two followed by rotation about axis one. Groups for which transformations performed in a different order give a different answer are called non-Abelian. The symmetry group underpinning QCD is a non-Abelian group.

The groups looked at so far have an obvious interpretation in terms of rotation angles. For many groups things are not quite so transparent. For example, the group $SO(4)$ is the group of rotations in four-dimensional space – in which case it needs a room having up, along, across, and some additional length in which to work – and its smallest representation is a four-by-four matrix. Yet these rotations demand six rotation angles in all, even though there are only four possible directions. For this reason, it's easier on the mind to view transformations not in terms of angles but simply in terms of independent parameters. For simple cases these are readily identified, perhaps as angles of rotation in the accepted way. In more complicated cases, they may be rotations in some abstract space, or simply "parameters of the transformation."

As a hint of where this is all leading, the symmetry group of electromagnetism has a single parameter, in parallel with the physical fact that there is only one type of photon. Correspondingly, there is also a single conserved quantity, the electric charge.

The groups $SO(2)$ and $SO(3)$, interesting and powerful though they might be in the study of rotations in classical mechanics, are not quite the groups needed to understand symmetry in theories of fundamental forces, though they emulate them and are very closely related. In fundamental theories of particle interactions, the description of the basic entities in the theory makes use of a more abstract space than that represented by a tabletop or the room it stands in. This space is based on one in which, for instance, the two independent directions of a tabletop (along and across) are welded together in a particular way. In this more abstract space, length-preserving orthogonal transformations are replaced by the corresponding length-preserving "size one" unitary transformations.

The symmetries of particle physics

The group that plays the equivalent role of the simple rotation in the plane is $U(1)$, the one-dimensional unitary group. This group and $SO(2)$ are mathematically equivalent. There's no "S" here since $U(1)$ is already "special" so an "S" would be redundant. It does appear in other cases, however, and once again is about eliminating jump transformations. The rotation group $SO(3)$ is replaced by the intimately related "special unitary group" $SU(2)$, likewise non-Abelian and having three parameters. It plays an essential role in understanding the partnership between protons and neutrons, and also in the rigorous description of the "spin" of electrons. The group $SU(2)$ has a bigger sister, called $SU(3)$, also non-Abelian and this time having eight parameters. The group $U(1)$ is the symmetry group of QED. Combined with the group $SU(2)$ it expresses the symmetry of the weak force. The group $SU(3)$ describes the symmetry of the strong force: it is the symmetry group of QCD.

According to QCD, the most fundamental strongly interacting particles carry any of three labels. The strong force between these particles is insensitive to the way in which these labels are mixed up, and physicists have discovered that this underlying symmetry is described by the group $SU(3)$. The "master equation" of QCD, *the* equation which really defines the theory, must remain unchanged by the action of this group, which transforms the labels carried by the fundamental particles of the theory into mixtures of one another.

Knowing the symmetry group of the theory, particle physicists can then use all the know-how mathematicians have accumulated concerning that group. The group $SU(3)$ has eight parameters, and this means that in QCD there will be a clutch of eight force-carrying particles that play a role equivalent to that performed by the photon in quantum electrodynamics.

That, in a nutshell, is how symmetry enters into QCD. The other main ingredient of the theory, besides symmetry, is the system of mechanics applicable to the world of the very small: quantum mechanics.

3 The quantum world

THE BEST THEORY EVER

In a magnetic field, electrons behave like tiny magnets. They have a north pole and a south pole, and when placed close by a magnet they suffer a turning force as the tiny electron-magnet attempts to align with the magnet's field. The size of that turning force is determined by an inherent property of electrons, their magnetic moment.

The magnetic moment of an electron, or a simple bar magnet for that matter, is just the ratio of the aligning force to the strength of the magnetic field causing the aligning. In other words, a larger magnetic moment makes for a bigger turning force for a given field. It turns out that the magnetic moment of electrons isn't what simple theory says it should be, however. The correction factor is called the anomalous magnetic moment, a dimensionless quantity denoted a. Experimenters have measured this quantity many times, achieving incredible accuracies. In 1987 Robert Van Dyck, Paul Schwinberg, and Hans Dehmelt at the University of Washington, Seattle, used single electrons trapped in a magnetic bottle, a Penning trap, to achieve a value of

$$a_{\text{expt}} = 0.001\,159\,652\,1884 \pm 0.000\,000\,000\,004\,3.$$

Compare this to a theoretical result of

$$a_{\text{theory}} = 0.001\,159\,652\,1535 \pm 0.000\,000\,000\,029,$$

based on a 1999 value given by Cornell University's Toichiro Kinoshita, who has spent decades teasing out and comparing values of this quantity a. Exactly what "a" is doesn't matter: the thing to notice and to be impressed by is the close match between the two numbers. The agreement is to eight significant figures, and as agreements between theoretical predictions and experimental measurements go, this is touted as the best there is. The theory used to give this result is quantum electrodynamics. It's this agreement that justifies the claim: quantum electrodynamics is the best theory ever.

The "quantum" in the name derives from the fact that this theory is directly applicable to the world on an atomic scale. The "electro" is because it is about electromagnetic fields, and "dynamics" is because it's about fields that change over time. Quantum electrodynamics, QED, is the theory of how charged particles such as electrons interact on the minutest scale. It is, in the words of the US physicist Richard Feynman, "the strange theory of light and matter."

QED is the most fundamental description we have for the way matter interacts via the electromagnetic force. Since all atoms and molecules are held together by electromagnetic forces, it underpins our understanding of them too. And since everything, including us, is composed of atoms and molecules, QED in a sense explains all of chemistry and a great deal more besides.

Though the electron anomalous magnetic moment is a stringent test of QED, experimenters have carefully assessed the theory's predictions in other ways. For instance, the electron has a cousin – an elementary particle called the muon – which also has an anomalous magnetic moment. Beginning in 1996 experimenters at a huge dedicated machine at New York's Brookhaven National Laboratory have watched billions of muons decay. Their measurement of the muon's anomalous magnetic moment agrees with the calculated value to six decimal places.

Electrical discharges in gas cause it to emit light and radio waves with a complicated but well-defined spectrum. An historically important case is hydrogen. Its spectrum contains one particular high-frequency radio contribution, the so-called Lamb shift, that simpler theory says should not be there. According to QED, however, it should. The measured value agrees with the QED calculated value to five significant figures. Other tests exploit different light emissions from hydrogen, or light emissions from other atoms, and all agree with the QED calculated value to various levels of accuracy.

The L3 experimental group at CERN used their enormous particle detector to study reactions in which two or three photons are produced from collisions between electrons and their opposite numbers, called positrons, in LEP, the Large Electron Positron collider. Their results, and many others from the world's accelerator laboratories, support the predictions of QED and in a much higher energy regime than the spectral and anomalous magnetic moment tests.

Physicists studying matter in the solid state provide still further support for QED, and in a completely different environment. Two exotic phenomena from condensed matter physics – the quantum Hall effect and the ac Josephson effect – give results for the strength parameter of QED that agree with that derived from electron anomalous magnetic moment measurements to one part in ten million.

That's all very edifying, but so what? None of these effects are particularly familiar to the man or woman in the street, so why are they important? The reason is that although QED is billed as a theory of almost everything, to see it in detail means looking in obscure places. The point is that anomalous magnetic moment measurements and the rest offer clean tests of the theory. The name of the game is to find something that can be measured and calculated to high accuracy, and which is sure to give effects that depend only on QED. Or if other effects creep in, they too can be estimated and the QED-only part extracted. There's no point in looking for definitive tests of QED in the bulk properties of coal or some biochemical reaction, even though QED is at the root of both, since these systems are too complicated and messy. Obscure effects as acid tests are here to stay.

And there's another question to deal with. Why so much about QED – where is QCD in all this? Quantum electrodynamics is tried, tested, and trusted, and it has an impeccable pedigree as the child of classical electricity and magnetism. It's (relatively) simple and very elegant as a theory. And it is the template for QCD. Quantum chromodynamics is very similar to QED, but QED lacks the details that confuse a discussion of the basics common to both. The details will follow soon enough, and flowing from them will come all the remarkable features of QCD. But basics first.

QUANTUM MECHANICS

The "quantum" of quantum mechanics is the same as the "quantum" of quantum chromodynamics. Quantum mechanics describes the behavior of the world at tiny distances, roughly speaking no bigger than the size of an atom. Since the field of play of the strong force is the atomic nucleus, any theory of that force *must* be a quantum mechanical theory. So really getting to grips with QCD, the theory of the strong force, means first exploring the surprising inner world where quantum mechanics reigns supreme.

The ancient fortified city of Grenoble, in south-east France, is home to a unique research facility. Set against a beautiful Alpine backdrop is the Institue Max von Laue–Paul Langevin, or ILL for short, one of the world's leading centers for research using neutrons.

That's research *using* neutrons, rather than research *into* neutrons. At the ILL, and at a number of other similar sites around the world, scientists are using the particular properties of neutrons to study everything from stress in railway rails to photosynthesis in plants. The facilities at ILL are centered on a nuclear reactor, inside which uranium undergoes nuclear fission to produce large numbers of neutrons that are piped off to several experimental areas. The ILL reactor is the most powerful neutron source in the world, and has the greatest number of experimental instruments attached to it.

Neutrons are particles. They have a minute but measurable mass, and other well-defined "particle-like" properties that experimenters have measured and tabulated.

When a neutron enters a material, it bounces off the nuclear hearts of the atoms inside – a consequence of the strong force. This is in contrast to electron and X-ray scattering, in which the scattering is from atomic electrons, and is one reason why neutron scattering provides a useful complementary experimental tool to other scattering methods. The surprising feature of neutron scattering is that the scattered neutrons, recorded as particles in detectors at various angles around the sample, give a neutron intensity pattern of a type peculiar to the *interference* of *waves*. Neutrons are waves.

The peaks and troughs of two waves traveling together need not tally exactly. The amount by which the peaks of one wave lead or lag those of the other is called the phase difference; if the phase difference is the same from one moment to the next then the waves are described as coherent. The phase itself expresses a wave's progress through its cycle, in much the same way that the angle hub-to-valve on a bicycle wheel labels the revolution of a spinning wheel. Indeed phase is an angle, and the possible values of that angle match the range of angles available in a circle. So phase has a value between $0°$ and $360°$. The phase part in the description of the wave doesn't contribute to the wave's amplitude, in other words its size, so it has the geometry of a circle.

Phase and coherence are key to understanding interference, and interference in turn is enjoying something of a golden age as the basis of many of the most accurate measurement techniques in physics. For example, the gravitational wave detectors in Louisiana and Washington use laser interference techniques to measure changes in length of less than a thousandth of the diameter of an atomic nucleus in the 4 km long arms of the detector.

If two coherent waves are brought together (for example, carefully prepared light waves arriving at a screen) then two peaks arriving in concert will combine to give twice the amplitude and *four* times the brightness. A peak arriving with a trough will cancel, and give a low-intensity region. This is interference – overlapping waves that either reinforce or cancel one another depending on phase difference.

Back in Isaac Newton's time, and largely as a result of his views, light was deemed to be little particles called corpuscles. This picture was undermined by the work of Thomas Young, another Englishman, who around 1802 performed a simple experiment that has become one of the best known in all of physics. Young shone a beam of light at two narrow parallel slits, and on a screen beyond saw a pattern of light and dark bands where naively he might have expected to see just two bright lines, projections of the two slits. Realizing that this was interference at work, Young surmised that light must be a wave. The pattern is caused by light from the two slits combining so that peaks add to give bright areas, and cancel with troughs to give dark areas.

Two centuries on and scientists are still performing variants on Young's double slit experiment. Austrian physicist Anton Zeilinger and collaborators performed the first clear demonstration of Young's double slit experiment for neutrons at ILL in 1988. They directed a neutron beam onto a fine boron wire positioned between jaws made of boron-enriched glass, boron being a strong neutron absorber. The gaps between the wire and the jaws either side constituted two slits. Their results, shown in Figure 3.1, revealed a clear neutron–wave interference pattern so good that the group claimed theirs was the most precise realization of "matter waves" up to that time. So, as if the fact that countless researchers at ILL have used the wave aspect of neutrons to glean their experimental results wasn't enough, the classic wave-proof experiment confirms that neutrons can behave like waves.

And that's not all. Double slit experiments have even been performed for whole atoms. First to achieve this were Oliver Carnal and Jürgen Mlynek at the University of Konstanz in Germany. In their 1991 experiment, they directed a stream of helium atoms onto a pair of 1 μm wide slits cut in gold foil, and saw atom–wave interference beyond the

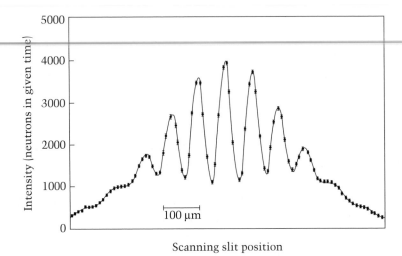

FIGURE 3.1 Neutrons passing through a double slit create an interference pattern, the peaks and troughs of the neutron count reminiscent of the light and dark bands formed by light passing through two slits. The solid curve shows the calculated value. Credit: Anton Zeilinger/*Reviews of Modern Physics.*

slits. And in 1995 David Pritchard's group at the Massachusetts Institute of Technology (MIT) in the USA saw interference in two-atom sodium molecules.

Now, this is getting serious. Atoms and molecules are supposed to be the building blocks from which the material world is made. Surely this means they must be particles? Yet these experiments showed interference: atoms and molecules can behave as *waves*.

So are neutrons, atoms, electrons, and other entities waves or particles? Probably the best response to this question is "the question is wrong." Experiments show there is an essential wave–particle duality in nature. Depending on how they are manipulated and observed neutrons, atoms, and the rest can exhibit wave-like properties or particle-like properties, though not both simultaneously. But both are intrinsic, and both apply across the board to all entities in the Universe.

The framework for understanding this duality is quantum mechanics. Though wave–particle duality applies to everything, it becomes more noticeable on a scale of atoms or less, and for this reason quantum mechanics is sometimes said to be the mechanics of the submicroscopic world, where the mechanical rules applicable to everyday objects go wrong.

The quantum world

The "quantum" in quantum mechanics originates from the recognition that light, proved wave-like by Young, only exists in multiples of some basic packet, or quantum. For a given light frequency, the energy of the light quantum is the product of the frequency and a constant, called Planck's constant. The light quantum has been given the name photon. Newton's corpuscles were on the right track after all, though it would have been hard for him to confirm photons directly because their individual energy is so tiny and their numbers so vast.

Planck's constant is named after Max Planck, discoverer of the quantum nature of the Universe. It was Planck who, in 1900, introduced the idea of "energy packets" – quanta – to resolve a long-standing light emission and absorption problem. Einstein introduced the quantum of light five years later. Denoted h, Planck's constant has the amazingly small value of $h = 6.625 \times 10^{-34}$ (in units of joule seconds denoted Js), and is now recognized as one of the fundamental constants of the Universe. Physicists these days often use $\hbar$, equal to h divided by 2π, as this combination appears so frequently in their equations.

The tiny value of Planck's constant reflects the fact that the wave-like nature of matter is not apparent in everyday objects. The wavelength of a "golf ball wave" is about 6×10^{-34} m, far too small to excuse poor putting. The wavelength associated with a car on a freeway is even less, about 2×10^{-38} m. These are incredibly small numbers, many orders of magnitude smaller than the size of the atomic nucleus, and make the wavelength of light, about 5×10^{-7} m for green light, seem huge by comparison. Electrons in a household television have a wavelength of about 7×10^{-12} m, so even this is pretty minute.

Golfers and freeway drivers will never experience the strangeness of the quantum world directly, since these wavelengths are so short. This is probably a good thing, since quantum golf and freeway cruising could be distinctly weird experiences. The double slit experiment can give some hints why, as it shows much more than simply the wave-like nature of "particles." In fact double slit experiments are held up as perhaps the most fundamental realization of quantum effects.

In a double slit experiment, for instance one with light, if one or other of the slits were to be blocked off then the interference pattern would vanish. This is because the interference pattern requires recombination of light that follows two paths, in this case one path through one slit, and the other path through the second slit. What happens if the light source is turned down so low that the incoming beam is so weak that only one photon at a time arrives at the slits? Common sense says that a light particle would go through one or other slit. But common sense is wrong: a single photon effectively passes through *both* slits. The final interference pattern is created once sufficient single photons have run the gauntlet of the slits and arrived at the screen to be recorded as single tiny dots. These individual dots eventually build up the final, distinctive intensity pattern.

Ever more sophisticated and persuasive experiments, starting in 1927 at the University of Chicago with Arthur Dempster and Harold Batho, who just turned down the light source intensity, confirm that single photons interfere with *themselves*. Doubters might think they can beat this by watching to see which hole the photon "really" goes through. If they do, they'll find that the mere act of watching in itself destroys the interference pattern. The quantum eraser, first suggested by the US physicist Marlan Scully in 1982, and implemented by Paul Kwiat, Aephraim Steinberg, and Raymond Chiao at the University of California, Berkeley, in 1992, shows this beautifully.

Figure 3.2 shows a schematic drawing of a two-slit quantum eraser experiment. Photons arrive one at a time at the double slit. Light that travels via the upper path passes through a "marker" that in some way marks the light as having passed via the upper route. The paths

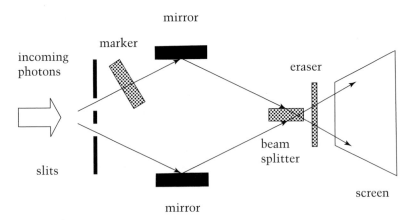

FIGURE 3.2 Schematic diagram of a quantum eraser. Arrows show the paths available to incoming photons. The paths converge onto a beam splitter. The marking and erasing can be achieved using devices called polarization rotators and polarization analyzers, respectively.

then converge on a beam splitter, a partially silvered mirror that reflects 50% of incident light, and allows the rest to pass through. The function of the splitter is simply to provide a place for the two different paths to recombine and overlap. Between the beam splitter and the screen is a device to remove any marks on light put there by the marker. This mark-removal device is the quantum eraser: so-called which-way information is stored in the polarization states of the photons, and erased using polarizing filters, but the details do not really matter here.

Without either marker or eraser this is a straight single-photon Young's double slit interference experiment. It gives an interference pattern. With the marker in place, but not the eraser, the interference pattern vanishes. With both the marker and eraser in place, an interference pattern *will* form.

What's happening is that, with the marker alone, the upper photon carries which-way information, that is something that says which route it took, which slit it came through. Once the paths are in any way distinguishable, interference disappears. With the eraser in place, although upper-path light is marked, the mark is removed before final detection, the paths are no longer distinguishable, and interference reappears.

With marker and no eraser, the photons behave like particles, passing through one slit or the other and showing no interference. With the eraser added, photons behave like waves, and pass through both slits. Just one of the mysteries of the quantum world is how the photon arriving at the slits "knows" if the eraser is in place, and so whether to behave like a particle or like a wave.

It's not reasonable to say that electrons, photons, and the rest are really particles or waves. The best that anyone can say is that there are times when, say, electrons exhibit particle-like behavior and others when they exhibit wave-like behavior. Or, in the words of Sir William Bragg, "Physicists use the wave theory on Mondays, Wednesdays and Fridays, and the particle theory on Tuesdays, Thursdays and Saturdays." For their part, particle physicists, as perhaps the name implies, habitually call everything particles, even though they know in their hearts that this is not naively true.

Chance and uncertainty
One important consequence of the wave-like nature of "particles" is that their precise position and momentum become a little blurred. If an

electron is described as a wave, how then can anybody say exactly where an electron is? There is an inherent uncertainty in a particle's position, likewise in it's momentum. The two are related through a fundamental quantum uncertainty due to the wave-like nature of "particles": the uncertainty in position combined with that in momentum at best equals Planck's constant.

This is Heisenberg's uncertainty principle, introduced by Werner Heisenberg in 1927. It's a statement of nature's intrinsic inscrutability, nothing to do with limited skills on the part of experimenters. It says that there's no way of measuring both position and momentum to perfect precision, even in principle. It also says that a high-precision measurement of one quantity may be at the expense of an increased uncertainty in the other, in order that the uncertainty relation is respected.

In particle physics, the uncertainty relation pops up everywhere, often in an alternative but exactly equivalent form. This version says that uncertainties in the measurement of a particle's energy and the time of that measurement are likewise wed through Heisenberg's principle. If the energy is known exactly, then the time at which the particle had that energy is known only imprecisely, and vice versa. In this form, the uncertainty relation is used to justify all kinds of extravagant things, right up to the creation of the Universe out of nothing.

Wave–particle duality also begs the question of how an electron or light wave, say, translates into numbers of electron "particles" or photons picked up by some detector. The two-slit experiment provides some clues.

The first clue comes from comparing the brightest parts of the two-slit pattern with the brightest part of the single-slit version. Two slits instead of one give not just double the brightness, but four times the brightness. Doubling the wave size, its amplitude, gives four times the intensity – evidence that intensity is the square of the amplitude. A second clue comes from the fact that the distinctive interference pattern vanishes if either one of the holes is closed up. The intensity pattern seen from two slits is not just the intensity pattern of one slit plus the intensity pattern of the other slit. The distinctive pattern arises only from addition of the wave *amplitudes* contributed by both slits; this total amplitude, multiplied by itself, gives the intensity.

The rule is this. Add the contributing amplitudes, then square the resultant amplitude to give the final intensity. This is how waves work. But this is only part of the answer. How does a wave description translate into particles counted on a screen or in some kind of electronic detector?

Each "particle" arrives at the detector screen, via the two slits, to be detected as a single dot. In time, a pattern of dots will build up to give the interference pattern. An individual dot, corresponding to an individual incoming single electron or photon, does not by itself create a particular intensity pattern. But there's a greater chance that the dot will be found in a region that will eventually be a bright region, and less chance that it will be found in a region that will eventually emerge as a darker region.

Chance. Chance, or probability, is an integral part of the Universe, and of quantum mechanics. Because of their wave-like nature, an electron or photon, say, with some exact momentum is never certainly here then later certainly there. Yet their particle-like nature means it is reasonable to ask after the whereabouts of electrons and photons. The way out is probability. Quantum theory does not calculate where an electron or photon will be located, but the probability of where it will be located. It does not provide a way of calculating the precise moment an electron will make a transition within an atom, merely the probability that such a transition will take place.

In quantum theory, the amplitudes of all contributing quantum waves are first added, and the resultant squared to give a probability. Experimenters only see the final probability in their measurements, the amplitude itself remaining hidden.

Schrödinger and hydrogen

Physicists discuss quantum theory in terms of quantum states. For example, a quantum state might be an electron at some point in space and time, and a second quantum state might be the same electron at a different point and a later time. Quantum states can also have other attributes, some of which will appear in due course.

Things happen, according to quantum theory, by transitions from one quantum state to another. A simple example is the movement of an electron from one place to another, a transition of a state at one point at one time to a state at a different point at a later time. Quantum theory provides rules for calculating the amplitude for transitions such as this. Squaring those same amplitudes gives *probabilities*, and it's these that are to be compared with experiment.

The simplest description of a changing quantum state is given by Schrödinger's equation, devised by Erwin Schrödinger in 1926. The classic application is to the hydrogen atom. Schrödinger's equation describes the influence of the atom's central positive charge on its single electron's quantum state, and predicts a number of distinct electron "energy levels," each having a well-defined energy. The electron in a hydrogen atom may jump from a low-energy level up the "energy ladder" to a higher energy level by absorbing a photon of just the right energy. The reverse process is also true, and the electron can fall from a higher energy level to a lower energy level, accompanied by the emission of a photon whose energy matches exactly the gap between the levels. Schrödinger's equation even provides the basis for calculating the probabilities of making such transitions.

Historically, the fact that quantum mechanics applied to the hydrogen atom gives energy levels between which electrons can jump, with emission or absorption of photons, was hugely important. Not only did it explain how atoms could actually exist – according to non-quantum electromagnetic theory alone they could not – it also gave a natural interpretation of the spectrum of hydrogen. This is the particular set of precise light frequencies, in effect colors, generated by an electrical discharge through hydrogen gas, which scientists were at a loss to explain prior to quantum theory. Understanding the details of such emission spectra in hydrogen, and in other gasses, or the absorption spectra, in which gases absorb specific frequencies from light passed through them, has been critical at several stages of the development of quantum theory.

Schrödinger's equation forms the basis of much of modern working quantum mechanics, for example that used by chemists and atomic physicists. It is enough of a quantum theory for their needs, but it is not the whole story. The simplest applications of Schrödinger's equation view the atom's nucleus merely as a source of electrostatic force, with the nucleus and electron obeying Coulomb's law of electrostatics. This says that, as the distance between two charges is doubled, the electrostatic force between them decreases by a factor of four, an example of what physicists call an inverse square law. Or, equivalently, the energy of the electrostatic interaction is proportional to the reciprocal of the distance separating the charges. So although the electron itself obeys quantum rules, its interaction with the atom core is treated using old-fashioned non-quantum, or classical, electrostatics.

This kind of treatment generates lots of powerful and useful results, but is not really complete, and there are other things besides that are demanded of a full quantum theory. Schrödinger's equation is merely the beginning.

The least of many paths

Surveyors use light rays to help them line up the walls of buildings and the edges of roads. In doing so, they are making two important assumptions: that light rays travel in straight lines, and that the path the light follows is unique. Luckily for surveyors, houses and roads are so large that wave and quantum effects are not a big issue, for the double slit experiment shows that neither assumption is quite true when the subatomic detail is important.

If there is a double slit between the starting point and the finish, then an electron or photon will pass via two paths. There's no reason to stop at two slits – if there were thousands of slits then there would equally be thousands of possible paths that all contribute. Why bother with arbitrary slits? Remove the spacers between the slits and now there are, in effect, an infinite number of paths connecting the start with the finish across the intervening space.

In the two-slit case, the final intensity pattern is controlled by the phase differences between the two contributing parts. And the same is true when there are millions and millions of possible paths. For paths well wide of a "reasonable" path, the phase differences are such that contributions from these paths tend to cancel, and so the widely off-track paths contribute little. For paths nearer a "reasonable" path, this is not the case, and these paths contribute much more. The sum over all the infinite number of tiny amplitudes, one for each path, gives the complete quantum amplitude for making the transition from the initial state to the final state.

This prescription is called the path integral. Richard Feynman, at the time a young Princeton graduate student, introduced the idea in the early 1940s. It's the most powerful formulation of quantum mechanics. Chemists and atomic physicists can be satisfied with the Schrödinger equation approach, which incidentally can be derived from the path integral, for it does what they need. But for theories of fundamental forces, there is really little alternative to Feynman's path integral.

The path integral combines all contributions from each of the infinitely many possible paths linking two quantum states, even crazy paths that do not gently swerve across space from start to finish. Just how much each path contributes to the final result depends on its phase, which ultimately depends on the "master equation" of the theory.

One of the magical things about the path integral is that the master equation used is a simple "classical" non-quantum one. Inserting the classical master equation into the path integral turns the theory into a quantum theory. This is because, once inside the path integral, the master equation controls the contributions from infinitely many paths, instead of just the one path of classical mechanics. Working with a classical master equation means the correspondence between the classical and quantum is therefore always easy to see. Many additional properties required of the final theory can be readily introduced via the master equation, and automatically follow through via the path integral into the final quantum theory. For the exotic symmetries of the theories of fundamental forces, the path integral alone is up to the job.

Given the path integral, the starting point for theorists building a quantum theory is to devise the appropriate master equation, resplendent with symmetries and other special properties. They can then plug it into the path integral, and follow a well-defined recipe to produce a set of calculational rules. With those rules they can then work out physically measurable quantities such as how fast unstable particles decay, or what happens when particles collide. It's easy with the path integral approach to see how changes in the master equation manifest themselves in the calculational rules.

In some respects, there's nothing terribly mysterious about the master equations that plug into the path integral: they express the difference between the total motion energy of the participants and the total energy of their interactions. This might mean the motion energy of both electrons and photons, minus the interaction energy of electrons with photons. Different master equations fed into the same path integral will give different quantum theories.

A bicycle ride helps to illustrate the underlying classical idea. There are several roads between your starting point and your final destination. The shortest route takes you straight up a mountain and down the other side. Freewheeling down the distant side of the mountain, converting that energy of position into energy of motion, might seem attractive, but the ride up is a killer. Then there's the route through the valleys, almost flat but it drags on for miles. You settle for a third option through the low foothills, not so long and tedious as the valley route, not so debilitating as the summit route, somewhere in between in terms of both pain and boredom. What you have done is to select an optimum path. But it is not an optimum path in the sense of being geographically shortest, nor the flattest – it's a compromise between flat (and long) and hilly (and tiring).

Nature also selects optimum paths. Nature doesn't know about pain and boredom, but she does know about energy. Where the cyclist uses a map of the terrain, nature uses energy horse-trading, keeping a running tally of motion or kinetic energy versus position-related or interaction energy, the potential energy. Kinetic energy over a long time is undesirable (like the valley route). Extremes of potential energy over short times (like the summit route) are also undesirable. The way that nature tracks this is through the difference between kinetic energy and potential energy, accumulated over time. This energy-trade tally is called the action. The optimum path, the one that nature takes, is the one that gives the lowest value of the action: this is called the principle of least action.

The action lies at the heart of mechanics. Feed the action into some scheme for finding its smallest value, something that mathematicians know how to do easily, and from the principle of least action you have the laws of classical mechanics. A baseball flying through the air is minimizing its action, and flies along a curved path and not a gravity-defying straight line as a consequence. A stripped down version of the least action principle says that the beam from the surveyor's laser follows a straight line, yet light rays from an object on the bottom of a swimming pool get bent as they break the water surface. In quantum theory, the classical least action principle is upgraded to the path integral: the master equation enters in the form of the action, which appears as the phase controlling the relative importance of each of the possible paths.

Feynman was first introduced to the magic of the least action principle while still at high school. His doctoral dissertation was entitled "The Principle of Least Action in Quantum Mechanics."

QUANTUM MECHANICS MEETS RELATIVITY

Two sentences are enough to encapsulate an idea that has changed forever our understanding of our Universe. Firstly, the results of experiments performed in two inertial reference frames, one moving at a constant velocity relative to the other, must always be the same. Secondly, the velocity of light has the same finite value for all.

What's an inertial reference frame? More on that in a moment. Whatever it is, simply saying that experimenters will get the same results working in two different places seems like a tedious detail. And so what if everybody gets the same answer when they measure the speed of light?

For such mundane statements, they turn out to say a great deal. They are the basis for Einstein's theory of special relativity, the precursor of the general relativity theory which, by including accelerating participants besides just acceleration-free ones, describes gravity.

Special relativity features in the story of QCD because it lays down rules that QCD must adhere to if it is ever to be taken seriously as a theory of the strong force. Secondly, much of the way in which high-energy physics is expressed is a direct consequence of relativity, since the particle energies and speeds involved are so high. Finally, special relativity is at the root of our understanding of some important particle properties.

Jiggle a magnet near a stationary wire and you can make electric current flow in the wire. Alternatively, leave the magnet alone and instead move the wire, and this likewise causes an electric current.

Thus, with a simple discussion of moving magnets and wires, did Einstein begin his famous 1905 paper in which he unleashed special relativity on the world. His point about wires and magnets was that only *relative* motion mattered. So far as producing a current is concerned, it doesn't matter if wires or magnets are moving or stationary, just so long as they move relative to one another. It was this observation that led Einstein to his principle of relativity, that all inertial reference frames are equivalent for all physical experiments.

There's no real magic about the meaning of the inertial reference frame. A reference frame is a set of measuring axes against which an experimenter might monitor an experiment, for example the movement of some object. Without a reference frame, there's no way of telling if something *is* moving. An inertial reference frame is simply one in which a freely moving object experiences no acceleration relative to that reference frame. An Earth-bound laboratory is *not* an inertial reference frame, since everything inside experiences a downward acceleration due to gravity. The classic example of an inertial reference frame is a space ship drifting through empty space.

In this sense, inertial frames are a bit specialized. In practice, they are needed only in theory to explain relativity, which is one reason why discussions of relativity so often wind up talking about observers O and O' whizzing round in space ships. An inertial frame really just means a laboratory where there's no gravity or acceleration to obscure what's going on. Take two such laboratories, moving relative to one another at constant velocity, and experimenters inside will discover the same laws of physics. This is the content of Einstein's principle of relativity.

It means there is no one definitive place to do experiments, no great laboratory in the sky that is at rest in some universal sense. There's no place where experimenters produce measurements and definitive physical laws against which all others must be compared. It's a very deep, fundamental yet simple idea.

The second principle – that the speed of light is the same for all – likewise owes its genesis to electromagnetism. Maxwell's equations of electromagnetism reveal how an electric charge, when stationary a mere source of *electric* field, becomes in addition a source of *magnetic* field the moment it moves. Relativity enters with the observation that if, instead, the charge stands still and it's the experimenter that moves, the electric and magnetic effects seen by the experimenter turn out to be the same.

A hugely important consequence of Maxwell's work was the prediction that an electromagnetic disturbance should propagate as a wave through empty space, moving at the velocity of light. These electromagnetic waves, first discovered as such in the form of radio waves, include a whole spectrum, ranging in frequency from radio waves through microwaves to visible light and up to X-rays and gamma rays. For all, the velocity is the same, that of light, and can be written down in terms of fundamental constants of electricity and magnetism. Since there is no reason at all to expect those constants to depend on who's doing the measuring, the implication is that the speed of light is similarly constant for everybody. That the speed of light is the same for all, an incredible 3×10^8 m s^{-1}, has some odd consequences.

Imagine a stationary footballer kicking a ball at an unfortunate oncoming opposition player. Neglecting friction and air resistance, the second player will suffer the impact of a ball traveling at a speed equal to the speed the ball left the foot of the motionless player, plus the speed at which he himself was running. This kind of simple velocity addition idea is how the everyday world works.

However, picture instead a motionless white-coated scientist flashing her torch at a colleague running towards her. The running scientist measures the speed of the light from the torch. The answer turns out to be just the same as when both parties were motionless. What the second scientist measures is simply the book value for the speed of light, the very speed at which the pulses left the torch. By everyday standards, it's an absurd outcome.

But this is how the world really works, and the experiment has been done. For instance, instead of white-coated scientists flashing torches, experimenters have used particles called neutral pions, also called π^0 mesons, which decay into two photons. Using π^0 mesons traveling at a significant fraction of the velocity of light, experimenters see photons emitted in the direction of meson travel moving at the usual velocity of light, and not the sum of the velocity of light and that of the meson. Similarly, photons emitted in the backward direction also travel at the velocity of light, and not at some lesser value that depends on the meson's movement. The mechanics of everyday objects has no explanation for this.

Together these two principles, that experiments performed in different inertial reference frames always yield the same answer and the constancy of light's velocity, allowed Einstein to develop a whole new set of mechanics whose startling effects only become obvious at very high speeds.

Two experimenters using light signals to compare results provide a useful introduction to relativity at work. The expressions linking distance and time quantities in the two experiments are called Lorentz transformations, after the Dutch physicist Hendrik Lorentz, who derived them some years before Einstein's theory of relativity. So if one experimenter had a meter-long rule, then that same rule would appear to the other experimenter moving past to be a bit shorter that one meter, by an amount given by the appropriate Lorentz transformation. Correspondingly, the first experimenter can look at the second experimenter's meter ruler and will *also* see a *shortened* ruler. This is one consequence of the idea that no one reference frame is special compared to any other, so there is always a mind-boggling symmetry between two experimenters.

This effect, again a little hard to reconcile with everyday experience, is called length contraction. The original 3 km long linear electron accelerator at SLAC, the Stanford Linear Accelerator Center in

California, accelerated electrons to within a whisker of the velocity of light. At this speed, the length contraction effect is dramatic: a three kilometer accelerator seems about a meter long to such fast electrons.

Time does the same kind of thing. If both experimenters have clocks, then they can each look at one another's and *both* will see the other's clock running more slowly than their own. This slowing of clocks is called time dilation, and is commonplace in experimental particle physics where decaying particles appear to live longer than they should simply because they are moving through the experiment very quickly.

Lorentz transformations also give a new rule for combining velocities, so that at high speeds, velocities do not add in the simple way that they do in the everyday world, for example with the footballers. The new rule combines two velocities in such a way that the result never exceeds the velocity of light.

These strange features of the relativistic world are more important in particle physics than in everyday life because in the ordinary world things move slowly. It's only at speeds close to the speed of light that relativistic effects become important. The mechanics of the everyday world is actually the low-speed version of relativity.

Blend together time and space . . .
A fly buzzing round a room is a pretty slow, non-relativistic and everyday kind of object. To describe the fly's position in the room you need three numbers, one for how far along the room the fly is, one saying how far across the room it is, and the third to give its height above the floor: in other words, the inside of a room is a three-dimensional space.

A little while later, the fly's position is given by three different numbers, assuming that it has moved in the interim. Time, the waiting between sets of measurements, is separate from space (along, across, and up) in the everyday world. Einstein was the first to understand that space and time are *not* separate in this simplistic way, but are intertwined. It's only by comparing clocks and meter rules for two experimenters traveling at high speed relative to one another that this intertwining really becomes obvious.

Non-relativistic experimenters can easily agree on the time interval between the fly being over there and over here. They can also agree on the distance between where the fly was and where it is now. What they agree on is the overall distance given by the "distance rule," rather than the three individual numbers themselves. The distance rule says to square each of the three distance numbers, add the results, and then take the square root of the answer. It's a bit like Pythagoras' theorem for triangles. There's no trickery here – the distance rule just gives what you would measure if you took a rule into your lounge and did it yourself.

In special relativity, experimenters can only agree on a modified "distance" that combines *both* time and space intervals in a clearly prescribed way; they cannot agree on distance and time measurements in the usual sense, as separately these are modified in a way expressed by Lorentz transformations. In fact, Lorentz transformations are simply the set of transformations that preserve intact this modified distance rule. And true to form, as a set of transformations leaving intact some quantity, they form a symmetry group, in this case the Lorentz group.

The lesson from this, and one of the main points of special relativity, is that in situations involving high velocities it is tidier to think in terms of a *four*-dimensional space-time, rather than the old style three-dimensional space plus a fourth but distinct dimension called time.

Back to the room with the fly. A straight line between the place where the fly was and the place where it is now can also be described using a set of three numbers, a little bit along, a little bit across, and maybe a lot up if it was trying to escape. The line also has a direction, pointing from where the fly was, to where it is now, as though to provide a hint on how to follow it. A line given by a bunch of numbers and with a direction specified is called a vector, in contrast to a quantity such as the fly's mass that has size but no direction, and which is an example of what's called a scalar. In the fly example, three numbers means it's a three-vector. In fact, because three-vectors have been in use longer than all other kinds, they are often just called vectors, dropping the descriptive "three."

In special relativity, to make use of the modified distance rule that two experimenters can agree on means using vectors given by *four* numbers plus a direction. These are examples of four-vectors. Four-vectors are a basic quantity for discussing what's going on both in special relativity and in particle physics. A four-vector is a set of four quantities that transforms relativistically (i.e., under a Lorentz transformation) in the same way as the three space labels and the time label of an event. To ensure this, the additional portion that appears in the position four-vector, but not in the three-vector, is a product of the time interval and the velocity of light. Different experimenters moving at very high relative speeds can then plug four pairs of numbers into the modified distance rule to give a space-time "distance" between the two events that they can then agree on.

Apart from a vector that relates two points, a position vector, another important vector is velocity. Back to the fly for a moment; it would help to know both how fast the fly is moving, and in what direction, to allow the fly hunter to achieve his or her goal. Speed plus direction give a velocity vector, in this case a three-vector again, since only three numbers are needed to specify the fly's velocity.

If it were the fly's collisions with other flying objects that mattered, then a different vector, called momentum, would be even more useful than velocity. Mathematically, momentum is the product of mass and velocity; physically, it says how much of something is going somewhere and how fast it's doing it. It's momentum, not velocity, that is conserved when flies collide or particles interact. So when describing physical processes momentum is invariably far more useful than velocity.

This is certainly the case in high-energy physics. There, with particles zipping through the insides of detectors at speeds close to that of light, special relativity says that the smart thing to use is an extension of regular momentum called four-momentum, another example of a four-vector. Four-momentum turns out to be conserved in particle collisions at very high velocities, and forms part of the language physicists use when they discuss and measure particle interactions.

Why bother using four-vectors? The reason is that any relation expressed using four-vectors will hold for everybody, since at the root of four-vectors is an agreed distance rule. The form of the equation is unchanged by Lorentz transformations, meaning that if one experimenter determines a physical law and writes it down using four-vectors, that same equation will hold for all other experimenters. This rather handy property is called form invariance, or covariance.

Though the relation between four-momenta is retained, the measured values of the four-momenta will depend on who does the measuring. However, some quantities retain their exact numerical value in all reference frames. For example, the mass of a stationary electron will always yield the same numerical value for all. Two four-vectors, combined in the same way that the modified distance rule combined two position four-vectors, gives a number that is always the same regardless of who does the measuring. A four-vector combined with itself in this way to give the so-called square of the four-vector is a special case of this. The square of a particle's four-momentum always equals the square of the mass of the particle at rest. These are Lorentz invariants – they remain totally unchanged by Lorentz transformations.

To demonstrate that this is not really getting too abstract, it's worth pointing out that the way particle accelerator energies are written down uses this idea. For example, the CERN proton–antiproton collider, or Spp̄S (Super proton–antiproton Synchrotron), where both the W and Z bosons were found, had an energy of 630 energy units (more on those "units" in just a moment). But to talk in terms of "energy" of an accelerator can be a bit misleading, since a beam of particles colliding head-on with a second similar beam has a very much greater "useful energy" than the same beam hitting a stationary target particle. For this reason, particle physicists use a quantity called invariant mass, denoted $\sqrt{s}$, which depends only on the colliding particle four-momenta, and which has the same numerical value for all observers – it's a Lorentz invariant. Experimentalists often publish plots of particle collision data with energy $\sqrt{s}$ along the horizontal axis. The invariant mass of those colliding beams at CERN's proton–antiproton collider was 630 energy units. But a single one of those 315 energy unit beams striking a *stationary* target would have given a much more modest invariant mass, the useful energy available for particle creation, of just 25 energy units.

Results of particle collision experiments, and theorists' efforts to figure out what those particle collisions should look like, must not depend on who is doing the measuring; they must be the same for all. Really, this is just writing things down in a form that makes sure it all matches Einstein's demand that the laws of physics be the same in all inertial reference frames.

What it means for a theory of the strong force, and for theories of other fundamental forces for that matter, is that *only* Lorentz invariant theories get a look in. This is the first of many tests. Any theory that fails it is destined for the scrap heap. Quantum chromodynamics is a Lorentz invariant theory, so passes on to the next hurdle.

. . . blend mass and energy . . .

In all the world, there's one equation that "everybody" knows, and that's Einstein's mass–energy relation, $E = mc^2$. Here, E is energy, m is mass, and c the velocity of light. This natural consequence of the ideas introduced above expresses the fact that all energy has mass, and this too has some surprising implications.

A moving object has energy of motion, the energy available to crumple bumper bars and more in an automobile accident. The mass appearing in the mass–energy relation expresses the mass equivalence of this motion energy, in addition to the universally agreed "rest mass" of the object, its mass as measured when it's stationary: things become effectively heavier as they go faster, another fact of relativistic daily life in particle physics. In fact, the m appearing in the mass–energy equation is not the mass of an object at rest, but the mass of the moving object, which is larger. For example, the electrons in the beam of a television set appear fractionally heavier than the stated "book value" for the electron's mass, which applies to stationary electrons. And electrons whizzing around a particle accelerator appear much heavier still. This is one way of seeing why nothing can be accelerated through the speed of light barrier, as an infinite amount of energy would be required to do the job.

The mass–energy relation takes on an almost literal form given the special units that high-energy physicists use. Anybody has the right to use systems of units that make life a little easier. For example, to discuss journey times in seconds, the agreed standard unit of time measurement, would be painful: travel agents give flight times in hours, which is much more useful for journeys of sufficient duration to warrant a plane. On the other hand, seconds (and fractions of seconds!) are a good unit for the Olympic 100 m race. In high-energy physics, masses and distances are very small indeed. On top of that, because everything is relativistic and quantum, the speed of light and Planck's constant appear liberally scattered all through the equations.

Physicists have fixed all this up to make life easier, and their equations neater, by assigning both the speed of light and Planck's constant $\hbar$ the value of one. This has the effect of making both quantities vanish from all equations. It also means that rest masses of particles are written in terms of MeV or mega electronvolts, a unit of *energy*. For example, in these so-called "natural units," the proton has a mass of 938.3 MeV, equivalent to 1.673×10^{-27} kg in more conventional units. An electron has a mass of 0.511 MeV, or 9.109×10^{-31} kg.

So what is an electronvolt? It is the unit of energy gained by accelerating a particle having a charge equal to the electron's charge through a voltage of 1 V, and it's written as eV. For example, a 12 V

car battery could be connected to two metal plates a few centime-
ters apart. Electrons introduced into the vacuum by the negative plate
would accelerate to the positive plate, and would have motion energy
of 12 eV when they got there. An electronvolt is a pretty small unit
of energy: raising the temperature of a liter of water by 1 °C consumes
about 2.6×10^{21} of them.

A few electronvolts is the sort of energy relevant to electrons jig-
gling around in atoms. Nuclear physicists are more interested in higher
energies, like thousands of electronvolts (keV, or kilo electronvolts) and
millions of electronvolts (MeV, or mega electronvolts). Particle physicists are more at home with thousands of millions of electronvolts (GeV,
or giga electronvolts) and these days even millions of millions of electronvolts (TeV, or tera electronvolts). To return to our "energy units"
of the CERN proton–antiproton collider example above, that machine
had an energy of 630 GeV.

. . . and add quantum mechanics

Special relativity was already twenty years old when Schrödinger wrote
down his equation of quantum mechanics. It was clear from the outset
that Schrödinger's equation, for all its successes, would not sit easily
alongside relativity. The challenge of devising an alternative version of
quantum theory that did was met by a young British physicist, Paul
Dirac, and the result was spectacular.

A relativistic quantum theory needs to be covariant under
Lorentz transformations, must reproduce the relativistic relationship
between energy and mass, and give only positive probabilities. This
third condition features because the most obvious way to tackle the
issue founders on just this point, as it implies negative probabilities, a
very dubious concept. An additional problem with the obvious route
is the emergence of particles having negative energies, which is almost
as dubious.

In 1928 Dirac wrote down a new quantum mechanical equation
for electrons that satisfied the necessary relativistic conditions. He was
guided by the notion of incorporating space and time on an equal basis,
which they are not in Schrödinger's equation. Dirac's scheme only
worked with some inventive interpretation of the mathematics, and if
electrons have an attribute called spin. Added to that, there is an entire
second solution that threatens to undermine Dirac's equation, since
it implies negative-energy particles, though this time at least there's
no issue with negative probability. With extraordinary audacity and

insight, Dirac turned the negative-energy problem into a triumph: the "negative-energy" solutions describe a whole new type of object, the antiparticles.

First the electron spin. Dirac's equation and special relativity are not how electron spin entered physics. That electrons have spin was first suggested in 1925, by Dutch physicists Samuel Goudsmit and George Uhlenbeck, to explain the pairing of lines in atomic spectra – another example of the importance of spectra to the development of quantum theory. Spin would mean that electrons, having electric charge, would also then behave like little magnets, since moving charge is the origin of magnetism. The electric charge on the atom's nucleus, seen from the moving electron, is a source of magnetic influence. So the spinning electron is a tiny magnet in a magnetic field, and its energy then depends on whether its spin is aligned along the direction of the magnetic field, or in opposition to it. This results in two closely spaced energy levels where otherwise there would be just one. And this splitting of the energy level is what causes the pairing of lines seen in spectra.

Sodium provides a well-known example of spectral pairing. A spark through sodium gas gives off a distinctive yellow light that is actually not one spectral line but two, the sodium D lines. The corresponding wavelengths are 589.6 nm and 589.0 nm. Without electron spin, there would be a single line, with a wavelength about halfway between these two.

Special relativity shows that spin is a child of the space and time we live in, with its special symmetries encapsulated in the group of Lorentz transformations. One hint of this special relationship comes from the fact that the Lorentz transformations are nothing more than stretching and rotating transformations. Applied to the motion of an electron, for example, should the electron's four-momentum remain unchanged by a Lorentz transformation, then what's left are transformations according to the group $SU(2)$, which describe simple rotations: spin. Spin arises from the most natural description of Lorentz transformations, and in fact Dirac's equation can actually be derived this way.

Electron spin is also the child of the quantum nature of the Universe. It is this quantum aspect that imposes limits on the measurable values of the electron's spin. Planet Earth spins about its axis, and so far as classical mechanics is concerned, it could in principle spin at pretty much any rate about any axis. Electrons don't enjoy this level of freedom. Sitting an electron in a magnetic field provides a reference direction, conventionally termed the z-direction, against which the electron spin is measured experimentally. The result is that the electron can spin clockwise (taking electron spin rather literally) or anticlockwise about this direction. Moreover, the magnitude of its spin has always a fixed size, $1/2\hbar$. Conventionally, the $\hbar$ is dropped and the electron is said simply to have spin $1/2$. The result of a measurement of electron spin is either $+1/2$, termed spin up, or $-1/2$, which is spin down. There is another useful way of describing spin. Spin up is also called clockwise or right-handed – a thumbs-up sign made with the right hand will give the thumb in the direction of motion for a moving electron, while the curled up fingers show the sense of rotation. Spin down is anticlockwise or left-handed. That there are just two *discrete* values corresponds to the splitting of single lines into double lines in atomic spectra.

The magnitude of $1/2$ for the electron spin corresponds to the experimentally observed value of the strength of the interaction between the electron and a magnetic field in which it resides. The splitting of levels into two indicates two spin states, and shows that spin does not take on continuous values, but only discrete values, for which reason spin is said to be *quantized*. The numbers associated with the electron's spin attribute are an example of quantum numbers, experimentally measurable labels of a quantum state.

The quantum measurement rules not only limit, to specific values, the outcome of measuring spin along a given direction, but they also limit the prospects of *simultaneously* measuring spin along some other direction. Having sat an electron in a magnetic field, and thereby giving it a reference (z) direction, and then measuring its spin as $+1/2$ or $-1/2$, means that, according to the rules, the option of precisely knowing the electron's spin along the x and y axes at the same time is lost.

A stream of silver atoms sent through a specially designed magnetic field clearly shows these quantum restrictions on the electron's spin, or rather the experimenters' knowledge of it. Silver atoms have a nucleus surrounded by 47 electrons, 46 of which form a symmetric spherical cloud about the nucleus. The spin of the 47th electron is what gives the silver atom its magnetic moment, and thus controls the atom's response to an applied magnetic field.

What all this means is that when a beam of silver atoms passes through a magnetic field that has a change in strength across the beam direction, i.e., a magnetic field gradient, there's a force on each atom

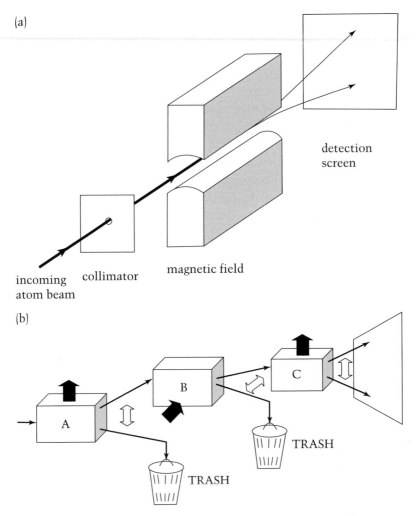

(a)

detection
screen

incoming   collimator   magnetic field
atom beam

(b)

A

B

C

TRASH

TRASH

FIGURE 3.3 (a) The Stern–Gerlach experiment. A beam of silver atoms
is split vertically into two parts by a magnetic field gradient. (b) A
sequence of three such experiments, the middle one having a magnetic
field oriented at right angles to those of the other two. Each box
represents a beam-splitting magnet assembly like that in (a).

proportional to the electron spin, Figure 3.3(a). With the magnetic field
arranged as depicted, this force depends only on the combination of the
magnetic field gradient, and that portion of the electron spin aligned
along the z-direction, called the z-component of spin. So the magnetic
interaction, being sensitive to direction, picks out only that portion of
the electron spin parallel (or antiparallel) to the field gradient.

What would an experimenter see? The spin of an electron that behaved classically would have any one of a complete range of possible z-aligned portions, so correspondingly the different atoms in the beam would experience any of a smoothly varying range of forces in the z-direction, and the output would be a single, smoothly varying peak at the detector. However, this experiment has been done, originally by Otto Stern and Walther Gerlach way back in 1922, and this is not the kind of thing they found. In their classic experiment using beams of silver atoms, they arranged for a single line of silver atoms to be deposited on a glass plate in the absence of a magnetic field. When they then switched on the magnetic field, in place of a single line they saw a double line of deposited atoms. Classically, the single line would have become a broad smudge: Stern and Gerlach showed that the magnetic field splits the atom beam into *exactly two* separate components. They were pretty smitten with their results, and fired off a postcard to Copenhagen's Neils Bohr, who that same year won the Nobel prize for his work on the structure of the atom. Their postcard showed an image of the single line of atoms minus field, and alongside it an image of the split line with a field.

The corresponding z-components of the electron spin are simply $+1/2$ or $-1/2$, in the usual units of $\hbar$, and not some arbitrary value between the two, as would be expected from classical mechanics. In other words, such a measurement finds that electrons have only spin up or spin down, and no in-between state. Prior to the measurement, a particular electron is in an indeterminate state of spin up and down. The act of measurement pushes it into one or other possible state. Once a particular electron has been found in, say, the spin up state, a second identical measurement soon after will find it still in the spin up state.

But what if that second measurement is not identical? This can be demonstrated using a sequence of three Stern–Gerlach experiments, Figure 3.3(b), where the output of one experiment is the input of the next. Each box is simply a representation of the set-up of Figure 3.3(a). The spin down electrons leaving box A are dumped, and the spin up electrons passed to box B. This has a field gradient oriented horizontally, in the y-direction. Leaving this box, electrons divide evenly between those having spin oriented along the y-direction, and those with spin oriented opposite to the y-direction. The "opposite" part is dumped, and the remainder fed to a final box C, which is identical to box A. The final output, from box C, shows equal numbers of electrons having spin up *and* spin down, measured relative to the z-axis.

The beam entering box B is 100% spin up, yet the beam leaving box B contains 50% spin *up* and 50% spin *down*, according to box C – some spin down seems to have appeared from nowhere, as if by magic. This demonstrates that, although box B gives a clear measurement of spin along the y-direction, inside box B spin up and spin down (relative to the z-axis) have become "confused" by virtue of the measurement made there: inside the one box, measurements of spin along the z-direction and along the y-direction cannot be made unambiguously at the same time.

In other words, the "measurement" performed by box B destroys all knowledge of the beam gained via box A and the trashing of a portion of its output. Then a further measurement by box C in turn destroys the knowledge of the beam that enters box C. What this shows is that a particular box throws the electron into one or other of just two possible quantum states, as determined by the field direction, independent of the history of the beam.

Only a single quantum number is needed to characterize the magnetic response of the silver atom – the outermost electron's spin as measured relative to the selected axis. The value is either $+1/2$ or $-1/2$. There is no second quantum label, corresponding to spin determined along a second direction. This property is mirrored in the underlying symmetry transformations associated with spin, the $SU(2)$ group. For this three-parameter group, there are transformations involving just one parameter that preserve the states labeled by the quantum number. It is this subset of transformations, in terms of a single parameter, that corresponds to the single quantum number – an indication of the deep link between quantum theory and group theory.

Picturing spin is quite tricky, since it is inherently a quantum property. While thinking of electrons as spinning tops is quite useful at times, such a simple "classical" picture is quite misleading. In fact, a classical mechanical "electron ball" would have to spin so fast to reproduce the electron's properties that parts of it would move faster than the speed of light, and that's against the rules. Nobody actually looks at electrons and sees them literally spinning clockwise or anticlockwise.

There's much more to spin than splitting of lines in atomic spectra – it's the basis for the great divide between particle types. Particles such as electrons and any others having a half-integer spin ($1/2, 3/2$,

5/2, and so on) are termed fermions, after the Italian physicist Enrico Fermi. Those having zero spin, or integer spins 1, 2, 3, and so on, are termed bosons, after the Indian physicist Satyendra Bose. In general, a particle of spin 1, such as W and Z bosons, has three possible discrete spin states, labeled +1, 0, and −1, again a feature of the quantum world. Photons, despite being so common, are in one sense a special case. They have zero rest mass; for this reason one of the possible spin states, the state labeled 0, is not available to them: photons, though spin one, have just two spin states.

Next, the disaster-turned-triumph in Dirac's equation. The preconditions Dirac set out to meet with his blend of relativity and quantum mechanics forced him to a second solution of his equation that appeared as a partner, by symmetry, to the first. It seemed to correspond to a particle identical to the electron but having a negative energy.

Dirac's stroke of genius was to reinterpret these negative-energy solutions as positive-energy *antiparticles*. The electron, according to Dirac's idea, should have an antiparticle partner of identical mass and identical spin but opposite electric charge. This object, the positron, was discovered in 1933 and Dirac was vindicated.

Because antiparticles, according to Dirac's theory, arise as a consequence of space-time features and the quantum nature of the Universe, they are not limited to partnering electrons. The neutrino, the special left-handed player in weak interactions, is another spin 1/2 particle that should and does have an antiparticle partner, the antineutrino. Likewise the proton has an antiparticle partner, called the antiproton: antiprotons, which have a negative electric charge and a mass identical to that of the proton, were discovered in 1955; antineutrinos were discovered in 1959.

The enormous success and importance of Dirac's equation is reflected by its presence on a plaque in London's Westminster Abbey, unveiled in 1995 in honor of Dirac, who died in 1984 and is buried in Florida. The plaque is set in the floor of the abbey alongside Newton's grave.

Quantum mechanics meets special relativity and the result is relativistic quantum mechanics. This leads to an understanding of particle spin and to antiparticles. Relativistic quantum mechanics also explains some fine detail of atomic spectra, and a constant relating the strength of the electron's magnetic moment to its spin. That constant has a value of two, which must be inserted "by hand" in non-relativistic quantum mechanics, but which emerges as a prediction from Dirac's theory.

Actually, the measured value of this constant is a shade more than two. Dirac's equation, for all its strengths, actually falters here. It also runs into trouble with the Lamb shift, which, according to the Dirac equation, should not exist. And there are other, deeper issues that limit the Dirac equation's usefulness as a full theory of fundamental forces. Dirac's equation is on the way, but it is *still* not enough.

QUANTUM BIRTHS AND DEATHS

Dirac's antiparticles have themselves become an aid to physicists hunting for new particles or exploring the interactions of known particles. For example, experimenters at KEK (Koh-Ene-Ken), Japan's National Laboratory for High Energy Physics at Tsukuba, 60 km north of Tokyo, smash 8 GeV electrons and 3.5 GeV positrons together in a 3 km-circumference energy-asymmetric collider, called KEK-B, creating particle showers in the process. And at Fermilab in Batavia, Illinois, protons and antiprotons collide head-on in the mighty Tevatron accelerator, likewise producing particle showers. In both examples, a particle and an antiparticle meet and die, and whole showers of other particles and antiparticles are born in their place.

Now this is awkward for "everyday" quantum mechanics, the kind of quantum mechanics used to describe the energy levels of hydrogen atoms and the like. The reason is that there is nothing in the simplest formalism that can accommodate the creation or destruction of particles. Yet these are basic features in so many high-energy particle interactions where two particles strike one another, are destroyed, and other particles emerge from the ashes.

Creation and destruction of particles is everywhere. For instance, a high-energy gamma ray photon, perhaps shed from a slowing charged particle in the Earth's atmosphere, itself a daughter of a "cosmic ray" from space, can spontaneously expire in favor of an electron–positron pair. This process, called pair production, is commonplace in particle physics, representing one of the standard mechanisms for the absorption of gamma rays in matter. And when an electron and positron meet they can annihilate one another, creating two photons in their place, a well-known process called pair annihilation.

Even the simplest atomic transitions contain the germ of the problem. When an electron absorbs a light photon and jumps to a higher level, where has the photon gone? And when an electron drops down an energy level and a photon is emitted, just where does that photon come from? In the former, a photon vanishes, in the latter it is somehow created, but where and how? This, loosely speaking, is the original particle creation and annihilation problem faced by the pioneers of quantum theory. They eventually resolved the enigma, and in so doing created quantum field theory: three of nature's four fundamental forces are now understood in terms of quantum field theory, with only gravity remaining outside the quantum field theoretic fold. The most successful quantum field theory is QED, which forms the template for the other two, QCD and the electroweak theory. So what is quantum field theory, and how does it explain quantum births and deaths?

Quantum fields

A "field" is some quantity or influence that exists over a region, rather than solely at a simple point. Weather maps, for example, illustrate the basic idea quite nicely. The temperature varies over the whole region covered by the weather map in a smooth and continuous way – a temperature field exists across the region. In principle the temperature at each and every point could be marked in. Likewise every spot on the weather map has wind, a second field spanning the same region, and depicted for a selection of locations by both numbers for the wind speed, and by arrows for the wind direction.

Wind and temperature fields are actually examples of two different kinds of fields. Temperature has a size, but no direction, so is a scalar, and a temperature field is an example of a scalar field. Wind, on the other hand, has both a size and a direction, so gives rise to a vector field. This is effectively represented as a region covered by little arrows whose direction represents the wind direction (direction of the vector) and length, the wind speed (size of the vector).

There are examples of fields everywhere. Planet Earth is, amongst other things, a giant bar magnet with quite literally a north and a south pole. The resulting magnetic field can be detected all around the Earth by a magnetic compass. Overhead power lines that convey electricity from power stations to consumers are sources of electric fields, as can be demonstrated (but should not be!) by holding a fluorescent light tube vertically under the lines – the light glows. Both magnetic and electric fields, the two faces of the electromagnetic field, are vector fields.

A basic theory of the interaction between an electron and the charge on an atom's nucleus is really a theory of the interaction between the electron and an electromagnetic field. Simple quantum theory explains atomic structure by obliging electrons to obey quantum rules, yet leaves the electromagnetic field as a simple classical field free from quantum impositions. Though this approach can get terrific results in chemistry and atomic physics, it's plainly not the whole story, since the electromagnetic field should also obey quantum rules. It was, after all, in electromagnetic radiation where Planck first noticed the quantum properties of nature.

How then is the electromagnetic field to be represented in a quantum world? Since electromagnetic energy comes in little discrete energy packets, or quanta, called photons, it's reasonable to expect that the field is a collection of photons. And since there's no reason to assume that all the photons are of the same frequency, all possible frequencies must be allowed. Sometimes some frequencies may be favored over others, meaning that the number of photons of different frequencies will be present in different numbers. A picture emerges of the electromagnetic field as made up of various numbers of photons of different energies.

There's a musical analogy. With all due respect to Beethoven and others, the output of an orchestra is just a pressure wave in air called sound. It's composed of contributions from a number of different instruments, present in different numbers, giving the sound a complicated make-up of different frequency waves. Soft, almost constant passages give a nearly even sound through the auditorium; a crashing chord is a pulse that blasts through it. In an analogous way, electromagnetic fields and pulses are the sum of contributions of electromagnetic waves of different frequencies. In this picture, the analogy of a single photon is the sound of a trumpet or some other one instrument that almost drowns out the rest of the players.

The big difference is that one is a quantum system and the other is not: a trumpeter can just play louder, but a light quantum cannot be made any brighter. In the quantum electromagnetic field, the different frequencies are photons of different energies, and a larger contribution of a particular frequency means more identical photons with that same energy. By contrast, in the sound wave case, a larger contribution of a particular frequency is simply a bigger wave – there's no notion that the contribution of a particular frequency is as multiples of some basic unit, the quantum.

Absorption and emission of photons then becomes easy to understand. When a photon is absorbed, it simply adds one unit to the assemblage of photons that make up the field. Count plus one for the number of photons of that energy in the field. When a photon is emitted, it is a photon lost from the assemblage, with transitions from different levels giving emission of different photons. This is how quantum field theory pictures particle creation and annihilation, at least for the electromagnetic case, and this is how Dirac introduced quantum field theory in 1927, even before his relativistic quantum mechanics.

Representing a field as a sum over quanta that obey quantum rules is termed field quantization, occasionally and less appropriately termed "second quantization." Whichever, it's a *field* that's being quantized, as distinct from the quantizing of the motion of single *particles* such as electrons in quantum mechanics.

Going full circle, since field quantization provides a natural way to understand the appearance and disappearance of photons, can the same idea be applied to electrons and positrons? They too are created and destroyed, for example in pair production. Another example is radioactive decays, in which an electron and antineutrino appear "from nowhere." So not only should quantum field theory explain these effects, it should also put all the players – electrons, positrons, and photons – on an equal theoretical footing.

Because quantum field theory applied to photons is based on Maxwell's equations of electromagnetism, the theory automatically satisfies the conditions of special relativity. Those same conditions must also be met by a *field* that replaces *point-like* electrons and positrons. Since point particles obeying the Dirac equation satisfy the constraints of relativity, quantum field theory builds on this by introducing a new field, the Dirac field, which satisfies the Dirac equation for fields. Then the Dirac field is an assemblage of electrons and positrons, the field quanta, just as the electromagnetic field is an assemblage of photons. The electromagnetic field satisfies Maxwell's equation; the Dirac field satisfies Dirac's equation. An electron and positron colliding and vanishing into a pair of gamma ray photons, for example, is an electron and positron vanishing from the Dirac field assemblage, and two photons appearing in the electromagnetic field assemblage.

These quantum fields are very different from classical fields. A classical field, for example wind or temperature, or electromagnetic fields treated classically, are smooth and continuous throughout space. There is a simple meaning to the question "what is the field strength at this point?" Not so for quantum fields. Though they too are spread throughout space, they are not smooth and continuous, but have a grainy structure when "viewed" close up, with smaller and smaller regions exhibiting greater fluctuations in field strength. The uncertainty principle reigns: measurements made in brief time intervals will give uncertain values for energy, and precise measurements of position will give uncertain values of momentum. At best the strength of a quantum field is an average value over some finite region of space and time.

A quantum field is an assemblage of quanta: a single particle is a fluctuation of the field, a region where the field is very strong, the probability high for creation of quanta with an energy in some narrow range dictated by the uncertainty principle. Loosely, a particle is equivalent to a localized swirl of wind in an otherwise relatively calm region, but a swirl that can only have a quantized energy. Stretching this picture to the limit gives a way of imagining pair production. A swirl of wind – the photon – makes contact with the surface of the ocean where it whips up the water into a pair of little whirlpools – the electron and positron. A "quantum" of one field species has died, and two "quanta" of a different field species have been created, with energy and momentum passed between the two.

The energy of the field changes by the addition or removal of quanta. Quanta of every possible energy contribute to the final field by an amount that is the probability of their being created or destroyed. So a high probability for creation of, say, photons of a particular energy means that the electromagnetic quantum field will be rich in photons of that energy.

One way of viewing quantization is the imposition of the special quantum rules on these probability factors that control the relative contributions of different quanta. This rule-based approach is called canonical quantization, where in this context the word "canon" means general principle, rule, or law. Proceeding via this route for QED, for example, means writing down a classical master equation of the theory from which action-minimizing recovers both the Dirac equation and Maxwell's equations for their respective fields. Both types of field are represented as sums of all possible quanta; the quantization step is the enforcement of the rules to be obeyed by the probability factors.

There are many technical problems with this rule-based approach when it comes to getting out some useful physical predictions, however. Luckily, there is an alternative – the path integral, Feynman's revolutionary innovation that he first introduced in the context of the quantum mechanics of particles. In quantum mechanics, the path integral gives transition probabilities for going from one particle quantum state to another. In field theory, this becomes the probability of a transition from one configuration of field quanta to another, via all possible intervening paths.

So there it is, quantum field theory, built on the twin pillars of special relativity and the quantum nature of the Universe. The basic entity is the field, a collection of quanta. Particles, regions where the field strongly fluctuates, correspond to localized configurations of quanta. There is a different species of field for each fundamental particle type, for example a photon field and an electron–positron field. The theory can accommodate other bosons in much the same way as it handles photons, and similarly other spin $1/2$ fermions in a way that parallels the treatment of electrons and positrons, so the formalism of quantum field theory can go further that just quantum electrodynamics. Quantum field theory begs us to believe that the tough, interactive material world all around us is actually a seething mass of fields of a few basic kinds that behave in roughly similar ways. It's a bizarre picture, but it seems to be true.

Quantum field theory is not so much one theory as a whole class of theories. Once the framework has been set up there's little room for maneuver. The only real remaining freedom is the internal symmetries of the fundamental fields involved. The prescription for quantum theories of fundamental forces is the same: take quantum field theory, add symmetry, and stir.

Particles united and divided

Pair annihilation is a good way of seeing why a quantum field theory must embrace both electrons and positrons, and indeed other particle and antiparticle pairs. If an electron could be created or destroyed alone, or in pairs, then electric charge could be created or destroyed. There's no evidence that this happens in nature, and electric charge is understood to be a conserved quantity. This means that the total charge of a system remains constant. So if an electron is destroyed somewhere, then to preserve charge conservation a positron must also be destroyed. Similarly, electrons and positrons can only be created as a couple. In fact, a workable quantum field theory featuring the Dirac equation for fields predicts both electric charge conservation and antiparticle partners to charged fermions.

And that's not all. Dirac's original relativistic quantum mechanics inspired his idea of antiparticles as a way of making sense of negative energies. In field theory, negative energies no longer make an appearance yet antiparticles remain – an altogether more satisfactory situation.

However, this in turn implies that fermions exhibit a strange and mysterious property: no two electrons, or any other fermions for that matter, may occupy the same state. It's as though an electron takes up a seat on a subatomic bus and thereby prevents any other identical electron from occupying that space. It's this rule, that electrons cannot all sit on one another's laps in the one front seat, which ensures the electrons in an atom inhabit different states, even though they might like to all fall into the state of least energy.

Mathematically, a combination of two fermions A and B is the negative of the same two fermions swapped over – two fermions form an antisymmetric state. In other words, A combined with B is equal to minus the combination of B with A. Fermions are antisymmetric under the interchange of two particles.

A little vignette might help. You're driving in your automobile, going on vacation. It's a long journey, and the twins in the backseat, one on either side, are crabby and fed up. You get them to swap places, and now mysteriously they are happy. To the casual observer nothing has happened to the location of the children – they are twins remember – though the final (happy) state is the "negative" of the initial (unhappy) one.

This antisymmetry property of electrons and other fermions means that two identical fermions combined in the same location would then imply the fermions are to be equal to the minus of themselves – A combined with A is equal to minus the combination of A with A, which doesn't make sense. This is a whole new kind of field, quite different from the scalar and vector fields of everyday experience.

This apparently bizarre behavior seems a high price to pay for the satisfaction of seeing negative energies vanish but antiparticles stay. In fact, what appears here in tandem with vanishing negative energies is a famous and powerful postulate due to the Austrian physicist Wolfgang Pauli, called the exclusion principle – no two electrons in an atom can occupy the same quantum state. Pauli introduced his principle in roughly these terms early in 1925, before the advent of modern quantum theory or the introduction of the idea of electron spin. His motivation was simple: there had to be something to prevent all the electrons in an atom collapsing down to a single lowest state. Pauli knew that even unexcited atoms have electrons in a series of different energy levels, and his exclusion principle was a way of explaining why they stay there. So Pauli's exclusion principle keeps electrons – and other fermions – from invading each other's space.

At the time, Pauli could not give more explanation of his rule than that. It was an ad hoc device to make sense of an atomic model that showed good agreement with experiments. Only much later did Pauli and others see that the exclusion principle appears through quantum field theory both as a result of insisting that energy remains always positive and as an explicit consequence of the quantum nature of the theory.

Besides its importance in understanding electrons in atoms and the whole of the table of different elements in nature, the exclusion principle is also vital for understanding the structure of matter at the finest levels. The exclusion principle signals the great divide of nature's particles into two camps, bosons and fermions. There *is* no exclusion principle for photons: in contrast to electrons, photons can be piled together without restriction. Photons on the subatomic bus *can* all sit together on the same front seat. On this basis, it's easy to see that, *en masse*, photons and electrons will behave very differently. For example in lasers, photons act together in a way that would be impossible for electrons, irrespective of effects due to electric charge. In line with their division into two camps, nature's fundamental particles obey two systems of "population statistics" for describing their assemblies: fermions obey Fermi–Dirac statistics and bosons obey Bose–Einstein statistics.

Fermi–Dirac statistics are the basis of our understanding of electrical conduction in metals, which is due to a "gas" of effectively free electrons floating in the metal. Bose–Einstein statistics reproduce Planck's law for the emission of light from an almost closed box of photons and are one of the main ingredients in understanding lasers.

What makes laser light so special is that laser light photons lose their individual identity and coalesce into a single state, something that only bosons can do. One consequence of this is that laser beams can be focused into much finer spots than light from ordinary sources, and it's this tight focusing that makes laser light so useful.

Lasers are not the only example of the special properties of bosons. Atoms of the most abundant variety of helium, called helium-4, are bosons. When helium-4 is cooled to about $-271\,°C$, it becomes super-fluid, meaning that it loses all resistance to flow. This is because the atoms coalesce, or condense, into the same quantum state. One conse-quence of reaching this state is that superfluid helium-4 can creep out of its beaker as if by magic. The very rare helium-3, its atoms fermions, has a markedly different behavior.

In 1995 Eric Cornell, Carl Wieman, and their colleagues at the Joint Institute for Laboratory Astrophysics in Boulder, Colorado, cooled a gas of rubidium atoms to within a few thousand millionths of a degree above absolute zero, the theoretical minimum temperature with a value of about $-273.15\,°C$. With the atoms trapped as a cloud using a system of magnetic fields, they witnessed a sudden and dramatic increase in the density of atoms in the middle of the cloud. What they were seeing was the very first true demonstration of Bose–Einstein condensation, free from any other interactions that tend to obscure the effects of con-densation in cases such as helium-4. Rubidium atoms formed into a blob of atom-matter, as distinct from a more normal collection of sep-arate atoms. This is a new form of matter in which individual atom identities are lost in much the same way that individual photon iden-tities are lost in laser light. And like laser light, it's a manifestation of Bose–Einstein statistics. Cornell and Wieman won the 2001 Nobel prize for physics for their feat, shared with MIT's Wolfgang Ketterle who achieved a similar result using sodium atoms the same year.

The divide of particles into two camps depending on spin, little more than an empirical fact without quantum field theory, runs even deeper than particle-counting statistics. Fermions are matter particles, the "stuff" of the Universe. Bosons are force-transmitting particles. For example, for a hydrogen atom, the proton at its core, a fermion, holds tight to an electron, another fermion, through the mediation of photons, which are bosons. The rule is quite general: bosons transmit force between fermions.

FEELING THE FORCE

How does a charged particle feel another nearby charged particle? How does a planet know to respond to the gravitational field of the Sun? How does one strongly interacting particle sense the presence of another?

In Newton's world, force was an instant business. A wobble in a planet's orbit would give an instant response in another. With Einstein, things changed. A wobble in a planet, or an electric charge, would give a response in another planet, or in another electric charge, only after the wobble had had time to propagate between the two at the velocity of light.

Where does this leave conservation laws? The energy and momentum transferred between the two planets or charges resides in the field between them for the duration of the transfer, a convincing way of seeing the need for fields. In a quantum world, these force-carrying fields comprise force-carrying quanta. The force between two objects becomes the exchange of force-carrying quanta between them. This is the basic idea of exchange forces.

If nature really permits particles to be created and destroyed, as quantum field theory suggests, does this not open the floodgates to chaotic particle vanishings and appearances? In short, yes. Amongst other things, it forms the basis of exchange forces.

Electrons, as carriers of charge, are sources of electromagnetic field. In quantum-speak, an electron is a source of photons, emitted and reabsorbed by the electron via particle creation and annihilation, and forming a seething cloud around it. Two electrons interact by the transfer of photons between the two "photon clouds" associated with the two electrons. This photon exchange replaces the classical notion of electric field permeating the empty space between two charges.

An exchanged photon changes the momentum of each charged participant. This is manifest as a tiny force between them. Lots of exchanged photons each contribute to the measurable push and pull between two charged particles.

However, there's something odd about the exchanged photons. Each photon emitted or absorbed should result in a change in the electrons' energy and three-momentum. But for regular "real" photons, like those from a light bulb with their zero mass, the numbers do not stack up correctly to guarantee energy and momentum conservation.

The resolution lies in the uncertainty principle. Energy conservation as a photon leaves or arrives at a charged particle is relaxed for a brief interval of time. The amount by which energy conservation is relaxed is linked, via the uncertainty principle, to the interval of time over which the relaxation can occur. The larger the energy borrowed, the less time the borrowing may last.

Similarly, momentum conservation is relaxed to a degree dictated by the momentum–distance version of the uncertainty relation. The greater the amount of momentum borrowed, the less distance over which the borrowing may occur.

Detecting actual quanta with relaxed conservation of energy and momentum would mean violating the sacrosanct energy and three-momentum conservation laws. Acknowledging this difficulty, particles participating in this "particle exchange" are not manifestly observable in some naive sense. Physicists call them virtual particles. Virtual particles are a totally quantum phenomenon.

Thus, quantum field theory offers a whole new interpretation of force as the exchange of force-carrying particles. Photons are exchanged between charged particles. Gravitons, the purported quantum of the gravitational field, are exchanged between objects tugging at one another through a mutual gravitational attraction. And weak interactions involve the exchange of W and Z particles.

Force-carrying candidates

The idea that a stream of exchanged photons could explain the force between charged particles was introduced by Fermi and the German–American physicist Hans Bethe in 1932. Three years later, the Japanese physicist Hideki Yukawa developed the idea into the first field theory of the strong force.

Yukawa reasoned that the strong force is short range since it is at play inside the atomic nucleus. Therefore, it should be carried by an exchange particle having a substantial mass. The particle's mass would mean that it could borrow energy for only a very short time, as dictated by the uncertainty relation, and in that short time could travel only a small distance. So the mass of the new exchange particle postulated by Yukawa should have a value directly related to the range of the strong force, roughly the radius of the proton. For a distance of a couple of fermi, where a fermi is set as 10^{-15} m, Yukawa's exchange particle should have a mass of about 100 MeV. So it's no surprise that a particle of mass 106 MeV, discovered two years later, was immediately seized on as Yukawa's heavy strong force carrier, and served to bring Yukawa's ideas into the limelight.

Yukawa was lucky. In fact, the object discovered by Carl Anderson in 1937 is now known as the muon, or μ, the electron's cousin introduced above, and which has no role in strong interactions. But in due course Yukawa's strong force carrier *was* found, and christened the pion, or π meson. Charge-free pions have a mass of about 135 MeV. Pions with positive or negative electric charge are heavier, at 140 MeV.

Pion exchange is still used to describe how the strong force works at distances a bit in excess of a fermi. That is, pion exchange is more part of nuclear physics, where the challenge is to understand how protons and neutrons combine into atomic nuclei. Physicists have come to realize that pions themselves are not actually elemental, but in turn are built of even smaller component parts. The way those components combine into pions is described by QCD.

Pions are not the only heavy carriers of force – weak interactions are also very short range, and are mediated by the heavy W and Z exchange particles. And QCD, the theory of the strong force, presumably also has heavy, fundamental force-carrying particles, since the strong force is short range. Right? Wrong! There are indeed fundamental force-carrying particles in QCD, but they turn out to be massless. There's something new and unique to QCD to account for the short range of the strong force.

Virtual particles

A classical or even relativistic quantum electron is a lonely, solitary object. However, quantum field theory says that the electron really has an entourage, a welter of virtual particles leaving the electron and then being absorbed. Dominating this virtual cloud are photons, living on borrowed time and energy, and giving rise to what is perceived classically as an electric field. Likewise a photon also has its cloud of virtual transients, electrons and positrons that in turn are sources of other, virtual, photons.

When an electron strikes a proton, for example, the two interact electromagnetically by the exchange of virtual photons. It's as though virtual photons from the cloud around the electron sniff out and find a new home, the charged proton or some charged object inside it. So this collision process is mediated by virtual particles having only a fleeting, unreal existence, and which obey only uncertainty-relaxed energy and momentum conservation rules.

What experimenters see when they collide an electron with a proton is a scattered electron, with a modified direction and possibly reduced energy with reduced four-momentum, plus a recoiling proton. The difference between the electron's initial and final four-momentum passes, by way of virtual photons, to the proton or the proton's constituents. The four-momentum carried by those photons is termed the momentum transfer.

Usually, the square of a particle's four-momentum is the rest mass of the particle concerned, in which case it's automatically positive. The square of the momentum transfer in the case of an electron colliding with a proton, however, comes out to be negative. So high-energy physicists use the *negative* of the momentum transfer squared to give a final positive quantity. This they denote Q^2; it is a standard quantity used throughout particle physics in virtually every discussion of collisions.

A real photon, one that pings across the room from a light bulb, has a four-momentum squared of zero, equal to its zero rest mass. A virtual photon, on the other hand, borrows energy for an instant and becomes massive, so its four-momentum squared is not zero. That the value is negative is related to its unphysical nature; nobody can "see" a virtual photon.

A virtual photon involved in the collision thus has a mass of sorts, but only for a short time. The greater the mass of a virtual particle, the shorter its lifetime and the smaller its range. For this reason, the larger the momentum transfer, the deeper the photon probes into the target's structure. The momentum transfer is, in effect, the resolving power of the electron scattering "microscope," and the higher the momentum transfer, the finer the detail revealed.

This is why Q^2 – sometimes simply called momentum transfer for brevity – appears so frequently in experimental plots of particle collisions, and in theorists' calculations: it's a measure of the magnifying power of particle accelerator microscopes. As the magnification increases, experimenters can see the structure of matter ever more clearly.

GETTING SOME ANSWERS

What do theorists do when the calculation is too hard, and they don't know how to do it? The ghastly truth is that this is the norm, not the exception. Real physical problems that can be worked through to give a neat, exact mathematical equation as a final solution, what mathematicians call a closed form analytic solution, are rare indeed. Most problems are just too difficult.

So what to do when the problem is too hard? There are three basic strategies, apart from giving up altogether or trying harder to solve the problem directly. One is to do a simpler problem, perhaps a cut-down version of the original. Another is to do the original problem on a computer, a strategy called simulation. The third is to use approximations.

Why bother doing a simpler problem? The answer is that the simpler case will, hopefully, give insights into the real problem, or perhaps can be extended to the real problem. Maybe it will show some aspect of the physics, which is the thing of ultimate interest; maybe it will show how the real problem can be solved, or at least illustrate why it can't.

A skill here is in picking the simpler problem, in addition to solving it. A simpler problem that is so far removed from the real problem as to stretch credibility is called a model. Models and simplified problems abound all through physics, and some of them take on a life of their own. A simplification, for example, might be the assumption that the atomic nucleus has an infinite mass, reasonable given that the nuclear mass is vast compared to that of an electron. An example of a model, important in the study of bulk matter, is the famous Ising model, a simplistic lattice of interacting spins for which theorists can derive exact solutions.

Massive computing power, now plentiful and cheap, opens up whole new vistas. The collective forces acting within a pile of sand grains are not so simple as one might think. One way to try to understand what's happening is to feed to a computer all the basic mechanics of how sand grains slip and slide and pressure one another, then let the machine churn through the thousands of equations a step at a time to model how a real sand pile behaves. This problem, one example of a computer simulation, is actively researched by people interested in piles of everything from grain to iron ore. Huge computer power also opens up the exciting prospect of simulations in QCD, called lattice QCD, the subject of Chapter 10.

Then there are the approximations. If the problem is too hard, it may be that some mathematical trickery will give an approximate answer that's still good enough. One such piece of trickery, which features in the story of QCD, is perturbation theory. Perturbation theory is not new, nor is it inherently part of QCD, or even of particle physics. In fact, it's pretty well as old and widespread as mathematics itself. It owes its origins to the study of planetary motion, and to people like Isaac Newton, Joseph Lagrange, William Hamilton, and others who long ago developed and perfected the basic mathematics.

Planetary motion is as good a place as any to see what perturbation theory is all about. For example, the motion of a satellite around the Earth can differ from a simple circular orbit on account of the Earth being slightly fatter round the equator. Perturbation theory involves breaking the description of the process, in this case a satellite around the Earth, into two portions. One portion contains the bulk of the interaction, in this case a satellite going round a perfectly spherical Earth, and this can be solved mathematically in some way. The other portion is much smaller, and represents the disturbance, or *perturbation*, of the simple case to give the real case. Mathematicians don't know how to solve this in some simple direct way. In the satellite example, the perturbation is the effect of the non-spherical aspect of the Earth on the satellite's orbit.

A classic case is the motion of the Moon around the Earth, with the Earth in turn orbiting the Sun. Writing down and solving the equations that describe in detail the motion of these three celestial bodies does not sound like such a tall order, and would be useful for predicting tides.

This problem is an example of a three-body problem, meaning one in which three objects all interact with one another. Such a problem cannot be solved directly, a state of affairs long suspected but only proved in 1994 by US mathematician Zhihong Xia. In fact Xia showed that three-body problems are chaotic: this means that given precise current positions for the Earth, Sun, and Moon, it's not possible to predict their exact future positions well into the future.

Such are the limits of mathematics, although at least in this case mathematics was powerful enough to explain its difficulty. Mathematicians now also understand why perturbation methods were so important in the old days of celestial mechanics – there was no chance of an exact solution, and the computer hadn't yet been invented. Ironically, Xia used a perturbation technique to arrive at his result.

The idea in perturbation theory, then, is to break the problem into two parts. One represents a simple version of the system that can be satisfactorily handled mathematically. The other is the small deviation from the simple case, which makes the problem real but also makes it more difficult to solve. This division is the first trick.

The second is to actually get an answer. Mathematicians usually do this using what they term a "series expansion." A series expansion is the representation of some mathematical expression as a sequence of powers. For example, $(1 + x)^{20}$ can be represented by the power series $1 + 20x + 190x^2 + \cdots$ ($\cdots$ means more terms). For some values of x, the first three terms give a fair approximation to the true value, but for others they do not. For instance, when x is 0.01, these first three terms of the series give a value of 1.219, when the true value is slightly in excess of 1.220. For an even smaller value of x, the agreement is better: when x is 0.001 then the true value and the value from the first three terms of the series agree to five decimal places. But for larger values things go badly wrong. When x is 0.1, then the true value is almost 7, but the first three terms of the series give just under 5. A simple way to get better agreement is to write down more terms of the series.

And that, given one important proviso, is one of the main rules of the game – more terms gives better agreement, at least in this and other select instances. The example series used above is very artificial, not least because the series terminates, meaning that it is possible to write out the whole thing and then stop. The result contains twenty-one terms, and is just the answer one would obtain by multiplying $(1 + x)$ with itself twenty times over and tidying up the algebra. Using all twenty-one terms means the answer for any value of x is exact.

But series exist for well-known mathematical expressions that typically have an infinite number of terms. Mathematicians have rules for generating the series representation required, and often the first few terms offer a fair approximation to the true value. You need more accuracy? No problem, just calculate a few more terms.

Perturbation theory typically makes use of a series expansion. For instance, in the satellite orbiting a misshapen Earth example, the final answer comes out as a series expansion. The first term is just the simple case of a satellite orbiting a spherical Earth. The second and subsequent terms give increasingly accurate adjustments to this orbit arising from the irregular shape of the Earth. The series is given in terms of ever-increasing powers of some parameter of the problem, which in the perturbation series expansion assumes the title of expansion parameter. The power of the expansion parameter is the order of the expansion. When physicists talk of higher order corrections, they mean they are working with terms of the perturbation series containing higher powers of the expansion parameter.

Each term in the series needs to be smaller than the term before it, loosely speaking – this is the important proviso alluded to earlier. If this is not the case then successive terms do not give increasingly fine detail in the answer, they change it dramatically. And because there is no end to the sequence, there is no way of extracting a meaningful answer. So the expansion parameter must be small, and larger powers of a small quantity are smaller still.

If successive terms grow very small very quickly, then just a few terms of the series yield an accurate answer. This is a good thing since those extra terms might be hard to calculate. Disaster can strike, however. If the expansion parameter is large, say approximately one, then higher powers are *not* smaller and successive terms of the series are of comparable size. When this happens perturbation theory, or at least that particular way of implementing perturbation theory, will not give useful answers. For QCD, this turns out to be a real issue.

In QED, electroweak theory, and QCD, perturbation theory is the workaday route to answers for setting alongside experiments. The strategy is to divide the master equation into two parts. One part expresses the freely moving particles of the theory, and can be solved directly. The other part, the perturbation, is the interaction between particles and this is the interesting bit. It's the source of a perturbation series.

For QED, the interaction part is simply the conjunction of a photon and an electron–positron, with the overall strength factor given by the electron's electric charge. There's no surprise there – it's perhaps intuitive that the electron's charge, as a measure of the electron's capacity to exert an electric influence, should control the strength of its interaction. The perturbation approach then yields a series of successively smaller terms that account for the interaction with increasing levels of accuracy. That said, as we shall discover shortly, there's a caveat that ultimately curtails the pursuit of ultimate accuracy.

Since the electron charge e turns out entering as even powers of two, four, six, and higher, for convenience physicists have introduced a new quantity, the coupling constant, which is proportional to the electron charge squared. It's the coupling constant, denoted α, that appears in the final answer. Successive orders of the perturbation series for the amplitude correspond to a series in successive powers of the coupling constant. The coupling constant, in particle physics' use of perturbation theory, is the expansion parameter. For QED, the perturbation series is an expression for the amplitude as α to the power one (the first order term) multiplied by other factors, then α to the power two (second order) with other factors, and so on.

Starting from the master equations of QED, QCD, or the electroweak theory and figuring out a workable perturbation theory solution is a big job. Luckily this was done long ago, and the results are listed as a bunch of rules in the back of most text books on particle physics. Theorists, instead of working from the master equations each time they want to perform calculations, work directly from these rules. These rules are the Feynman rules, given in terms of Feynman diagrams.

A baker's name immortalized

Experimentalists like to set theorist's predictions alongside their measurements, looking for places where the two agree – and places where they don't. Their experiments characteristically involve directing a beam of particles onto target particles, then counting the number, direction, and properties of particles emerging from the collision point.

What experimenters are typically out to measure, and the thing theorists correspondingly need to calculate, are cross-sections. A cross-section, which has units of area, expresses the strength of an interaction between particles.

There's three parts to a cross-section calculation. One tracks the numbers and speeds of colliding particles – it's a kind of particle flow contribution. The second part constrains the way energy and momentum can be partitioned amongst the collision debris. But these are really just bookkeeping – the real business, the real work, lies in the third and final part, the likelihood that the interaction in question actually occurs.

In quantum field theory, particles can be spawned from the void, the "empty space" that physicists call the vacuum, and vanish back into it. If a particle is created, then vanishes again but at a later time and at a different point in space, then the particle has traversed the interval between the two points. The quantum amplitude for doing this has a name – it's a Green's function. Actually it's a two-point Green's function, since it links a particle creation at one space-time point to its destruction at a different point. Since this effectively describes a particle propagating from one point to the other, the two-point Green's function is called a propagator. In QED there's a propagator describing the amplitude for the creation of either a positron or an electron at one place, and its ultimate destruction at a different place. A different propagator does this same job for photons. In pair annihilation, an electron and positron collide and destroy one another, and two photons emerge in their place. The quantum amplitude for doing this is described by a four-point Green's function, since there are now four players embroiled.

It's this entity called a Green's function that describes the core physics in particle scattering. To calculate cross-sections ultimately means calculating Green's functions. And the best way to calculate Green's functions is in terms of Feynman diagrams, themselves extracted from the Feynman path integral.

Since Green's functions lie at the heart of field theory, they were presumably named after some giant of quantum field theory? Not so: Green's functions are named after George Green, a miller and baker from Nottingham, UK, who was born in 1793 and died in 1841, nearly a century before the genesis of quantum field theory. His windmill still works, and is open to visitors. Green was a self-taught mathematician who, amongst other things, contributed a piece of mathematical machinery, now called Green's functions, useful for tackling problems in electromagnetism and heat flow.

A simple Green's function, for instance a propagator that describes a non-interacting particle going from place to place, is easy to figure out. And a four-point Green's function describing a system without interactions is just a bunch of two-point functions for particles going from start to finish without noticing one another.

That's not so very interesting. Interactions are interesting, and interactions make for a cross-section that experimentalists can look for, but incorporating interactions makes calculations far more complicated. In fact calculating two- and four-point Green's functions is no longer trivial, because they cannot be isolated on one side of an equation but always finish up expressed in terms of themselves. This is why in quantum field theory exact solutions are not possible: nobody knows how to do the mathematics. The only way anybody knows of calculating interesting Green's functions is to use an approximation technique in the form of perturbation theory.

In the case of QED, the result is sets of (bare) electron–positron and photon propagators, linked at junctions associated with the electron charge. For example, the complete two-point Green's function for a transiting electron–positron, subject to interactions, is a string of terms in increasing powers of the coupling constant: the first and simplest corresponds to an electron going from point to point as a non-interacting particle. The next in line, having two junctions (so the coupling constant has power one), corresponds to an electron that travels from A to B, then an electron from B to C in tandem with a photon from B to C, and an electron only from C to D. The third portion has the coupling constant raised to the power two, and a correspondingly more complicated interpretation.

These terms of the perturbation series can be drawn as diagrams, and the result of all this effort suddenly becomes much clearer. Some of the contributions to the complete fermion two-point Green's function in QED are shown as diagrams in Figure 3.4. These are examples of the famous Feynman diagrams.

Figure 3.5 shows a second example of Feynman diagrams, those for electron–positron scattering, $e^+ + e^- \rightarrow e^+ + e^-$, so this is a four-point QED Green's function depicted to first order in the coupling constant. There are just two fermion–photon junctions so the coupling constant appears with power one in the amplitude, making these the simplest terms of the four-point function that allow for electrons and photons to interact. Figure 3.6 shows second order diagrams for this same process.

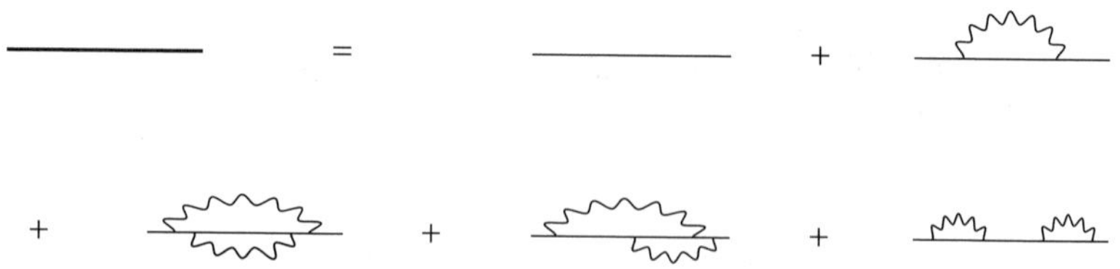

FIGURE 3.4 Two-point Green's function for the electron. The complete electron propagator, thick solid line, is the sum of an infinite number of terms of a perturbation series, shown here as Feynman diagrams.

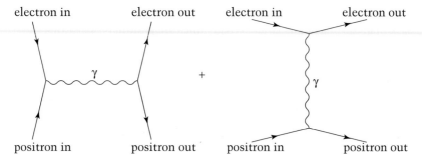

FIGURE 3.5 First order diagrams contributing to electron–positron scattering.

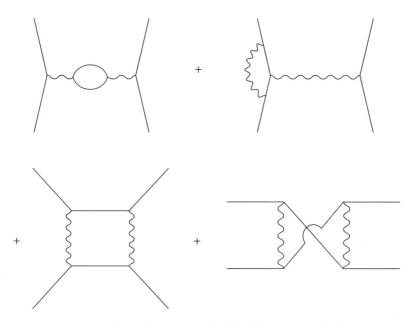

FIGURE 3.6 One-loop, or second order, diagrams contributing to electron–positron scattering.

Diagrammatica

Feynman diagrams, which made their first appearance in Feynman's seminal paper on quantum electrodynamics in 1949, are a representation of the terms of the perturbation series expansion of the theory in question. They are *not* simply quaint thumbnail sketches of particle collisions. Each line and blob in a diagram corresponds to a precise mathematical expression. By convention, straight lines are fermion propagators, which in QED means they correspond to the passage of electrons or positrons. Wiggly lines are photon propagators; they meet the fermion lines at a junction, called a vertex. At each vertex, four-momentum is conserved, meaning that the total four-momentum flowing into the vertex matches that flowing out. Strictly speaking the fermion lines should have a directional arrow. Diagrams without loops are termed tree-level diagrams since they simply branch; diagrams involving loops turn out to have a sting in the tail that tree-level diagrams do not.

All possible diagrams required for a real calculation can be created using simple subdiagrams. In the case of QED, these subdiagrams total just three in number. They are the electron propagator, the photon propagator, and the three-pronged electron–photon vertex. These, the basic rule that all possible distinct diagrams having a given number of vertices are to be included, plus a few other technicalities, are enough to allow realistic calculations of particle scattering processes.

Feynman's diagrammatic approach so simplifies perturbation theory calculations that they are almost mechanical. Graduate students can perform meaningful calculations without ever having heard of Green's functions and path integrals. Now anybody can do it!

Well, almost. The truth is that for diagrams involving more than a handful of vertices the number of diagrams to work out explodes. This, and other complications, means that in practice only the simplest levels are readily calculated, and nothing more than about the four-loop level has ever been worked through fully for any realistic theory. For example, calculating the electron's anomalous magnetic moment to three loops involves 72 diagrams, to four loops 891 diagrams, and to five loops an incredible 12 672 diagrams. To put this in perspective, the *three-loop* diagrams only succumbed in 1996 after almost thirty years of effort. Calculating the four-loop diagrams is an ongoing problem – the theoretical value a_{theory} quoted at the start of the chapter includes merely an *estimation* of the four-loop contribution.

There is also an interpretational hazard riding on the diagrams. They are almost too clear – for example, one diagram in Figure 3.5 really looks as though the electron and positron are scattering by first combining into a photon then re-emerging, via pair creation, as the photon bursts into an electron and positron. Though there are instances when a low-order diagram may closely approximate the physical process, in general they do not. The diagrams remain terms of a series expansion, and simple tree-level diagrams are just the first terms in that series. In reality, the physical process equates to a cloud of photons, electrons, and positrons that collectively mediate the interaction.

Even a humble electron going from point to point indicates this, with the true amplitude for an electron to propagate from one point to another involving a succession of ever more complex contributions (Figure 3.4) and not just a single "bare" electron line. The totality is the dressed propagator, as distinct from the bare one, and the loop diagrams are called electron self-energy diagrams. This, then, represents in Feynman diagrams the physical idea of quantum field theory that a particle such as an electron has associated with it a cloud of photons. The "true" situation is the dressed propagator, not the bare one.

Likewise for photons traveling from place to place. A dressed photon propagator includes the bare propagator, then the first order term in which the photon creates a fermion loop, and higher order diagrams containing more loops. The loop diagrams are called photon self-energy diagrams, or vacuum polarization diagrams after the way a transiting photon disturbs the space through which is passes. This is a pure quantum effect with no counterpart in classical physics, and can be seen by experiment.

Slippery series

Perturbation theory *à la* QCD closely parallels that for QED. However there is one major difference, which in some ways is the Achilles heel of QCD. Though in electrodynamics perturbation theory, exercised through Feynman diagrams, can be freely used in the knowledge that it works just fine, the same isn't always true for QCD. In QCD there is a problem. There are times when "little" corrections turn out to be bigger than the thing they are supposed to correct, like ripples on an ocean wave turning out to be larger than the wave itself. Mathematically, what happens is that the expansion parameter, which appears in increasing powers in ever-finer calculations, just isn't small enough. When this happens theorists are in deep trouble. Some turn to exotic mathematics, some to computers, but without the means of exactly solving the theory this issue, the lack of applicability of perturbation theory in some instances, remains one of the greatest challenges in particle physics.

As if this wasn't bad enough, there is an underlying problem with perturbation theory itself. This issue doesn't get in the way of day-to-day calculating, unlike the problem of the applicability of perturbation theory to QCD outlined above, but it chews at the theoretical roots of the entire field theory edifice.

Perturbation theory applied to field theory gives a series of terms of diminishing size, and theorists once assumed that this series would exhibit a property called convergence. But in 1951 British theorist Freeman Dyson showed that in general this is not true, an apparently technical point that has deep significance.

Mathematicians know about many series, not just those arising in perturbation theory. They use the term convergent to denote a series for which the terms get successively smaller, so that they can be added together to give a single number, the sum of the series. For instance, the series $1 + 1/2 + 1/4 + 1/8 + 1/16 + \cdots$ (more terms) is convergent, and the infinite sequence of terms sums up to give two.

Perturbation theory applied to quantum field theory at best gives a type of series termed asymptotic, and this is a nasty problem. To illustrate the difficulty, the first few terms of a sample asymptotic series (for which the general term is given by $(-1)^{n-1}(n-1)!/x^n$, with $x = 2$) are $1/2 - 1/4 + 2/8 - 6/16 + \cdots$ and, since these are getting successively smaller, all seems well. The trouble starts when a few more terms are included: $+24/32 - 120/64 + 720/128 - \cdots$. The sum over all terms of such a sequence gives an infinite answer, and the series is said to diverge instead of converging to a fixed value when all the terms are summed.

More formally, mathematicians describe a series as asymptotic if the difference between the quantity represented by the series and the sum of the first few terms of the series reduces to zero if the expansion parameter reduces to zero. For values of the expansion parameter larger than zero, eventually things will go wrong, meaning that at some point in the series terms will suddenly start to increase in size and the series will diverge.

In quantum field theory, the expansion parameter is the coupling parameter of the interaction, and this is clearly not zero, since that would mean a non-interacting theory. The finite value of the coupling means that, at some point in the perturbation series, a term of the series is *larger*, not smaller, than the one before it. And the next one is larger still.

So perturbation theory for realistic theories is not infinitely accurate. Even with the largest computers and biggest teams of theorists calculating vast numbers of Feynman diagrams, there's no chance of calculating an answer that is arbitrarily accurate. Perturbation theory has a maximum level of accuracy, a direct consequence of the fact that the perturbation series is asymptotic but not convergent. For a theory such as QCD, even in a regime where perturbation theory works, it can only provide an approximate answer.

QED, the best theory known to science, from a formal mathematical point of view makes no sense. And the same weakness is true also of QCD and the electroweak theory. *C'est la vie!*

There is yet another problem. For some theories, notably QED and "toy" theories that researchers work with, though not QCD, perturbation theory fails in an additional way. The coupling constant grows slowly stronger with increasing energy until at sufficiently large energies it's no longer small, and perturbation theory no longer holds. Luckily for QED, this is well beyond energies obtainable in any experiment. The key to understanding this problem, and at the same time a key property of QCD, is one of the great ideas of twentieth-century mathematical physics: the renormalization group.

TAMING THE INFINITE

Feynman once described it as "hocus pocus" and "a dippy process." Some have labeled it a mathematical fudge. It has been lauded as a guiding principle for theorists building new visions of the Universe. And more recently theorists have come to link it with shadows of a world beyond that which experimenters have already glimpsed. At the crudest level, it's a way of pulling out sensible answers from Feynman diagrams that explode. It is renormalization.

Here's the problem. There is no concrete limit on the energy of virtual particles buzzing around a transiting electron or photon. Provided they respect the Heisenberg uncertainty relation linking borrowed energy to the duration of that loan, virtual particles can be as energetic as they like. This permits the spawning of potentially incredibly heavy beasts, way heavier than everyday particles such as protons. And heavy particles that exist over short times embrace length scales that are minute even by the standard of the atomic nucleus. So, as given, quantum field theory demands that virtual particles of unlimited energy operating over infinitely small distances be included in calculations. When theorists do this, they get infinite answers.

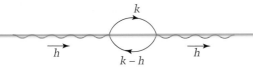

FIGURE 3.7 Photon momentum h passes through the loop, where there's nothing to restrain the electron's momentum.

The problem is easy to see in terms of Feynman diagrams. In all but the most basic calculations in quantum field theory, theorists find they encounter Feynman diagrams having closed loops. For example in electron–positron scattering $e^+ + e^- \rightarrow e^+ + e^-$, the simplest calculation involves only tree diagrams (Figure 3.5), but the next order diagrams involve loops (Figure 3.6).

At every vertex of a Feynman diagram the four-momentum must add up – what goes into the vertex must come out. When the vertex opens up into a loop, following round the four-momentum something odd happens. The only way of making the momenta add up is to introduce momentum that appears only within the closed loops and never outside. This in turn means that the internal loop momentum can assume any value, so long as the numbers tally at the vertices. Figure 3.7 illustrates this for a diagram with one loop on a photon line.

The rules for calculating the contributions from each diagram insist that the contribution from each value of this loop momentum be incorporated, and this is where disaster strikes: as the loop momentum roars off to infinity, these closed loops give infinite answers. The final answer of a full calculation is a mess of infinities. Yet only finite answers, having some sensible maximum numerical value, are of any use for comparison with experiment. As we shall see shortly, a theoretical U-turn underpins a modern perspective in which theorists see these infinities as reflecting the fact that all known field theories are just an approximation to something more fundamental. So infinities are no longer perceived as pathological. But until the 1970s they were a source of much anguish on the part of the world's theorists.

This kind of running off to infinity is called a divergence. In fact this is an instance of an ultraviolet divergence, where the "ultraviolet" recognizes that the fountainhead of the divergence is very large momentum, or correspondingly very short wavelengths – in other words the potentially massive virtual daughters operating at infinitesimal length scales. The ultraviolet appellation is handy as a means of distinguishing these divergences from slightly less nasty cases that arise in a low-energy context, called infrared divergences for that reason.

The problem of how to deal with the infinities arising in diagrams with loops is what held up QED for so many years, and almost led to the scrapping of quantum field theory. The same divergence problem is present in QCD and in the electroweak theory, and for all the fix is the same.

That fix is renormalization. Renormalization hides the infinities away, and gives a prescription for getting meaningful answers to compare with experiment. The physical picture of renormalization parallels that of the motion of an electron in a solid. Take one electron, measure its mass, and confirm that the answer agrees with the accepted book value. That same electron in a solid interacts with both the lattice of positive ions and other electrons, so behaves as though it were of a mass different from the book value. In a similar way, the electron's effective electric charge in a solid is modified by its environment, and will be different from the book value. When the electron is immersed in a solid the environment modifies, or *renormalizes*, the book values for its charge and mass.

In quantum field theory the idea is that an electron, for example, interacts with the vacuum in which it resides – implying that the vacuum is to be taken as rather more than inert empty space. The electron is then enveloped in a cloud of virtual photons and other particles so that the "bare" electron is never actually seen. A real experimental measurement of the electron's mass or electric charge actually yields a value equal to the value modified by the presence of the particle cloud.

There are two key differences between this and the case of an electron moving inside in a solid. Firstly, the ions and other electrons in the solid modify the mass and charge by a finite amount, but in the vacuum the bare charge and mass must be modified by an *infinite* amount. Secondly, though an electron can be removed from a solid and tested in isolation, it can never be isolated from the vacuum: the modifications to bare quantities arising from the vacuum can never be switched off.

The renormalization program has two parts. Firstly, the divergences must be written down in some way so that they can be compared, added and subtracted, and otherwise manipulated even though they are really infinite. There's work to be done here, since generally mathematicians are nervous about the meaning of infinity plus infinity, or infinity minus infinity. Secondly, the various divergences must be gathered together and extracted from the rest of the calculation, leaving a portion that gives physically meaningful answers for comparison with experiment. The first part of this program is called regularization. The second part is the renormalization proper. There are infinitely many ways of doing either, but at the end of it all the answers must come out the same. After all, they are supposed to match the one real world!

Pruning space-time

First, then, the regularization. The root of the problem is the possibility of creating particles of colossal energies, or equivalently the properties of the theory at very small distances. In terms of Feynman diagrams, the problem is that the momentum in a loop can run away to infinity. The simplest way to fix the problem is to simply do the obvious thing, and impose a limit on the allowed short distances to be included in the calculation, or equivalently impose an upper limit of the heaviest virtual particle that can be included. This is the way Feynman tackled divergences in electrodynamics in the late 1940s. He introduced into the equations a symbol for the largest possible value of momentum. No value will ever actually be selected; the effect of this strategy, which is described a momentum "cut-off," is to give answers in terms of this symbol. If it were to be allowed to assume its true value, that answer would once again be infinite.

Sometimes this strategy works well enough. In makes better sense when the underlying problem has a built-in cut-off, for example in solid-state theory where the natural crystal lattice spacing between atoms translates into a momentum cut-off. The smooth space-time of theories such as QCD amounts to a lattice spacing that goes to zero, equivalent to the cut-off running away to infinity. In practice, and certainly for QCD calculations, the cut-off strategy is not used, primarily because it does not properly respect the symmetry properties of the theory.

For QCD, there is one regularization scheme that offers advantages over all others. It respects both the symmetries of space-time, and also the underlying symmetries of the theory itself. It's called dimensional regularization, and even by the standards of quantum theory at first sight it's a pretty bizarre idea.

Some of those errant contributions that become infinite would not misbehave if the theory were formulated for a world in which everything happened on a string or a tabletop. That is, for toy theories in which the total number of dimensions is two or three, rather than four, the summations sometimes turn out to be finite rather than infinite. This is a hint: since the summations may work out fine in a different dimension, the idea is to allow the dimension to vary from four to a value just a little less than four. The dimension is carried along through the calculation: if at the end it were reset equal to its true value of four, then again the answer would be infinite.

What's weird about dimensional regularization is that this number, the dimension, does not have to be a whole number. Thinking in terms of a three-dimensional world is easy enough, and four is not too bad with practice. But what is the meaning of a dimension four minus a little bit? Luckily, this is of no real concern to physicists, since the use of dimension as a continuous quantity is a means to an end, and never appears in any experimental prediction.

The important consequence of using dimensional regularization is that troublesome parts, the infinite bits, are separated out, quarantined, in calculations. Once there, they can be readily manipulated, with the exploding divergence permanently on view in an especially easy-to-see form.

The perhaps unfamiliar idea of tinkering with the number of dimensions actually has a long history in theoretical physics. In particular in the 1920s Polish physicist Theodor Kaluza and the Swedish physicist Oskar Klein worked up a general relativity–electromagnetism combo that involved an extra dimension, making five in all. At the National University of La Plata, Argentina, Carlos Bollini and Juan Giambiagi, experienced in manipulating infinites in errant diagrams, picked up Kaluza and Klein's model as a way of exploring what would happen to infinite diagrams in an odd number of dimensions. They soon realized that the number of dimensions itself could be used as a handle for manipulating infinities. Dimensional regularization was born, but not before it became the subject of publication leapfrog. Bollini and Giambiagi's first paper outlining their idea reached the journal *Physics Letters* in October 1971. But it fell foul of a referee who didn't care for higher dimensional theories on the grounds that the real world is four-dimensional. Their second paper discussing dimensional regularization arrived at a second journal on February 8 1972; days later a third journal received an independent account of dimensional regularization from Gerard 't Hooft and Martinus Veltman in Utrecht, The Netherlands. Yet the 't Hooft–Veltman account appeared in print *before either* of the Argentinean's papers, for which reason dimensional regularization is sometimes attributed only to 't Hooft and Veltman. Bollini observed "it takes much longer for a paper to arrive to [sic] the Editor's desk from La Plata, Argentina, than from any city in Europe."

The advantages of using dimensional regularization are considerable, so it's worth setting aside anxieties about the meaning of those non-whole dimensions. For even after the application of dimensional regularization, the mathematics of the Feynman diagrams themselves is relatively undamaged, the diagrams retaining a recognizable form. In addition, dimensional regularization preserves all necessary symmetries. Finally, a common simplification in high-energy physics is to set the masses of some participating particles to zero, since they are small relative to the energies of big accelerators. Theoretically, this can cause trouble, however, since setting masses to zero in calculations can provoke divergences of the infrared variety. Dimensional regularization can take these in its stride by allowing the dimension to *increase*, to four *plus* a little bit.

In all this, there's a detail missing that turns out to be rather more than a detail. In any mathematical equation, any two portions added or subtracted must have the same dimensions. For instance, a distance in meters plus a second distance in meters gives a total distance, also in meters. But add a distance in meters to a volume in cubic meters and the result just doesn't make sense.

Of course the equations of QED and QCD also have well-defined dimensions. But when the space-time dimension is fiddled with, as in dimensional regularization, then in QED for example the description of how an electron and photon interact is messed up.

In fact for QED the bit that's messed up involves the coupling constant, whose dimension changes rather unexpectedly when the space-time dimension changes. The way this is tackled is to introduce what appears to be a compensating factor, apparently plucked from thin air, with dimensions to counter those of the coupling constant, so that when the two are multiplied together the combination has trivial dimensions, and is rendered harmless.

Followed through, this new parameter is still there in the finite part of an expression corresponding to an infinite diagram. It looks as though there is a penalty for using the trick of dimensional regularization, which is the introduction of a parameter about which we know nothing, except that it entered the game with the right dimensions to fix up a technical problem, and that it does not trivially vanish.

Is this a disaster? On the contrary, the fact that this parameter has no assigned value, and can take any value, will lead to some amazing new physics. For that reason, it's worthy of a name and a symbol: it's called the renormalization scale and is denoted μ. Actually, this apparently mysterious quantity is in effect the momentum cut-off re-emerging in a different guise in a different regularization scheme. The quantity μ marks the boundary between "short distances" and "long distances." Long distance is supposed to embrace all the players we care about. Excluding short distances is to admit that we don't know what goes on there, but that we assume it doesn't really matter anyway, at least for the time being.

In pursuit of cancelations
Dimensional regularization thus provides a way of reworking a loop divergence into a simple, infinite part, plus a useful, finite part. Now for the renormalization itself. The challenge appears pretty daunting, since for QED, QCD, and the electroweak theory there are infinitely many Feynman diagrams, with nearly all of them containing divergences caused by loops and their runaway momentum summations. It turns out that things are not quite that grim, and theorists have risen to the challenge.

Working again with QED, for QED displays the central issues without confusing details, there are just five basic loop diagrams to be investigated, depicted in Figure 3.8. All other Feynman diagrams, with even the most complicated of loops, can be resolved in terms of these five.

For the electron self-energy diagram, for example, depicted in Figure 3.8(a), dimensional regularization gives a well-defined separated-out infinity. Renormalization works by adding a simple, extra portion to the master equation of QED, then feeding the new master equation into the machinery of perturbation theory. The new portion gives another infinite contribution, of the same mathematical form but of opposite sign to the infinite piece arising from the electron self-energy diagram, which is therefore cancelled. The new additional portion in the master equation is called a counter-term, since its effect is to counter the infinities that would otherwise appear in the answer. Two more

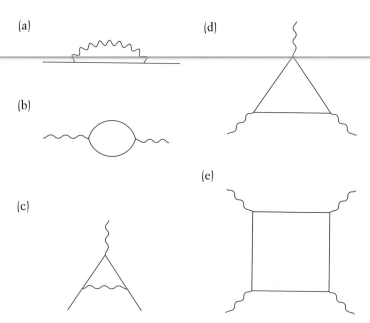

FIGURE 3.8 The basic loop diagrams of QED: (a) is electron self-energy; (b) is photon self-energy, or vacuum polarization; (c) is the electron–photon vertex; (d) is the three-photon vertex; (e) is light–light scattering. In practice, diagram (d) does not contribute and diagram (e) is finite. All the divergent diagrams of QED can be handled in terms of diagrams (a), (b), and (c).

counter-terms added to the master equation in a similar way neutralize the infinities due to the photon self-energy diagram and the vertex diagram.

The magic starts here, though to begin with it looks more like just a touch of good luck. The good luck is that nothing need be done to fix up the three-photon vertex and light–light scattering diagrams, (d) and (e) in Figure 3.8, respectively. The three-photon vertex diagram always cancels with a matching diagram in which the electron loop has a reversed direction (the direction is not shown on the diagram), so vanishes from the plot – literally. The final diagram is the light–light scattering "box" diagram, which turns out to be finite, a consequence of having so many photons attached to the box.

This is not the end of the story, however, since the next level of calculation of some real physical process involves more diagrams, with more loops. How will these higher order terms of the perturbation series

(a)

FIGURE 3.9 Two-loop
contributions to the photon
self-energy.

(b)

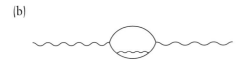

(c)

fare? It could be that these diagrams will contain more complicated loops savage enough to destroy what's been achieved for single loops using counter-terms.

Next in line are diagrams with two loops. For example, the vacuum polarization case with two loops corresponds to the diagrams in Figure 3.9. Each of these three diagrams contribute infinite summations since they have closed loops. Those loops are pretty complicated, raising the ghastly prospect that things might go badly wrong.

Not so, and this is the real magic. Recall that a counter-term added to the master equation provided a cancellation for the infinity arising from the single-loop vacuum polarization diagram. That cancellation showed up after the master equation, counter-terms and all, had been fed through the perturbation theory engine. Turn the handle one more time to find out what the counter-terms produce to this higher order. The remarkable thing is that the complete set of all diagrams, up to and including those with two loops, including the extra parts due to the counter-terms, gives all sorts of terms that cancel in unexpected ways. All that remains is the sensible (i.e., not infinite) answer, to the appropriate level of accuracy, and one last divergence. This in turn is cancelled by a counter-term from the next level up.

This is the heart of the matter. For renormalizable field theories such as QED, QCD, and the electroweak theory, a small number of simple counter-terms added to the master equation are enough to see *all* infinities cancel to *all* orders. This is the magic.

It's magic because, by rights, new counter-terms could be needed to neutralize each new level of divergence as it appears. But this is not necessary, and in QED the original three is enough. Another way of viewing this is that in QED and other renormalizable theories, as more and more terms of the perturbation series are calculated, the divergences are actually no more or less savage than those encountered at the simple loop diagram level. Complex multi-loop diagrams are just more of the same.

There are theories that contain no divergences at all, in which case they are termed finite. Chief amongst these are a class of theories based on an exotic symmetry between fermions and bosons, called supersymmetry. It is their finiteness that makes these theories so popular amongst theorists, despite the absence of direct experimental support for the idea of supersymmetry.

For some toy theories, the higher order diagrams eventually cease to contain divergences, so the theory has in total a fixed number of divergent diagrams. These theories are super-renormalizable, but no theory of the real world has this property. For other toy theories, and for quantum gravity, the divergences become increasingly vicious for higher order multi-loop diagrams. The first set of counter-terms is not enough, and at each level new counter-terms must be added to neutralize fresh divergences. This never-ending string of counter-terms is the signature of a non-renormalizable theory.

Needless to say the trick lies in selecting the counter-term, the portion added to the master equation whose job is to see off the infinities. And this is where dimensional regularization comes in, since it leaves the troublesome divergent part high and dry as a simple additive contribution. So conjuring a useful counter-term becomes easy. Even then, there's some latitude in the exact choice of the counter-term. No matter, the physical predictions come out the same – but more on that in a moment.

All this leaves a couple of nagging questions, in addition to a slight headache, and perhaps a sensation of having been swindled. For example, what has become of the idea that renormalization is about reinterpreting physical quantities such as charge and mass? And how does it make sense to go around sticking bits into the master equation of some theory such as QED or QCD, and yet expect that it is unchanged in terms of its physical predictions?

In fact, bits are not really added to the master equation; it's more that the master equation has been rewritten, in tandem with some careful and perhaps surprising interpretation. To start with, the master equation is given in terms of bare, immeasurable quantities, which in the case of electrodynamics include bare electric charge and bare electron mass. These bare quantities are understood to be *infinite*. This interpretation causes little anguish since bare quantities are taken to be invisible to experimenters. So the unrenormalized bare quantities start out infinite in the first place, and the divergences arising from loops then cancel these to leave finite physical renormalized quantities. The purpose of the renormalization program is to make this happen.

A renormalizable theory is one in which there are only a limited number of different divergences, such that curing them cures all possible divergences. In electrodynamics, there are three basic divergent diagrams, three counter-terms to match, and three infinite constants. The electron self-energy diagram transforms the bare electron mass into the physical mass. The vertex diagram links bare electric charge to physical electric charge, and finally the photon self-energy is related to the transformation of bare electric field strength to physical field strength. The photon itself has zero mass both before and after renormalization.

To make physical predictions with a renormalizable theory, experimenters need to make enough measurements to fix the parameters in the theory, which in the case of electrodynamics, for example, means measuring electron mass and charge. Only then can the theory be used to predict the outcome of some other process.

This gives further insight into what it means to be non-renormalizable. A non-renormalizable theory needs new counter-terms added at every new level of complexity of the perturbation expansion. Not only does this imply an infinite number of experiments to fix all their values, it implies an infinite number of "physical" parameters. This is at the core of the lack of predictive power of non-renormalizable theories such as quantum gravity. For renormalizable theories such as QED, QCD, and the electroweak theory, all divergences can be removed by the infinite renormalization of the physical quantities in the theory.

The "dippy process" actually works!
Theorists can calculate whole sets of diagrams, but for higher orders, meaning more loops and more diagrams, the calculations are just too

big. To simply go on calculating increasingly complicated diagrams is not a realistic way of checking that the renormalization program makes sense.

Besides, mathematics doesn't normally work this way. That is, mathematical proofs do not work by straight enumeration of all possible cases. Instead, mathematicians set to work to devise a general expression that holds for all cases. In the context of renormalization, this means that somebody needs to sit down and prove a general result showing that counter-terms make for cancellation of all divergences, even as more and more diagrams of ever-increasing complexity are incorporated.

Dyson managed to do just this in 1949, while at the Institute of Advanced Study in Princeton. In a classic paper, Dyson set out the basic terminology for analyzing Feynman diagrams, and proved that renormalization of electrodynamics works "to all orders," that is for all possible diagrams, given that the simpler ones correspond to a renormalizable theory. His strategy was to use proof by induction, meaning that he assumed that renormalization works for diagrams of a given order, then proved that the procedure works for the next highest order.

Dyson was aided in his task by the fact that all diagrams with loops can be analyzed in terms of a small number of basic diagrams, three in the case of QED. But some diagrams have overlapping loops, and these turn out to be especially hard to analyze. The task of understanding these overlapping divergences, and therefore of filling a gap in Dyson's work, fell to a Cambridge graduate student by the name of Abdus Salam, who untangled overlapping loops in 1950. Another assumption in Dyson's paper was proved correct by an up and coming researcher at Columbia University in New York, one Steven Weinberg (Figure 3.10). Weinberg's "convergence theorem," published in 1960, showed that, in general, after renormalization the summations really are well-behaved, even when they come from Feynman diagrams involving complicated loops.

Ironically, Weinberg and Salam themselves eventually fell victim to renormalization problems in their theory of the weak force, the work for which they are most well-known. Building on the efforts of US physicist Sheldon Glashow, Weinberg in 1967 and Salam in 1968 proposed a theory that was hamstrung until 't Hooft and Veltman proved that it was renormalizable.

FIGURE 3.10 Steven Weinberg in 1992. Weinberg is best known for his Nobel prize winning work on electroweak theory, and for setting out what's now called the Standard Model, embracing strong, weak, and electromagnetic interactions. Credit: Louise Weinberg.

That same year, 1971, 't Hooft and his graduate supervisor Veltman proved in addition that *all* theories of the QCD type are likewise renormalizable. Veltman and 't Hooft (Figures 3.11 and 3.12) were awarded the 1999 Nobel prize in physics for their work on taming infinities. Given that divergences were already causing headaches for Dirac, J. Robert Oppenheimer, Heisenberg and others in 1930, beating them into submission in a form useful to modern theories of particle physics took four decades, and the efforts of some of the best theoretical physicists around.

One last word on the latitude in choosing counter-terms, the choice open to theorists in selecting a renormalization scheme, alluded to earlier. This latitude amounts to a choice in the way the master equation is split into physically meaningful portions and infinite counter-terms. Yet the final answer cannot depend on any such choice, since that choice is arbitrary. This independence of choice is an invariance, an invariance that implies a symmetry. The symmetry finds mathematical expression in the renormalization group, an entity with a special role to play in QCD.

A renormalization epilogue

Four men are credited with fixing the divergence problem of electrodynamics and giving the world renormalization pretty much in its current form, but none of them can really be described as ardent fans. Feynman didn't like renormalization, as demonstrated by his "dippy process" comment. Julian Schwinger worked hard for many years to get round

FIGURE 3.11 Sartorially elegant renormalization: Gerard 't Hooft, standing, defends his graduate thesis at Utrecht University, 1 March 1972. Credit: A. van der Sloot, courtesy of Gerard 't Hooft.

FIGURE 3.12 Martinus Veltman in John Bell's office at CERN, 1973. Veltman's conviction that the infinities in emerging theories of fundamental forces could be tamed drove him and his student, Gerard 't Hooft, to seek a way through the renormalization maze. The pair shared the 1999 Nobel prize in physics "for elucidating the quantum structure of electroweak interactions in physics." Credit: CERN.

it, and Sin-itiro Tomonaga viewed it as a temporary fix. A dismayed Dyson moved on to other things.

Renormalization started out looking like a trick, the pragmatic fudge we had to have a get sensible answers out of an apparently beautiful but somewhat recalcitrant theory. Following the impressive successes of QED, the requirement of renormalization became hugely important as a regulatory principle, a fundamental test that any realistic theory must pass. As such it in no small way helped in the construction of the electroweak theory describing the weak force, and the quantum chromodynamics that describes the strong interaction.

More recently though, inspired in part by the problems of trying to build a renormalizable quantum theory of gravity, theorists have remolded their attitude to renormalization. There's a new flexibility in the air: perhaps, under the right conditions, it makes sense to hang on to what looks like a good theory even if it doesn't make the renormalization grade. Let's learn to live with incomplete theories that may nevertheless allow us to tackle interesting questions, they argue.

There's a hierarchy of length scales in our universe, reflected in different phenomena and in the different theories describing those phenomena. The range extends from the subnuclear world at one end, through atoms, then molecules, then "everyday solids" right through to planets, then galaxies and ultimately the Universe as a whole. Issues that enthrall atomic physicists have little to do with the details of what goes on inside the atomic nucleus. And engineers building bridges are about as interested in the atomic details of steel as astronomers are in the niceties of bridge construction.

Each group has its own range of length scales to deal with, and its own description of how the world is supposed to work within that range. An atomic physicist does not need to worry about the size of the galaxy in which his atom of interest resides – it may as well be infinite. And an engineer does not need to think about the size of an atom – it might just as well be zero and the bridge steel a truly continuous material.

This in fact is the big idea: focus on a range of length scales of interest, set everything smaller to zero, and everything larger to infinity. The result is an "effective theory." The fact that the small quantities are not really zero, nor the large quantities really infinite, can then if necessary be included as perturbations about this simple starting point.

When chemists discuss atoms combining into molecules through shared or donated electrons they are really using an effective theory. The mechanics used by the bridge engineer is another.

In the context of quantum field theory, an effective theory approach usually means looking long and hard at what happens at the very shortest length scales. Imagine having a quantum field theory that works just fine for some type of process and some range of energies. It's not hard to envisage that at much higher energies, in other words much shorter length scales, some new effect that's not part of the original picture comes into play, just as an engineer will find his notion of steel as a smooth substance will fall apart if he looks at the steel under a powerful microscope. The goal of effective quantum field theory is to build a theory that allows theorists to tackle questions over some energy range without having to know the full story, the full picture of how the world works at the highest energies.

The idea of effective field theory, which was introduced by Weinberg in 1979 and which has been steadily gaining support ever since, might seem like so much back tracking. But then again a zero-tolerance enforcement of the renormalization requirement might be at the risk of ditching a good idea. There is one classic example that theorists highlight in support of their modified position. Back in 1933 Enrico Fermi proposed a theory of weak radioactive disintegration of neutrons to protons. Applied to a decaying muon, for example, Fermi's theory would imply an electron, neutrino, and antineutrino emerging from precisely the spot where the muon vanished. Now, Fermi's theory gives some pretty sensible answers sometimes. For instance, it's easy enough to calculate the lifetime of decaying muons in a way that shows respectable agreement with the data. That's not bad – Fermi's theory can't be all wrong, surely?

For the simplest calculations and at low energies, Fermi's theory looks good. But set against refined measurements, and used to calculate higher energy cross-sections, things go wrong. Fermi's theory is not renormalizable – crucially the force-transmitting W and Z particles, an idea still in the future when Fermi wrote down his theory, are missing. So judged by the criteria of renormalization, Fermi's theory doesn't make the grade. But now theorists would call it an *effective* quantum field theory: it's a non-renormalizable theory that works fine up to a certain energy. That certain energy must be modest compared to the mass of the W and Z particles. So according to the new effective theory paradigm, rather than throw Fermi's theory away, use it where it works well and leave a space for the new physics yet to be discovered. In the case of weak interactions, of course, physicists now know what that new physics is – it's the W and Z weak force transmitters. And the new theory is the electroweak theory.

So where does that leave a successful theory such as quantum electrodynamics? It's the most successful theory ever, and it is renormalizable, suggesting that to be successful all quantum field theories likewise have to be renormalizable. According to the effective field theory perspective, however, the success of QED to date merely hides the fact that physicists have not pushed it hard enough. This puts a fresh spin on experimenters' measurements of electron and muon magnetic moments, conventionally held up as instances of quantities the theory can accurately account for. Now those measurements are seen to provide limits on the mass and length scale at which some new physical process *beyond* conventional QED cuts in. At the level of calculating the electron and muon magnetic moments to a few decimal places, the interaction of light with matter can be described successfully by a theory where the main mechanisms are described by a renormalizable theory – conventional QED. Adding in any new mechanisms for the interaction of light with matter will ultimately mean adding *non-renormalizable* parts that come into play only at extreme energies and distances. Realistic theories in general are likely to contain both renormalizable and non-renormalizable parts: renormalization has lost its status as a basic requirement of a theory of fundamental forces, which is good news for theories of gravitation. In place of the renormalization demand are constraints on the structure of effective quantum field theories. Effective field theory is a conceptual breakthrough.

So now QCD and the electroweak theory must be viewed as incomplete. These theories will work up to certain energies and correspondingly fine length scales: new physics will emerge either as tiny discrepancies between calculations and precision measurements, or at huge collision energies. There is a blank space reserved on theorists' whiteboards for what comes next. For now, however, we will run with "conventional" QCD.

ADDING SYMMETRY

Here comes the final piece of the puzzle. Quantum field theory provides a framework incorporating the quantum and relativistic features of the Universe. Symmetry under space and time transformations, including rotations, is built into this framework. What's missing is symmetry intrinsic to the fundamental quantum players. When this is included in a particular way, the result is called a gauge theory.

The key player in this game is phase. The phase of a wave is an angle expressing the wave's progress through its cycle, with the phase difference between two waves determining their relative displacement. Gauge theories involve symmetry that appears through phase. They rely on the special relationship between external influences and phase in quantum systems, a relationship that has been experimentally demonstrated clearly only recently.

Slits revisited

More double slits. Recall that light through a double slit gives a pattern of light and dark bands as a result of interference. The two slits are, in effect, separate in-step sources of light waves each described by some amplitude and phase. Where two peaks overlap on the detection screen, the amplitude is doubled and the brightness increases four times. The central bright band corresponds to overlapping peaks both in step and the same distance from the slits: for the central bright band, there's no phase difference between the two contributions due to the two slits. The first bright peaks either side correspond to light from one slit being one complete cycle ahead or behind the other, the difference due to the different slit-to-detector distances. Moving outwards, there's another matching pair of bright bands, one either side, and these correspond to light from one slit being two complete cycles ahead or behind the other. Peaks correspond either to no phase difference, or to a phase differences of a whole number of cycles. Troughs correspond to phase differences of half a cycle, one and a half cycles, and so on.

So far, this is nothing more than an elaboration of an experiment that made its debut earlier in the chapter. Now to vary things a little. If a "light retarder," which did nothing but slow the light passing through it, was to be inserted after one or other of the slits, what would happen? The interference pattern still forms, but because light from one slit takes a little longer to arrive, the phase differences are all altered a little. The result is that the entire pattern shifts left or right, but the pattern itself is otherwise unchanged. By symmetry, if the retarder were put over the other slit instead, the pattern would shift the opposite way. Incidentally the retarder does nothing that allows one path to be distinguished from the other, so there is no which-way information inserted that would destroy the interference.

What if the retarder were placed over both slits? There would be no observable difference: the pattern would not be altered in any way. This is because interference depends on phase *differences*, but not on the phase itself. So retarding light from both slits alters the phases equally, but the phase difference is unaffected so nothing appears to happen. Phase differences determine the position of the light and dark bands and their relative brightness, and the amplitude of the waves determines the overall brightness of the pattern. There is no way of seeing the phase itself.

Basic double slit experiments using light are easy enough to perform in a high school laboratory. In contrast a real double slit experiment using single electrons passing through mechanical slits has never been done. That's because it's too difficult, the apparatus required being impossibly small. What has been done, however, is a version of the two-slit experiment for single electrons in which the "slits" are spaces between a thin positively charged wire and two conducting plates between which the wire is suspended. The electric field between the wire and each plate focuses the electrons towards the central region in a way analogous to what's called a biprism in optics. In 1989, a team led by Akira Tonomura, of Hitachi's Tokyo research laboratories, inserted an electron biprism into a device called a field emission microscope, and passed single electrons through it and then onto a position-sensitive electron counter beyond. Electrons appeared at the counter one at a time, gradually building up the distinctive banded interference pattern as expected.

If Tonomura and his colleagues had put an electron retarder downstream of one or other slits they would have seen the interference pattern shift. And a retarder over both slits would have made no impact. It's possible to see electron wave phase differences, impossible to see the absolute phase in this or any other experiment. The message is the same for both photons and electrons: modify the phase of one or other arms of a split beam and the resultant interference pattern changes in a well-understood way.

Driving a car along a straight road provides an analogy for the sensitivity of experiments to phase *difference* rather than actual phase. Just as a wave repeats itself periodically, so too a car's wheels go round and round. All four wheels turn in unison, and the driver has no way of knowing where the wheels are in their revolutionary cycle – no way of knowing when the manufacturer's logo on the hubcaps is the right way up on any one wheel, for instance. But driving though a deep puddle on the passenger side will slow that side of the car, momentarily slowing the passenger side wheels. If by some quirk the hubcap logos *had* all been aligned before the puddle, they won't be afterwards; the puddle "retarder" will have introduced a phase difference between the wheels on the two sides of the car. This the driver will notice, since without some corrective measure he'll wind up in a ditch.

"Whodunnit" addicts may be able to sniff where the plot goes from here. A retarder at one or other slits clearly has an effect, yet a retarder at both slits seems to do nothing. The interpretation of this is that the experiment is invariant with respect to overall changes in the phase of the electron wave. Quantum theory expects as much, and correspondingly experiments may be sensitive to phase differences, but not to the absolute phase. Any deviation from this would mean rewriting all the books: it's a central feature of quantum theory. Phase can't be

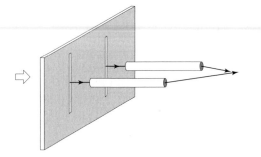

FIGURE 3.13 The Aharonov–Bohm effect using two electrically charge cylinders. Electrons enter from the left. Downstream of each of the two slits is a hollow metal tube that can be electrically charged.

seen directly, so physical effects are invariant with respect to overall changes in it, and this means there are conservation laws somehow related to phase symmetry.

Here's another experiment that has never been done. It hasn't been done because nobody has figured out how to do it, even though it was proposed way back in 1959. It's another double slit experiment for electrons, only this time downstream of each slit is a hollow metal tube, represented schematically in Figure 3.13. The idea of the experiment is to direct single electrons onto the slits, wait a moment until they have had time to reach a point a little inside the tubes, then apply an electric pulse to the tubes that finishes before the electrons have had time to reach the distant end of the tubes.

There is no electric field inside a closed conductor that's connected to some electricity supply, even though there's an electric field outside. So in this electron two-slit experiment, there's no electric field *inside* the tubes, and the electricity supply is switched off when the electrons are outside the tubes. So one might expect nothing much to happen, since the electrons have not been subjected to any electric field. So one might reasonably expect that, even when different sized pulses are applied to the two tubes, the interference pattern will not change. Certainly this is what a *classical* physicist would predict, since according to *classical* physics a charged particle will respond only to electric or magnetic *fields*.

When this experiment is done, there is no doubt in experimenters' minds about what they will see. When the pulses are different, the interference pattern will shift. When the pulses are the same, the pattern will not move. They know that some electric influence other than an electric field will impinge on the interference pattern.

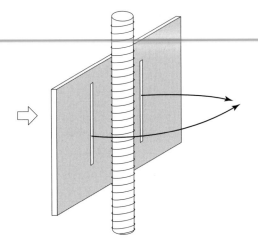

FIGURE 3.14 The magnetic Aharonov–Bohm effect. Electrons enter from the left. Downstream of the two slits is a solenoid. There is a magnetic field within the solenoid, but not outside it.

An experimental classic

One reason physicists are so confident of the outcome is that this effect is a variant of something that *has* already been seen experimentally. The experiment depicted in Figure 3.13 is one version of what's called the Aharonov–Bohm effect, first described by Yakir Aharonov and David Bohm back in 1959, while they were still at the University of Bristol in the UK. The other version of the Aharonov–Bohm effect, the one for which the experiment *has* been done, is represented schematically in Figure 3.14. Again it's an electron double slit experiment, but this time there's a long skinny electric coil, a solenoid, inserted downstream from the slits and placed between and parallel to them. When current flows in the coil it generates a magnetic field inside *but not outside,* a known feature of solenoids. So electrons moving past the coil, on their way from the slits to the detector, are moving in a magnetic field-free region. Yet when the coil is activated the interference pattern shifts. Again, classically this makes no sense since, because there's no field, there's no reason to expect the electrons to respond in any way. But there's something other than magnetic field at play here.

This experiment needed to be done several times before everybody believed the results, as the experiment is so hard. All doubts evaporated following a careful demonstration in 1986, also by Tonomura and his group, using a tiny doughnut-shaped superconductor-coated magnet just a few microns across. Thus the long solenoid is closed up into a ring in this version of the experiment, and the magnetic field is trapped within the material of the doughnut. Electrons are free to pass both through the hole in the center of the doughnut and round the outside, to interfere beyond. As expected the interference pattern shifted when current flowed in the tiny magnet, even though there could be no magnetic field in the space around it.

A charge-free particle that can behave as a tiny magnet presents the opportunity of doing an alternative version of these experiments, first pointed out by Aharonov and Aharon Casher in 1984. Once again there are two possibilities, paralleling the two cases of the Aharonov–Bohm effect, and recently both have been realized using neutrons.

Neutrons have no charge but, for reasons that will become clear later, they do behave like tiny magnets. A 1989 experiment by a collaboration from the University of Melbourne, Australia, and the University of Missouri in the US used the latter's research nuclear reactor to demonstrate one of these possibilities. Neutrons with their magnetic orientation upwards were passed either side of a charged wire. Given their orientation, they could feel no force from the electric field, yet again the interference pattern shifted as expected, showing the presence of an influence even without a force.

In 1993 the same group demonstrated the final variant, using a layout similar to that of Figure 3.13 but with two long solenoids in place of the two cylinders. Neutrons, magnetically oriented along their direction of flight, passed through the solenoids as electrical pulses were applied to them, generating magnetic fields within. The magnetic fields exerted no force on the neutrons, a consequence of the uniformity of the fields and the particular alignment of the neutrons. But despite the absence of a force, the experimenters saw tell-tale shifts in the interference pattern.

These apparently mysterious influences at work here are not really mysterious at all, and in fact they have a name. They are called potentials. Potentials are well-known in classical physics. In the electric case, the difference between two potentials – the potential difference – is familiar enough to have a household name: voltage. The change of electric potential from place to place is the electric field, which gives the usual force on a charge such as an electron that finds itself in a region of varying potential. The magnetic case is less familiar, but the idea is much the same. In the electric case, potential due to a charge drops away evenly in all directions as a simple number – electric potential is a scalar. The corresponding electric field therefore points radially towards or away from the charge, and since it has direction it is a vector. The magnetic potential surrounding a solenoid, on the other hand, while also diminishing in magnitude with distance from the coil, has a direction, and that direction is *around* the coil. Having direction, the magnetic *potential* is a therefore a vector. Sometimes it's pictured using concentric circles centered on the solenoid axis. Arrows on the circles then indicate the direction of the vector potential, which is the same as the direction of the current flowing in the coil.

The way the magnetic potential "circles" the coil provides a clue to understanding the Aharonov–Bohm experiment. Imagine two canoeists, one passing one side of a whirlpool and the other to the other side. One will be with the current, the other against: when they meet beyond, the one with the current will arrive first. For electrons passing a solenoid, to one side of the coil they run with the direction of the potential, and the other side they run against the tide. Here is the source of the phase difference suggested by the shift in the interference pattern that results from turning on the field inside the coil. To one side of the coil, the phase is advanced by the potential. To the other it is retarded. The result is a phase shift, and that phase shift depends on potential, not the field. This is how the world works at the quantum level – electrons (and other charged particles) experience electromagnetic influences through modifications to the phase of their quantum wave.

Potential: changing here, there, or everywhere

Birds sitting on high-tension overhead power lines emphasize the fact that electric potential itself makes no difference to the outcome of experiments. The bird is at a high potential, but there is no potential difference, and so no electric field, between different parts of its body; the bird continues to function in the normal way. However, if the bird used its beak to grab a cable that connected it to the ground, then a potential *difference* would exist between the bird's beak and its feet. Over that short distance the electric field within the bird would be large and the inner workings of the bird would no longer remain unchanged. The bird would fry.

Like the bird sitting on a power line, a quantum mechanical two-slit experiment performed at a raised electric potential, perhaps in a laboratory perched on the same power line, would likewise remain unchanged. In fact, the experimenters inside the laboratory on the power line might be performing the version of the Aharonov–Bohm experiment with two thin conducting tubes downstream of the slits, Figure 3.13. Whatever they see on the ground in an ordinary laboratory would be repeated exactly in the overhead laboratory; from the outcome of the experiment they would not be able to tell which laboratory they were in.

The point is that the potential of the complete experiment has been changed, yet there's no change in the result, just as changing the potential of only the two tubes by the same amount gives no change in the result. In fact, so far as the electrons in the experiment are concerned, these are the same thing, a change in potential that is the same everywhere. A change in anything that is the same change everywhere is dubbed a *global* change. Global changes in potential leave birds unharmed on power lines and the outcome of two-slit experiments unaffected. By contrast, local changes in potential, where local means that the change varies from place to place, are enough to fry birds on power lines and change the outcome of two-slit experiments. Altering the potential of the two tubes of Figure 3.13 by different amounts is a simple example of a *local* change in potential.

A global change of potential in the Aharonov–Bohm experiment of Figure 3.13 must influence the electron wave, yet the pattern is unchanged. In particular, the brightness of the pattern is unaltered, meaning that the number of electrons impinging on the screen is unchanged. This is an invariance, and with it there is a conservation law.

In fact what's important is not so much that the particles going through the apparatus are electrons, but that they carry electric charge, as it's this that permits them to experience the electric potential. So the conservation law is not one of electrons, it's the conservation of the electric charge they carry. Another way of seeing this would be to repeat the complete experiment with some other charged particle. The result would be the same.

Since potentials influence the electron wave by changing its phase, a *change in potential* is compensated for by a *change in phase*. A global change of potential gives a global change in the phase of the electron wave, just as a local change in potential gives a local change in electron phase. Then the conservation law above is this: global conservation of electric charge equates to the invariance of the system under a global change of phase of the electron wave.

Gauge transformations

The German mathematician Hermann Weyl was the first person to make the link between global phase changes of the electron wave and charge conservation. He reached this conclusion in 1928 without the benefit of the Aharonov–Bohm effect and the experiments demonstrating it, and via a totally different route. Years previously, Weyl had been attempting to unify gravitation with electromagnetism, work that led him to devising ways of comparing standards of lengths at different points in space and time. He introduced a word for length scales that wound up translated as "gauge." So the word "gauge," permanently etched into the mind of every particle physicist, originally meant something more akin to "gauge" as used by railway engineers discussing track spacings. Weyl's original unification theory and the length scale meaning of "gauge" are both consigned to the history books. What lives on is the word gauge itself and the form of the prescription Weyl used.

These days, the term "gauge transformation" is used, for example, to discuss changes of electric and magnetic potential that leave the corresponding electric and magnetic fields of classical electromagnetism unchanged. The electric and magnetic potentials are not unique: there is a freedom to redefine them, a kind of shifting of the goalposts, and it's this recasting that bears the appellation "gauge transformation." Significantly, these transformations leave Maxwell's equations unaltered.

In fact, local changes in the electric potential, as for example when charges move about relative to one another, that preserve Maxwell's equations are only possible if corresponding local changes are made to the magnetic potential too. These two potentials turn out to be the two faces of the photon in quantum electrodynamics.

In quantum theory, changes in potential are compensated for by phase changes, so "gauge transformation" is taken to mean a change in the phase of the quantum field. The theory is then invariant with respect to such phase changes.

Global charge conservation may be related to global phase transformations, but global transformations are just the start of the story. Countless experiments confirm that electric charge is conserved, but they surreptitiously say much more; if charge was only conserved globally, then one charge could vanish here and another could be created over there. In this case charge would be overall conserved, but the "over there" could be very distant. Perhaps even on a different planet. Not only is this at odds with special relativity, since it implies faster-than-light communications, it means that charge conservation would not be visible unless experimenters could watch all of the Universe at once. Then if a charge vanished on Earth, they would have a detector waiting on some distant planet ready to detect the creation of the charge that guarantees conservation. In reality, if charge really were only globally conserved, charge conservation would not appear as a fact of real experiments.

A reasonable solution is to restrict charge conservation so that it applies locally, meaning that the destruction of a charge must be matched by the creation of a similar charge somewhere close by. That "close by" is to be constrained by special relativity, so that there's no chance of linking two processes faster than the speed of light. Given local charge conservation, this then implies local gauge transformations. In quantum theory this means phase changes in which the change in phase of the quantum field depends on position, in contrast to a global change in phase that is the same everywhere.

Thus there are two types of gauge transformation, local and global, sometimes called gauge transformations of the first and second kind, respectively. One way of visualizing the difference between the two is to imagine a crowd in a stand by the finishing post watching the end of a horse race. As the horses approach the finish, all heads will turn in unison to see which horse crosses the line first – a global transformation, in this case a rotation, since all heads are transformed in the same way. After the last horse has passed the post, heads will turn as people consult their neighbors to discuss the outcome, or look for the popcorn seller, or check their form guide: the transformation is still a rotation but now it varies across the grandstand, a local transformation.

That experiments and experience point to charge being locally conserved suggests that the phase transformations of electron–positron quantum fields should also be *local*. When such a change is forced on them, and the requirement made that the theory describing freely moving particles remain overall unchanged, something very remarkable happens.

The only way that symmetry may be preserved under local changes in the phase of an electron–positron field is if a new field is introduced. Changes in this new field then absorb those contributions arising from the local phase change of the electron–positron field that would otherwise destroy the symmetry properties. To do this compensating, this new field must be linked to the electron field in some way – the formal expression is to couple, and the strength of this coupling is expressed in the master equation by a coupling constant.

By some incredible economy of nature this new field, called in general a gauge field, is not "new" at all: it is the photon. The photon is an instance of a so-called "gauge boson," a force-transmitting particle having integer spin, one in the case of the photon, which emerges in gauge theory as the preserver of local gauge symmetry. To complete this display of nature's economy, the relevant coupling constant is just the electric charge of the electron.

To enforce local phase symmetry, and hence local charge conservation, in a theory of free electrons plus positrons this means introducing photons, and a particular form of interaction with those photons. There is no choice. They must be there, and they appear naturally. The idea that the interactions of the theory are determined by forcing it to respect local phase symmetry, or gauge symmetry, is termed the gauge principle.

Physicists find this powerful link between symmetry and interactions hugely exciting. Quantum field theory gives the basic structure, leaving the phase of the quantum field unspecified. If the phase is allowed to assume different values at different points of space and time, and the condition imposed that the master equation has a symmetry with respect to the phase change, then the interaction appears *automatically*. So start with a quantum field theory having no interactions, add symmetry via the phase, enforce the gauge principle, and the result is a theory with interactions, a so-called gauge theory. The interaction is mediated by a new field that emerges to fulfil this role, a gauge field. In QED, the gauge field is the photon. Besides QED, this is how both the electroweak theory and QCD work. And, miraculously, gauge invariance, and the special features that follow from it, survives renormalization. If this were not the case, then gauge theories would not be useful as a basis for theories of physical processes.

It almost goes without saying that this is not the route by which QED was created, though elements of gauge theory have been around from long ago. In QED the symmetry itself is so simple that things had to get more complicated before theorists really understood the full significance of the mathematical structure of which QED is but the simplest case.

So what is the symmetry of QED? Quantum electrodynamics can be constructed by enforcing symmetry under the simplest possible local phase changes on a theory containing only free quantum electron–positron fields. "Simplest possible" means that the phase is an ordinary angle, ranging from 0° to 360°, with phases combining with one another in the same way that rotations describing a circle in a plane combine. The latter form a group $SO(2)$ of rotations in a simple plane, and phase factors form the mathematically equivalent group, $U(1)$, the group of rotations in an internal abstract space. The symmetry group of QED is the rotational group $U(1)$.

This group has a single group generator that drives changes in its single parameter, the rotation angle. For a global $U(1)$ symmetry, that single generator equates to a single conserved charge, the electric charge. Forcing the $U(1)$ symmetry to become a local symmetry, and applying the gauge principle, means that the "angle" of rotation appears in the transformation properties of the single gauge field of QED, the photon.

It's important to appreciate that the "rotation" corresponding to the $U(1)$ symmetry is not in the four-dimensional space-time in which the electrons and photons move and live. The rotations of the $U(1)$ group are an internal rotation in an abstract space associated with the field. It's a rotation more because its mathematics is the mathematics of rotations, not because there really is some little knobble on an electron that physically turns in real space, and which is somehow related to the electromagnetic force experienced by the electron.

Photons do not interact with one another – physically a consequence of the fact they carry no electric charge and mathematically a consequence of the fact that the group $U(1)$ is Abelian. This observation may seem a little trite at this stage, but for QCD things are different: the group is not Abelian and the objects that appear in place of photons *do* interact with one another.

Symmetry and conservation rules

Much of the structure of a gauge theory is classical rather than quantum. Gauge transformations as changes to the electric and magnetic potentials are a clear example. The link between global symmetry and charge conservation is also visible in classical electromagnetism.

A powerful result applicable to classical physics exposes this link, and it's a result that appears elsewhere relating less familiar conserved "charges" to symmetry properties. The theorem is called Noether's theorem after the mathematician who developed it in 1918, Emmy Noether of the University of Göttingen in Germany. Noether's theorem takes as input the premise that nature is governed by the principle of least action. It also uses Gauss' theorem, named after Carl Friedrich Gauss, that relates the appearance or disappearance of something in a region to the total flow of that same something through the boundary of the creation region. Then with the introduction of a symmetry group, out pop conserved scalar quantities called charges, whose rate of change in time matches the flow of current towards or away from the region inhabited by the charges. Electrical charge corresponds to $U(1)$ symmetry and the matching current is electrical current. With other symmetry groups, "charge" and "current" will mean charges and currents of different types that have little to do with the electrical versions.

It's Noether's theorem that really shows clearly how symmetry under translations in space equates to momentum conservation, and symmetry under time displacement equates to energy conservation. Other less well-known examples will follow.

In quantum field theory, gauge symmetry takes on significance much deeper than anything visible in simple quantum mechanics. Global phase invariance, via Noether's theorem, gives a conserved charge as is to be expected. But in quantum field theory this "charge" has a very clear interpretation as the number of particles minus the number of antiparticles; in QED the number of electrons minus the number of positrons. It's this particle number which is conserved, and which in QED is identified with electrical charge. What quantum field theory does not do, however, is explain why electric charge is quantized. Charge always comes in multiples of a basic charge unit, and the only explanation so far of why this is the case comes from a "grand unification" of electrodynamics with theories of both the electroweak and strong forces, a theoretical framework that is not adequately supported by experiment.

The fact that the mathematics of gauge transformations applies to classical equations is significant, since the master equation fed into the path integral is classical. So the resultant quantum theory lies with its symmetry properties clearly exposed, as though it were a classical theory. Strictly speaking "gauge theories" could refer to non-quantum creations, but in practice gauge theories are quantum theories with a symmetry under a local phase change.

For QED, the phase symmetry is symmetry under a single rotation and the quantum field is the electron–positron field. In other theories, more exotic symmetries mean more exotic symmetry groups and rotations, and different quantum fields. Each rotation angle corresponds to a gauge field, which in QED is the photon, and for every rotation there is a generator of that rotation which in turn gives a conserved charge; in QED, electric charge.

Since scientists' theories of electrodynamics, weak forces, and strong forces are gauge theories, there are a number of features common to all three. Naively at least their gauge fields should all be massless, known to be the case at least for photons, since if the master equation of any of the three theories contains a simple mass term for the gauge field, then the special gauge symmetry properties are destroyed. In all cases the gauge principle, the imposition of a local phase symmetry, defines the interaction and, for those theories with more exotic symmetries, imposes important constraints on the coupling constant of the interaction.

In all three cases, the fundamental matter particles are spin 1/2 fermions, just as electrons and positrons are spin 1/2 fermions. Gauge theory shows how the gauge fields that allow these matter fermions to interact with one another must be vector fields. This is easy to see in the case of QED, because the gauge field is composed of a scalar electric potential and a three-vector magnetic potential, so is itself a familiar four-vector field. Another way of seeing why a vector field appears is to view a collision involving charged particles in terms of currents. A moving, charged electron or positron is just an electrical current, and current is a four-vector. Upon interaction, the electron current will be modified in some way via an electromagnetic field, which must therefore have a vector nature to do the job. Since a four-vector field has the rotational transformation properties of a spin 1 field, gauge theories of weak, strong, and electromagnetic forces explain interactions as arising through the exchange of spin 1 gauge fields. Gauge fields having higher spins are possible: a gauge theory of gravitation should have a spin 2 gauge boson, the graviton, mediating the purely attractive gravitational force – two gravitating objects never repel.

In quantum field theory, particles having spin 0, spin 1, or other integer spins *must* obey quantum counting rules according to Bose–Einstein statistics, and hence are bosons. This is a consequence of enforcing the causality of special relativity: two events can only depend one on the other if a light signal is able to communicate between the two. Similarly particles with half-integer spins *must* obey Fermi–Dirac statistics and so are fermions. This in essence is the spin–statistics theorem, largely attributable to Pauli who published a key paper on the link between the two in 1940. Quantum field theory provides the only known explanation of the relationship between spin and statistics, a result that rates as one of its triumphs.

So, in gauge theory, force-carrying particles are all bosons. It is via these force carriers that the fermions interact. Fermions are the "other" component of the Universe, the "matter." Only as a consequence of their special statistics can they combine to give composites with their particular, familiar properties. A universe with no interactions would be a dark place, populated only by fermions unable to combine to give the structure we see all around us.

QED

Beautiful QED! The best theory ever, the epitome of theoretical elegance, despite the intrusion of renormalization and perturbation theory.

Expressed in modern terms, QED is a quantum field theory with a local Abelian gauge symmetry. The gauge group is the $U(1)$ group, and invariance of the theory under local gauge transformations leads to the introduction of a single massless gauge boson, in this case the photon, matched by a single conserved charge called electric charge. This charge is resident on a fermion particle–antiparticle pair, the electron and positron. Quantum electrodynamics is a quantum theory of the interaction of light with matter, consistent with special relativity, and catering for the creation and destruction of both fermions and photons.

Dirac's original 1927 paper on quantum electrodynamics not only laid the first stone of QED, it also marks the introduction of quantum field theory. Over the next few years, this would be used by Fermi in his theory of the weak force, and by Yukawa in his attempts at understanding the strong force.

But throughout the period up to the start of World War II, at which point physicists were distracted from their endeavors, calculations of real physical processes were beset by divergences. The divergence problem was beaten in the years immediately following the War, inspired in part by two significant experimental results.

In 1947 Columbia University's Isidor Rabi, together with two graduate students John Nafe and Edward Nelson, performed experiments on some very fine detail of the spectra of hydrogen. They studied two particular energy levels for which, significantly, the gap between levels is attributable to differences in the interaction between the magnetic moments of the electron and proton. Rabi and his team found a discrepancy between their measured energy level spacings and those calculated from existing theory. One possible explanation of their result, proposed by Yale University theorist Gregory Breit, was that the assumed value of the electron magnetic moment was wrong.

Also at Columbia University, and also in 1947, Willis Lamb and his graduate student Robert Retherford performed other experiments on a detail of the hydrogen spectrum. This was the first measurement of the Lamb shift, a feature of the spectrum which, according to Dirac theory, should not be present.

Both results were announced at the Shelter Island conference of 2–4 June 1947, held at the Ram's Head Inn on Shelter Island in New York State, a theory-dominated gathering of twenty-four leading US physicists that has taken on an almost legendary status. It heralded a new chapter in quantum electrodynamics, with Lamb and Rabi's results provoking new interest, and providing theorists with both fresh challenges and precision results with which they could compare their calculations.

Over the next couple of years Schwinger, of Harvard University, and, independently, Tomonaga, at the Tokyo University of Education, devised almost identical renormalization schemes based on redefinitions of mass and charge as a means of eliminating the dreaded divergences of QED, at least to low order. In 1948, Schwinger was able to calculate a field theory correction to the electron's magnetic moment that gave a value in line with new experimental values, and that same year Tomonaga and his colleagues published the first correct field theory calculation of the Lamb shift, just ahead of Schwinger's. At Cornell University, Feynman got to the same end point but by using a totally different approach based on path integrals and divergences limited by cutoffs, which he published in a series of papers between 1947 and 1950. Feynman's approach was shown to be equivalent to that of Schwinger, and Tomonaga by Dyson in 1949, who that same year also demonstrated that renormalization held to all orders. Feynman, Schwinger, and Tomonaga shared the 1965 Nobel prize for their efforts.

The Lamb shift is due in part to a modification of Coulomb's law by photon self-energy (or vacuum polarization), Figure 3.8(b). Electrons are held in place in atoms by photon exchange: according to quantum electrodynamics, a photon can spontaneously create a temporary virtual electron–positron pair, the vacuum polarization, which has the effect of shielding the electron from the nucleus; it's this shielding that modifies Coulomb's law. Higher order diagrams correspond to additional shielding. The Lamb shift comes about from two particular electron states that, naively, should have identical energies, but in reality do not. Quantum electrodynamic corrections show that the energy of each state is shifted slightly, with part of the energy shift due to the vacuum polarization. However, the shift is different for each of the two states, so they no longer have identical energies. According to QED, what was a single energy level has become two separate energy levels. And the QED-corrected version turns out to be the correct one: it's the tiny energy gap between these shifted levels that Lamb and Retherford measured in their experiment. Lamb was awarded the 1955 Nobel prize, shared with fellow Columbia physicist Polykarp Kusch.

The anomalous magnetic moment of the electron, on the other hand, corresponds to the vertex interaction diagrams, the simplest of which is depicted in Figure 3.8(c). The value of the magnetic moment predicted from Dirac theory is consistent with a structureless, point-like electron. An anomalous magnetic moment, a difference between this value and the actual measured value, would indicate that electron mass and charge do have some kind of distribution. QED explains this in terms of a cloud of virtual photons surrounding the electron, continually being emitted and reabsorbed. Then at any one moment some of the mass–energy of the electron is on loan to a virtual photon, in effect altering the distribution of the electron mass relative to its charge, and giving rise to a magnetic moment contribution. Sure enough, the single diagram of Figure 3.8(c), the first "radiative correction" to the electron magnetic moment, contributes a factor $\alpha/2\pi$. This has a numerical value of 0.001 161 4, to be compared to the experimental value of the electron's anomalous magnetic moment of about 0.001 159 6, based on the value quoted at the start of the chapter. The $\alpha/2\pi$ portion was what Schwinger calculated in 1948. Higher order corrections, including contributions from virtual photons that in turn create virtual electron–positron pairs, modify this result. The fact that experiment and theory agree to around the four-loop level, apart from providing the best available test of QED, also strongly supports the renormalization program.

It was Kusch, together with Henry Foley, who measured the magnetic moment of the electron in 1948 and confirmed, as Breit had suggested, that it does indeed differ from the value expected from Dirac theory. The anomalous magnetic moment of the electron accounts for Rabi's spectral results.

The Lamb shift and the electron's anomalous magnetic moment are both examples of physical properties modified (in field theory language) by the spontaneous emission and reabsorption of a photon from an electron, or an electron–positron pair from a photon. Capturing the emission aspect, these are termed collectively "radiative corrections" in the sense that, via particle emission, they modify the process that would otherwise be obtained had the electron stayed simply an electron, or the photon a photon.

Some processes are nothing but radiative corrections, and field theory predicts numerous exotic processes that just cannot happen according to simpler quantum theory lacking particle creation and destruction powers.

One such process, which will rear its head again in due course, is the collision and scattering between photons. At first sight, this doesn't seem possible, since each photon lacks charge, so how can two photons interact? The answer is via the conversion of each photon into an electron–positron pair, as depicted by the light–light scattering diagram of Figure 3.8(e). This process, through which two photons collide to give two photons, symbolically $\gamma + \gamma \to \gamma + \gamma$, has been observed numerous times.

For example, in the 1980s, experimenters at DESY, the Deutsches Elektronen-Synchrotron in Hamburg, northern Germany, studied it using the Crystal Ball detector at the DORIS electron–positron collider. Electrons and positrons interact via photon emission: in a suitable mode, each becomes a source of a real photon, so the true collision process studied is photon–photon collisions. The Crystal Ball is a sphere of sodium iodide crystals covering almost the entire interaction region, and its function as a detector also depends on a radiative process. A photon entering a crystal gives rise to an electron–positron pair. As these pass through matter, they throw off further photons via a radiative process called bremsstrahlung or "braking radiation." These photons in turn create more electron–positron pairs via pair production and so on, producing an electromagnetic shower that causes the sodium iodide to give off a measurable flash of light that signals the arrival of the original photon.

An exotic QED radiative process underlies the first ever observation of antiatoms, announced early in 1996. Using LEAR, the Low Energy Antiproton Ring at CERN, Walter Oelert of the Institute for Nuclear Physics at the Research Center in Jülich, Germany, and his colleagues produced a handful of antihydrogen atoms. Where hydrogen atoms comprise a proton and an electron, antihydrogen atoms are made up of an antiproton and a positron, presumably in an exactly matching way. That's only a presumption, because one of the reasons for creating antihydrogen is to check that antihydrogen atoms really work exactly like regular hydrogen atoms.

Stan Brodsky and Charles Munger of SLAC, and Ivan Schmidt of the University of Federico Santa María in Valparaíso, Chile, had already proposed a way of creating antihydrogen atoms two years previously. Putting their scheme to work at LEAR, Oelert and his group passed antiprotons through a stream of xenon, selected because it is a gas with a heavy nucleus. In the presence of a nucleus, an antiproton can emit a photon that in turn becomes an electron–positron pair. Occasionally, the positron can be captured by the original antiproton to produce an antihydrogen atom, Figure 3.15.

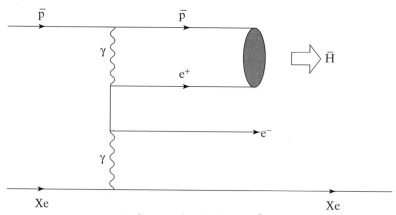

FIGURE 3.15 Production of antihydrogen, H̄, via quantum electrodynamic interaction between an antiproton, p̄, and a xenon nucleus, Xe.

Which all goes to show how much QED sets the scene for QCD, since analogous processes are believed to take place in QCD. For example, this mechanism, in the QCD context, predicts that collisions between a π^+ and a nuclear target will lead to the creation of another particle called the D^+ meson. So there are radiative processes in QCD paralleling those of QED, but the latter are tested more exactly, and are more easily understood.

In addition, QED establishes the utility of the field theory formalism, the power of the renormalization program to allow sensible predictions from perturbation theory, and the validity of the gauge principle, all of which carries over directly to QCD.

For its part, QED has undergone one more evolutionary jump that has so far been sidestepped. The electroweak theory does something rather more than build a gauge field theory of the weak force: it is a theory of both the weak *and* electromagnetic forces. Where QED alone is based on the group $U(1)$, the electroweak theory is based on the product of two groups, written $SU(2) \times U(1)$, where the $U(1)$ part is the QED group, and the $SU(2)$ part comes from the weak force. However, a theory of the weak force based on $SU(2)$ alone will not work. The electroweak theory only works if the weak *and* electromagnetic forces are unified. Quantum electrodynamics remains valid in isolation as the low-energy manifestation of a broader theory. But there is no corresponding, isolated theory of the weak force alone.

The electroweak theory of the unified weak force, based on $SU(2) \times U(1)$, and the QCD theory of the strong force based on the group $SU(3)$, stand together unchallenged in the minds of particle physicists as the explanation of three of nature's four forces. As if to reinforce the widespread acceptance of this view, it is called the Standard Model. There's a gut feeling amongst theorists that a further stage of unification will see the electroweak theory and QCD combined. So far, this has not been convincingly achieved.

For now, though, QED has served its purpose of showing how gauge field theory works, and it's now time to move on. Where does the symmetry of QCD come from, and what are the key players in that game?

4 Towards QCD

THE STRONG FORCE

Medieval alchemists would have done much better had they first discovered nuclear physics. The chemical properties of an element are dictated by the number and configuration of the electrons in its atoms, and the number of electrons must equal the number of protons in the nucleus. To turn one element into another therefore means changing the number of protons. No amount of chemistry can turn lead into gold, but chemistry was all the early alchemists had. They needed something else.

The world's first successful alchemist was the great New Zealand experimentalist Ernest Rutherford (Figure 4.1). By alchemists' standards he began modestly, turning nitrogen into oxygen. This he achieved by bombarding nitrogen with naturally occurring alpha particles from radioactive radium, in experiments performed in Manchester, UK, and reported in 1919. This was the first ever artificial transmutation of an element and the first induced nuclear reaction. It's a little ironic that the nuclear alchemist Rutherford had already been awarded the 1908 Nobel prize in *chemistry*, "for his investigations into the disintegration of the elements, and the chemistry of radioactive substances."

The long-range radiation produced in Rutherford's reaction he identified as hydrogen nuclei. By 1920 he had taken to calling them protons, and had reached what he termed the "natural conclusion" that protons were a component of the atomic nucleus.

Rutherford's results on bombarding nitrogen with alpha particles were part of a wider program of irradiating a range of lighter elements. His intention was to get alpha particles as close as possible to the nucleus to study the validity of Coulomb's law of electrostatics near the atom's core, and to look for the presence, or otherwise, of electrons there. Coulomb's law describes the classical electrostatic force between two charged objects, and says that as their separation is doubled, the electrostatic force between them decreases by a factor four. It's an example of what physicists call an inverse square law.

FIGURE 4.1 Talk softly. New Zealand-born Ernest Rutherford (facing the camera), pictured here at the Cavendish Laboratory in Cambridge in 1932, was the first to appreciate that atoms have a positively charged nuclear core. He also achieved the first artificial transformation of one element into another, having already won the 1908 Nobel prize in chemistry for his radioactivity work. Rutherford realized that protons were part of the nuclear core, and predicted the existence of neutrons. The "talk softly" sign is in deference to the sensitivity of nearby equipment: Rutherford was well-known for his loud voice. Credit: Courtesy Cavendish Laboratory, University of Cambridge.

Since lighter elements have less charge on the nucleus, an incoming alpha particle is able to approach more closely. The lightest element, and from this point of view the best, is hydrogen. But when Rutherford directed alpha particles onto hydrogen, he found the scattering did not match simple theory in which both alpha particle and hydrogen nucleus were point charges interacting via the electrostatic inverse square law. He saw evidence that the interaction force undergoes a dramatic change in magnitude and direction at very short distances. Crucially, he chose to interpret his results in terms of electrostatics, explaining the effects as due to the complex structure of alpha particles, and deformation of the actual electrons – an idea very much at odds with today's view of electrons as point particles – which he thought were present in the nucleus itself.

After Rutherford moved to Cambridge, he set James Chadwick (Figure 4.2) and McGill University's Etienne Bieler (Figure 4.3) to work on a careful study of the scattering of alpha particles by hydrogen. Chadwick and Bieler were looking for details of the size and shape of the alpha particle, and the field of force around it. As in Rutherford's

FIGURE 4.2 Chadwick used alpha particles scattered from hydrogen to reveal the strong force at work. In 1913 Chadwick had moved to Berlin to work with Hans Geiger, but following the outbreak of World War I he spent four years interned in a racecourse stable. As early as 1914 Chadwick had done important work on beta-decay. He went on to discover the neutron in 1932, winning the Nobel prize in 1935. Credit: Courtesy Cavendish Laboratory, University of Cambridge.

FIGURE 4.3 Etienne Bieler, pictured here at the Cavendish Laboratory, Cambridge, in 1921. He and Chadwick were the first to glimpse the strong force. Credit: Courtesy Cavendish Laboratory, University of Cambridge.

experiment, they directed alpha particles from a radium source onto a hydrogen target, but used an improved microscope to count the flashes produced by the emerging "H-particles" on a zinc sulfide screen. They made a systematic study of the how H particle production varied with alpha particle velocity, the angular distribution of the H particles, and their range at known scattering angles.

Their results, reported in *Philosophical Magazine* in 1921, showed clearly that, for high-energy alpha particles, the simple inverse square law of electrostatics failed, though it still held approximately for low-energy alpha particles. They considered several possible models of the alpha particle, eventually deciding that it behaves like an elastic "oblate spheroid" of semi-axes 8×10^{-15} m and 4×10^{-15} m. They wrote:

On this view, an H-particle projected towards an α particle would move under the ordinary electrostatic forces governed by the inverse square law, until it reached a spheroidal surface of the above dimensions. Here it would encounter an extremely powerful field of force and recoil as if from a hard elastic body.

They had glimpsed the strong force. The work of Chadwick and Bieler is taken to be the first evidence of the strong interaction, the adhesive holding together the components of the atomic nucleus.

Protons and neutrons unveiled

As early as 1920, Rutherford had speculated on the existence of an "atom" of mass one unit but with no electrical charge, which he envisaged as a kind of "neutral doublet" of an electron and a hydrogen nucleus. In due course he called this object the neutron, discovered after twelve years of searching by Chadwick by bombarding beryllium with alpha particles. All that was known about the neutron in the early days was that its mass was approximately the same as that of a proton, its charge was zero, and in collisions with atoms such as hydrogen or nitrogen it caused a recoil effect. Some speculated that the neutron was a proton–electron state, and others that it might be a fundamental particle in its own right.

Right up until the neutron's discovery, the nucleus had been viewed as a collection of protons and electrons, despite the increasing difficulties of justifying such a view. It was these difficulties, which included apparent inconsistencies between the electron–proton model and both special relativity theory and quantum mechanics, that motivated Heisenberg's interest in nuclear structure. Heisenberg seized on Chadwick's neutron as the key to the next step in understanding the nucleus, in 1932 proposing a model of the nucleus based on neutrons and protons.

Heisenberg's three-part paper applying simple quantum mechanics to the structure of the atomic nucleus marks the beginning of modern ideas of nuclear structure. Though he still entertained the idea that the neutron was an electron–proton pair rather that a particle in its own right, an offbeat notion from a modern standpoint, it was a view that was to prove crucial for later developments. ·

Important to Heisenberg's efforts was the discovery of deuterium, an isotope of hydrogen, by Harold Urey and co-workers early in 1932. Isotopes of an element all have the same number of protons (and hence electrons, and hence chemical properties) but different numbers of *neutrons* in the nucleus, as phrased in modern language. Deuterium has one proton and one atomic electron, as has "ordinary" hydrogen, but in addition the nucleus also contains one neutron. Incidentally, unlike deuterium, most isotopes are not honored with their own name.

For Heisenberg, the neutron was in some way a proton–electron pair. So his idea was that for a neutron in combination with a proton, the electron from the neutron could be shared between the two protons. Thus it would bind them together via an "exchange force," in analogy with the way atoms can bind together as molecules. In this way, a proton and neutron could bind together to form the deuterium nucleus. And likewise larger numbers of protons and neutrons could stick together to form heavier nuclei. He had little idea of exactly what form the force between proton and neutron should take, and his first guess spawned an industry as others subsequently tried all kinds of mathematical forms for the force relation.

In terms of subsequent developments, though, one of the most important points of Heisenberg's work was the notion that the proton and neutron are essentially the same object, and part of his formalism included a mathematical description for turning protons and neutrons into one another. He introduced a new variable that could take two values – one value corresponding to the proton and the other to the neutron. In modern language, what Heisenberg had introduced was the concept that the neutron and proton are two states of a single underlying object, which several years later was to be given the name nucleon: today, the term nucleon is a collective term meaning *both* neutron and proton.

Heisenberg's model was geared toward binding together a proton and a neutron. On the issue of how two neutrons or two protons might hold together it was rather less clear, and indeed at first it was not certain that they did. However, by 1936 physicists realized that not only can two neutrons and two protons bind together, but that the strength of the binding was equal to that of the neutron–proton binding, ignoring electromagnetic forces. In short, the force holding nucleons together exhibits charge independence.

By way of analogy, people are a bit like nucleons: a person is a single object having one of two possible states, male or female. The claim in many countries is that men and women are treated equally in law, in which case the law acts only on people, and not on men and women in different ways – there is a symmetry here. However, in matters of biology and, some might say, in other matters too, men and women are *not* identical, and respond differently, just as the two states of the nucleon respond differently in electromagnetic matters.

The charge independence of the nuclear binding could be represented naturally using the methods and the new mathematical quantity that Heisenberg introduced in 1932. This new quantity was christened isotopic spin, contracted in 1937 to isospin by the man who introduced group theory to quantum physics, Eugene Wigner. For Heisenberg's proton–neutron converter mathematics was straight group theory: quantum theory and group theory applied to the atomic nucleus had given the world a wholly new property of matter, isospin. Years later, isospin symmetry would prove crucial to both understanding strong interactions and attempts to classify elementary particles.

Heisenberg's proton–neutron model had usurped the proton–electron model of the nucleus that had held sway for years. But it by no means resolved all the difficulties facing scientists attempting to understand the workings of the atomic nucleus. One of those difficulties was to understand beta-decay, the spontaneous radioactive decay that results in the emission of electrons and antineutrinos.

A weak interaction interlude

This problem was tackled in spectacular style by Enrico Fermi in 1933. Fermi's theory introduced two key ideas that were important not only for understanding beta-decay, but also to the development of nuclear and particle physics as a whole. These were the ideas that the decay was due to an entirely new force, and that particles such as electrons and neutrinos could be *created*, as opposed to simply being dislodged from an existing home. Fermi's new force is now called the weak force, and his theory was the forerunner of modern electroweak theory.

Fermi allowed for the creation of particles such as electrons by making his theory a quantum field theory, so that electrons could be created in an analogous way to the creation of photons in electrodynamics. This mechanism for electron creation obviated the need for somehow hiding electrons inside the nucleus or inside neutrons, until that time the supposed source of electrons in beta-decay.

Particle creation also catered for a speculative particle called the neutrino. A long-standing riddle in the study of beta-decays had been the energy of the electrons produced, which has a continuous range rather than a single fixed value. This implied violation of the law of energy conservation, a prospect taken seriously by some such as Bohr. In 1930, Pauli proposed an equally radical idea that a new and invisible particle was given off at the same time as the electron, and because it too carried away some energy then the energy balance could be restored. The difficult choice was to deny energy conservation in the inner workings of the nucleus, or retain energy conservation and introduce a mystical new particle.

Fermi ran with the latter option, indeed his theory only hangs together if the emitted electron is accompanied by a partner particle of a related kind. In modern terms, Fermi's theory explained the beta-decay of a nucleus as the weak decay of a neutron into a proton, electron, and an antineutrino. Electrons and antineutrinos are not sitting in the nucleus waiting to escape, but are actually created in the decay process.

The (anti)neutrino is certainly hard to see, though, which is one reason why nobody had seen it by the time Fermi devised his theory, despite the huge experimental effort devoted to the study of beta-decay. Frederick Reines and Clyde Cowan, using a nuclear reactor as a neutrino source, finally found the neutrino in 1956. Reines shared the 1995 Nobel prize for his efforts.

As for Fermi's theory, despite its successes and enormous impact, it eventually foundered. Recall that at low energies, just a few hundred mega electronvolts, Fermi's theory worked fine. But at higher energies the calculations started to go transparently wrong, giving impossibly large cross-sections. The trouble was that Fermi's theory violated a key principle called unitarity, which says that the total probability for all possible final states should be unity. And there was a second closely related problem – Fermi's theory was not renormalizable. Both of these problems were to be fixed in due course by the electroweak theory.

But the challenges facing Fermi's theory were in the future when it was seized on as *the* theory of nuclear force, including that responsible for binding neutrons to protons. Soon it became clear that Fermi's theory could not fulfil this second role, though, since as a force it was just not strong enough. Yet even as Heisenberg and many others wrestled with this new dilemma, an essential clue lay waiting in the literature.

That clue was Yukawa's 1935 suggestion that the force responsible for holding neutrons and protons together in the nucleus was mediated by some heavy exchange particle. This was a fresh contribution to the still youthful notion of exchange forces, one that would eventually inspire a sea change in the understanding of fundamental forces. Yukawa's massive force carrier particle, the pion, made its experimental debut in 1947. The second key feature of Yukawa's work was the recognition that the neutron–proton force was a new and distinct force many times stronger than the electromagnetic interaction. Thus with Yukawa the strong force, first spotted experimentally in 1921, finally took a seat at the table (Figure 4.4).

FIGURE 4.4 Sin-itiro Tomonaga (left), Hideki Yukawa (center), and Shoichi Sakata (right). Tomonaga won the Nobel prize for his work on quantum electrodynamics, and Yukawa won his for his meson exchange theory of the strong force. Sakata proposed a model of baryons and mesons using the proton, neutron, and lambda as building blocks. Credit: University of Tsukuba, Tomonaga Memorial Room, courtesy AIP Emilio Segrè Visual Archives.

The dawn of accelerators

The 1930s not only saw the first appreciation of both the weak and strong forces and many key theoretical ideas, but they also witnessed the introduction of particle accelerators. Their history effectively begins in 1932, when two Cambridge physicists, John Cockcroft and Ernest Walton, won the race to split the atom. Their particle accelerator used a system of capacitors and rectifiers to accelerate protons to 700 keV. With their beam of fast protons they bombarded lithium to give pairs of alpha particles, the world's first nuclear reaction using artificially accelerated particles.

Many other scientists were also building machines to accelerate charged particles. At Princeton University, Robert van de Graaff developed the electrostatic generator that still bears his name. Van de Graaff's original 1931 machine could produce electrostatic voltages of about 1.5 million volts, and like the Cockcroft and Walton voltage multiplier machine worked by accelerating particles through a simple voltage drop.

Over in California they do things differently. Contemporaneous with Cockcroft and Walton, at the University of California, Ernest Lawrence and his student, Stanley Livingston, were developing a radically new machine, the cyclotron, for accelerating charged particles. Rather than using a single large voltage drop à la Cockcroft and Walton, the cyclotron relied instead on the then-new principle of resonance acceleration, in which particles are accelerated by passing them many times through the same modest voltage.

The first stab at a resonance accelerator was a device build by Norwegian Rolf Wideröe in Aachen, Germany. Wideröe's accelerator comprised three cylindrical electrodes in a line, with a space between each. He applied an alternating voltage between the middle electrode and those on each side. At the start of the cycle, the middle electrode was negative with respect to the other two, so a positive particle accelerated from the first electrode towards the middle electrode. While it passed through the middle electrode, the voltage switched, so that the middle electrode became positive with respect to the final electrode, and the particle was again accelerated as it left the middle electrode and headed for the third and final electrode. So the same accelerating voltage effectively kicked each particle twice. The trick is fixing the frequency of the alternating voltage to match the transit time of the particle in the middle electrode. Wideröe's humble machine was the ancestor of all modern particle accelerators.

Lawrence, having stumbled on Wideröe's paper, realized that a magnetic field could bend the path of the charged particle so that it re-entered the first electrode, and so could go through the complete cycle again and again. The result was the cyclotron.

In the cyclotron, magnets guide charged particles in a spiral path inside two hollow D-shaped electrodes, placed flat sides facing one another with only a narrow gap between them. An alternating voltage applied across the two electrodes provides the electric field responsi-ble for accelerating particles. The magnetic field is adjusted so that the time required for a particle to travel along a semicircular path inside one of the D-electrodes is equal to the time taken for the voltage between the D-electrodes to flip. This means that each time a particle reaches the gap between the electrodes, it is accelerated again. The first work-ing cyclotron, demonstrated in January 1931, was a mere four inches across and would fit neatly in the palm of your hand. A later version, built in 1932, was a little larger, eleven inches in diameter: particles received about 300 voltage kicks as they spiraled towards the outer edge of the machine, reaching a final energy of about 1.2 MeV. With this instrument Lawrence, Livingston, and Milton White repeated the proton bombardment of lithium to give alpha particles, already seen by Cockcroft and Walton.

Cyclotrons grew in size and rapidly became a major force in nuclear physics. However, they have a built-in limitation. For high energies, the mass of the accelerated particle increases according to the rules of special relativity, and for protons with energies of 25 MeV or more the increase becomes significant. For the cyclotron this is disas-trous, since it ruins the resonance condition on which the machine relies: as the particles gain in mass, they start to take too long to move through the D-electrode, and so arrive too late at the gap, missing the accelerating pulse. The frequency of revolution of the particles in the machine is no longer in step with the applied accelerating voltage, and particles will no longer be accelerated. For simple proton cyclotrons the realistic upper energy limit is therefore about 25 MeV.

The fix came right after World War II with the introduction of a principle fundamental to modern accelerator design. The idea, arrived at independently by Vladimir Veksler in 1944 and Edward McMillan in 1945, is called phase stability. As the mass of the particles inside the accelerator increases, the accelerating voltage frequency and the guiding magnetic field can be juggled so that the frequency of particles around the machine remains in step with the accelerating voltage. Particles of the correct energy will continually arrive at the gap in step with the accelerating voltage. Those that arrive too soon will be accelerated more, find themselves on a larger orbit, and so will arrive relatively late the next time round. Those arriving a little too late, on the other hand, will be accelerated less, follow a smaller orbit, and arrive a touch too soon next time round. This way, the accelerating particles tend to get more in step. It's this preservation of the relative timing between the accelerating particles and the voltage doing the accelerating that's called phase stability.

For a cyclotron, the way to reach higher energies, then, is to keep the gap width and magnetic field fixed, and gradually decrease the frequency of the accelerating voltage applied across the gap. The result is a new machine with a new name, the synchrocyclotron.

There is a penalty for the increased energy, however. Particles no longer spew from the accelerator in a continuous beam but arrive in bunches, one bunch per complete cycle of frequency. This has quite serious implications, since for most of the machine's time there is no particle beam, and the number of particles available for experiments is low, about 1% of that of a cyclotron.

The first synchrocyclotron went into operation at Berkeley in 1946, accelerating protons to 350 MeV. Many synchrocyclotrons were built, including the large 680 MeV machine at the Joint Institute for Nuclear Research (JINR) at Dubna, near Moscow, that came into operation in 1954. It's still in use today, modified and rechristened Phasotron, for nuclear physics research. A 600 MeV machine became operational at CERN – the acronym comes from the full original title Centre Européen pour la Recherche Nucléaire – in 1958. Like the JINR machine, this is still active, though it's undergone several modifications.

But these larger machines represent the upper energy limit of this acceleration method. At still higher energies, the beam wobbles about the main path in a way that becomes impossible to control. Not only that, the size of the magnet needed to steer the particles on their spiral path becomes unrealistic.

Thankfully, synchrocyclotrons are not the only way of exploiting the principle of phase stability. An alternative is to increase the frequency over the acceleration cycle, and at the same time crank up the magnetic field. A huge advantage of this arrangement, apart from phase stability, is that the particles can be steered on a circular rather than a spiral path. Then, only a small doughnut-shaped region needs to be subjected to the guiding magnetic field, resulting in huge savings on magnet costs. In fact, since only a doughnut region is needed, the entire accelerator can be built as a closed-up circle of pipe surrounded by guiding magnets, with accelerating kicks provided by high-frequency electric fields supplied at one or more points along the pipe. The result is an accelerator called the synchrotron. The very first synchrotron was built by McMillan at the University of California, and began operation in 1949, accelerating electrons around an orbit of a modest 1.3 m radius up to an energy of 320 MeV.

The downside of synchrotrons is that an initial fast particle beam has somehow to be introduced, or injected, into the machine, and the accelerated beam extracted – two operations that add complexity to the machine. Again, the particles arrive in bunches; to get a bunch from its injection velocity up to full speed means both the accelerating frequency and the magnetic field have to run through from low to high values.

But the energy! The first proton synchrotron was the Cosmotron (Figure 4.5), which went into operation at Brookhaven in 1952 with a 3 GeV proton beam. That's at least four times the energy of the big synchrocyclotrons and, significantly, is enough energy to create heavy particles such as protons and other new and exciting objects that had by then begun to reveal themselves.

In the same year that the Cosmotron began operation, the Brookhaven trio of Ernest Courant, Stanley Livingston, and Hartland Snyder published details of an important beam-control scheme called alternating gradient focusing, or just AG focusing. This was to fix a problem that had already become apparent in synchrotrons, namely that the beam wobbles about its central path in a way that's harder to control the higher the beam energy. These wobbles threaten the weak focusing effect of the magnetic field constraining the particles on their circular orbit, imposing an upper limit on the beam energy.

Alternating gradient focusing is a clever use of magnets to stabilize and focus the beam. The idea is to use magnetic fields that vary across the region of space occupied by the beam – this is the magnetic field "gradient." Such fields are used to alternately focus and

FIGURE 4.5 The Cosmotron. The world's first proton synchrotron, the Brookhaven Cosmotron, came into operation in 1952. Credit: Courtesy of Brookhaven National Laboratory.

defocus the beam to provide an overall net focusing effect. In practice quadrupole magnets perform this task. They have four poles, two norths and two souths. In a given magnet, both pairs might contribute to focusing in the horizontal plane while defocusing the beam in the vertical plane. The next magnet down the line is configured with the poles reversed to yield the opposite effect. It seems an unlikely way to focus a particle beam, but it works. It's possible to demonstrate net focusing of a light beam by using suitable lenses to focus and defocus the beam in a similar way.

Alternating gradient focusing, sometimes called strong focusing, was first put to the test in a small electron synchrotron at Cornell University in 1954. The first of the big machines to rely on AG focusing was the CERN proton synchrotron, which commenced operation in 1959. All modern ring accelerators use both simple two-pole magnets (called dipole magnets) to guide the beam using even magnetic fields and quadrupole magnets that provide focusing via their uneven fields. With the synchrotron principle and the focusing sorted out, the next energy-limiting factor comes from the economic, political, and civil engineering problems associated with building huge accelerators.

Modern particle physics could not begin without accelerators able to accelerate protons to several giga electronvolts. Such accelerators couldn't be built until the basic principles of their operation had been discovered. Those principles, the technology necessary to build the machine, and the political and financial will, would not have been present without the forty years of effort that led from Cockcroft and Walton's machine to the synchrotron. Only with the synchrotron would terrestrial particle physics begin. Though small cyclotrons are still built and used today, the accelerators that hog the particle physics limelight are large and expensive variations of the basic synchrotron. As a spin-off there are plenty of small machines about: around 15 000 particle accelerators are hard at work in places such as hospitals, electronics factories, and research centers across the globe.

With the advent of the particle accelerator, the study of the very small assumed the budget of the very large. The need for accelerators and big money is simple; nature has hidden two of her four forces inside the atom, with only radioactive decays and some artificial transmutations to help figure out how they work. Scientists need a microscope powerful enough to peer inside the atom and see what the nucleus is like and how it works. That microscope is the particle accelerator, and ever-increasing energies give ever better magnifications. And sometimes just looking isn't enough; accelerators are also a good way of smashing things up in the hope of being able to piece together the jigsaw afterwards.

It came from outer space

So why *terrestrial* particle physics? In fact, nature has thoughtfully provided one natural particle accelerator to aid particle physicists throughout the Universe. This is the great particle accelerator in the sky. Cosmic rays, mostly protons and other nuclei that have been accelerated to high energy in space, constantly bombard Earth. Their origin and the way they are accelerated remain uncertain, though exploding stars called supernovae may play a role. What is clear is that when so-called primary rays blast atoms of the Earth's atmosphere, many thousands of meters above sea level, they initiate "secondary rays" which ultimately become showers of particles detected at sea level. About three quarters of all cosmic rays arriving at sea level are muons, with an average energy of about 2 GeV. Other cosmic ray particles detected at sea level are neutrinos, electrons, and positrons; about 180 particles pass through each square meter of the Earth's surface every second. They are hitting us all the time.

Some cosmic rays have incredible energies. The High Resolution Fly's Eye detector (HiRes) out in the Utah desert and the Akeno Giant Air Shower Array (AGASA), situated rather more appealingly in a Japanese vineyard, have both picked up monster rays with energies exceeding 10^8 TeV. In other words, some of the rays zapping Earth have energies more than a hundred million times the energy of each beam in the Tevatron, currently the world's highest energy accelerator.

In fact, to think of cosmic rays as a consequence of a giant particle accelerator in the sky is to prejudge one of the great mysteries of science. Where do cosmic rays come from? How is it they have such vast energies? Are they born as low-energy particles that are then accelerated, or are they created in some process with a huge energy from the very beginning, in which case they don't need accelerating? Nobody really knows the complete story.

Following the discovery of cosmic rays by Austrian physicist Victor Hess in 1912, experimenters charted the interaction of these rays with matter, often taking their experiments to high altitudes to capture the higher energy cosmic ray debris. Their simple detectors were photographic plates, in which charged particle tracks appear as thin lines of exposure, or cloud chambers revealing particle tracks as vapor trails. British physicist Donald Perkins used photographic emulsions flown in an aircraft at 30 000 feet to glean the first hint of Yukawa's pion. His single event was a four-pronged star, the faint wispy tracks spanning just a couple of hundredths of a centimeter. That same year, 1947, a group led by another British physicist, Cecil Powell, both confirmed the pion's existence and showed that it decayed into a muon. Powell's group used photographic emulsions exposed to cosmic rays 5500 m up in the Bolivian Alps, at Chacaltaya, and at the Observatory of the Pic du Midi 2800 m up in the French Pyrenees.

Several other key particle discoveries, including the positron, muon, and a particle called the K meson, came from cosmic ray experiments. The 1930s, 1940s, and early 1950s were the glory days of cosmic ray experiments. But cosmic rays are uncontrolled, their arrival time and their energy in the hands of higher powers. Experimenters prefer to rely on accelerator engineers, though cosmic ray research is currently enjoying a renaissance.

The discovery of the pion in 1947 traditionally marks the birth of particle physics. The stage was set. The world was at (relative) peace following World War II, and scientists were resuming their peacetime research. Feynman, Schwinger, Tomonaga, and others were closing in on the theory of electrodynamics that would eventually form a template for theories of particle physics. Plans for both the Cosmotron and Berkeley's Bevatron, proton accelerators that would transform the study of fundamental particles, were on the drawing board. And scientists now had the pion, not only the carrier of a strong inter-nucleon force, but the only known force-carrying particle besides the photon.

Like all new subjects, particle physics quickly developed its own vocabulary, some of which already existed in 1947, though much more was to follow. Certain of these "new" words have a Greek origin, rather in keeping with a penchant for naming particles after Greek letters, and using Greek symbols in equations. In some cases, the original meaning of the Greek names has been overtaken by events, so is best ignored – it's simpler to stick with the modern meaning.

So there are mesons, such as the pions. All mesons have a spin of 0 or a positive whole number. They not only experience the strong force, they are transmitters of it. The strongly interacting matter particles among which mesons carry force are termed baryons, for example protons and neutrons. All baryons experience the strong force, all are heavy, and all have spins that are half integers such as 1/2 and 5/2. Mesons are bosons, baryons are fermions.

Collectively, mesons and baryons are called hadrons. Put the other way round, a hadron is simply any particle that experiences the strong force, though they may also feel other forces – a charged pion, for instance, responds to the electromagnetic force. From hadron comes the adjective hadronic, for example hadronic collision meaning collisions between strongly interacting particles.

In stark contrast are leptons, matter particles that do *not* experience the strong force. Examples are electrons, muons, and neutrinos. All known leptons have spin of 1/2 and all are considered today to be truly fundamental, meaning they cannot be subdivided and have no internal structure.

By the start of the 1950s, the fledgling venture of particle physics knew of three strongly interacting particles, hadrons, in terms of which physicists hoped to understand the strong force. But they would quickly find many more particles before they had any sort of understanding of what was going on.

Scattering experiments

Physicists try to understand the strong force using scattering experiments. The archetypal hadron–hadron scattering, that is scattering involving two strongly interacting particles, is proton–proton scattering, and the subject of experimenters' interest is the cross-section. Cross-section measurements are set alongside theorists' calculations in an effort to unravel the inner workings of the nucleus.

The most all-embracing cross-section, and the traditional goal of many nuclear and particle physics experiments, is the total cross-section, written σ or sometimes σ_T, where the T stands for total. The total cross-section measures the overall strength of an interaction. Roughly speaking, it is the geometrical size of the colliding particles.

For example, a 1990 measurement of the total cross-section for proton–antiproton collisions at an energy of 1.8 TeV, made by the E-710 collaboration at the Fermilab Tevatron, gave a value of 72.1 mb. The mb stands for millibarns, a standard unit of cross-section measure in high-energy physics: one barn is 10^{-28} m^2; a millibarn, or mb, is 10^{-31} m^2. A disk 72.1 mb in size has a radius of about 1.5×10^{-15} m, or 1.5 fermi, so this is roughly the "size" of the proton and the range of the strong force. However, the total cross-section changes with energy, one reason for caution when equating geometrical size with cross-section. For comparison, scattering of electrons from protons gives a proton radius measure of about 0.8 fermi.

The E-710 group also measured the elastic cross-section, in which the two colliding particles "bounce" off one another. For this they found a value of 16.6 mb, or about 23% of the total cross-section. The difference between the total and elastic cross-sections is the inelastic cross-section, in this case 77% of the total, which measures particle creating processes.

Inelastic collisions can yield spectacular showers of exotic particles. With large numbers of collisions, and the prospect that many particles can be produced in each, measuring all the particles produced is difficult. Instead, experimentalists measure what they call inclusive cross-sections. These can be represented by a + b → c + X where a and b are the colliding particles, c is a specific particle the experimenters are looking for, and X is everything else. One advantage of this is that experimenters can set their detector equipment to "trigger" for a given particle c with perhaps a specific energy. In contrast to inclusive cross-section measurements, experiments to measure exclusive cross-sections, represented by a + b → c + d, measure and identify every particle.

Experimenters discover the strong force

The earliest hadronic scattering experiments, in which strongly interacting particles are scattered one off another, were the alpha particle on nucleus experiments of Rutherford, Geiger, Marsden, and others in the early 1900s. But the precursors of modern hadron scattering experiments, exemplified by proton–proton scattering, didn't take place until the 1930s.

It was a decidedly shaky beginning, however. Between 1929 and 1931 Christian Gerthsen at Tübingen University performed proton–proton scattering experiments using protons with energies of up to 100 keV. But at such low energy the smallest distance of approach is around 10^{-13} m, too large to feel the onset of the new force revealed by Chadwick and Bieler's scattering of alpha particles off hydrogen. Indeed Gerthsen was able to verify only that proton–proton scattering at these energies matched the scattering expected from a straightforward electrostatic force.

Over in the Department of Terrestrial Magnetism at the Carnegie Institution of Washington, Breit and his team knew they would need at least four times this much energy if they were to get accelerated protons close enough to the target protons to see beyond electrostatic effects and experience new physics. From 1926 Breit headed a group at Carnegie whose objective was to use "high-voltage research" to study the properties of the nucleus, a pretty tall order since there were no particle accelerators at that time.

The first "high-voltage" proton–proton scattering experiment was reported in 1935 by Warren Wells as part of that program, "high-voltage" being a rather subjective term. Though accelerators had been invented, Wells didn't use one, preferring instead high-energy protons knocked out of cellophane bombarded with alpha particles from a polonium source. Though the proton energy ranged up to 1.5 MeV, the experiment was too crude to yield much conclusive data. Wells found just

FIGURE 4.6 The cyclotroneers. Ernest Lawrence (center) with his fellow cyclotroneers in front of the 27 inch cyclotron, 1933. Stanley Livingston is the left-most seated figure. Milton White is to the right of Lawrence, wearing a light-colored top. Credit: Lawrence Berkeley National Laboratory.

33 so-called "intimate" proton–proton collisions from 200 000 proton cloud chamber tracks viewed using 15 000 stereographic photographs.

The next attempt faired a little better, though was still rather rough. Starting in late 1934, Milton White at the University of California used one of Lawrence and Livingston's small cyclotrons to fire 750 keV protons onto a hydrogen target. About 250 000 cloud chamber tracks were photographed yielding around 160 collisions, reported in a brief letter in *Physical Review* in 1935. "The data, though still very meager, are sufficiently at variance with the concept of a 'point-charge' proton to make a brief report of interest to theoretical physicists," wrote White. White's was the first high-energy accelerator proton–proton scattering experiment (Figures 4.6 and 4.7).

Merle Tuve, Norman Heydenberg, and Lawrence Hafstad at Carnegie reported a much more refined experiment in 1936. Daunted by the prospect of examining vast numbers of tracks on photographs necessary to get sufficient data, Tuve and his collaborators waited until

FIGURE 4.7 The eleven inch cyclotron that Milton White used first to bombard lithium with protons, and later to perform the first high-energy accelerator proton–proton scattering experiment. The accelerator itself was housed in a brass "pizza box" roughly eleven inches square and an inch deep. This was set between the two poles of a large magnet, the dark cylindrical objects right of center in the picture. Credit: Lawrence Berkeley National Laboratory.

they had the technology to count scattered protons electrically. Using a 1.2 MeV electrostatic accelerator, they measured thousands of protons scattered at different angles and various energies, instead of the handfuls of collisions in the earlier experiments. They too witnessed the breakdown of simple electrostatic scattering and the onset of a new proton–proton interaction at a distance less than 5×10^{-15} m.

Following their paper in *Physical Review* was a second paper, a detailed analysis of their results by Breit, Edward Condon, and Richard Present. This paper compared the proton–proton experiment to the results of neutron–proton scattering, particularly the work of Fermi's group in Rome, concluding that the p–p interaction was the same as the p–n interaction. This killed the notion that electromagnetic forces held together the particles of the nucleus. It was also the support and inspiration for charge independence, dressed in modern mathematical clothing by Condon and Benedict Cassen, using Heisenberg's 1932 formalism, in a paper immediately following that of Breit, Condon, and Present.

The year 1936 was therefore highly important for the strong force, which gained a secure foothold both experimentally and theoretically. Protons and neutrons look the same to this force, which binds them together in a grasp that's far stronger than the electrostatic force, and independent of it. Furthermore, the evidence had come from some of the earliest accelerator experiments, a foretaste of what was to come from the mighty Tevatron and other accelerators around the world.

Working mesons

Tourist guide books describe the Canadian city of Vancouver as one of the world's most beautiful, its inhabitants hedonistic, and the lifestyle relaxed. One thing that they don't tell you is that Vancouver is also home to TRIUMF, Canada's national meson facility.

TRIUMF is a meson factory, meaning it mass-produces pions with a reasonable energy of their own. It is one of just three meson factories in the world, the others being LAMPF, the Los Alamos meson production facility, and the PSI, or Paul Scherrer Institute, in Zurich.

TRIUMF is based on an 18 m diameter cyclotron, the world's largest. Negatively charged hydrogen atoms, hydrogen atoms with an extra electron attached, are guided in a spiral path inside the machine by an emblematic six-segment pinwheel-shaped magnet. Accelerating particles receive voltage "kicks" twenty-three million times a second. Within a fraction of a second, the particles are traveling at 75% of the velocity of light, and have a final energy of 520 MeV.

The beam passes through one of three foil strips that remove both electrons, leaving bare protons that are then extracted from the accelerator. Because the stripping foil can be moved within the accelerator, the energy of the final particle can be adjusted to any value up to 520 MeV; three separate foils allow three separate beams of different energies to be extracted from the one machine.

Beam number 2 is used for the manufacture of radioisotopes for use in medicine. Beam number 4 is guided to the Proton Hall for proton collision experiments. Finally beam 1, which like the other two beams leaves the cyclotron as protons, is guided to the Meson Hall where it strikes a carbon or beryllium target to spawn pions. In about twenty billionths of a second, the positive and negative pions break up to give muons and (anti)neutrinos. The muons themselves disintegrate into electrons, a neutrino and an antineutrino after a further two millionths of a second. The single beam has become a beam firstly of protons, then of pions, and finally of muons. Experiments using these different particle types are located at increasing distances along the Meson Hall from the cyclotron.

Towards the distant end of the Meson Hall, sufferers from cancer benefit from scientists' taming of the pion. Pion radiotherapy is an experimental form of cancer treatment for attacking tumors in the brain and pelvic areas. As a pion moves through living tissue, it does little damage until it has almost ceased moving, at which point it breaks up in a small blast of radiation that kills surrounding cells. In this way, radiation is delivered by pion "depth charges" to the errant cells, with little damage to other material, potentially adding to the armory of radiotherapy techniques, though pions as a weapon against cancer appear not to live up to earliest expectations.

Most TRIUMF experimenters working with pions direct their beams onto less animate targets, as the facility is primarily a nuclear physics research center. Many experiments focus on scattering pions from protons and neutrons. At TRIUMF pion energies, the cross-sections for these kinds of experiments show some very distinctive features. In particular, pion–proton scattering has a cross-section that first rises steeply as energy increases, reaching a maximum for pions having kinetic energy of 195 MeV. As the energy rises still further, the cross-section drops away again just as fast to give a bell-shaped cross-section versus energy plot.

Though TRIUMF's pion beam is of modest energy by today's standards, it was more than the early pioneers had when they studied pion–proton scattering. Using the University of Chicago cyclotron, in 1952 Fermi, Herbert Anderson, and others first saw signs of this same peak, even though the maximum available pion energy was a mere 136 MeV. Subsequent experiments confirmed their suspicion that the rise in cross-section with increasing energy was indeed one half of a symmetric peak. This was something experimenters recognized from their earliest transmutation experiments with neutrons almost twenty years before: the peak was a resonance.

True transients

Resonance is achieved when the natural frequency of some system matches the frequency of whatever is attempting to vibrate it. When this happens, the system draws energy and vibrates more vigorously. For example, piano tuners rely on establishing resonance between a tuning fork and a piano string. If a tuning fork is set humming and held alongside the string, the string tension can be adjusted until the string hums in unison. At this tension, the string is in tune, and absorbs energy from the sound wave coming from the fork and passing over it. A microphone placed beyond the string would register a small dip in volume as the string tension was increased until resonance and then over tightened. This dip matches the sound energy absorbed by the string and re-radiated in all directions at resonance; it's this re-radiation that a bystander hears and the piano tuner uses.

Resonance effects are common, ranging from radio tuners to the sound from organ pipes to annoying buzzes in cars. Particle accelerators exploit a resonance between the accelerating voltage frequency and the transit times of charged particles.

There's resonance in sunlight too. Sunlight is not of even intensity right across its spectrum. Spread out by passing it through a prism to show its range of colors, sunlight shows narrow dark stripes. These are due to absorption of certain frequencies of the light generated in the Sun's core by the solar atmosphere, and are called Fraunhofer lines. The strong sodium D lines in the yellow part of the Sun's spectrum are a classic example, and demonstrate the presence of sodium in the Sun's atmosphere.

The lines correspond to energy jumps in atoms of the solar atmosphere. When the energy of the light matches that required for an electron to step up from one energy level to another within an atom, the atom will absorb light. Because that light is re-radiated in any direction a dark line results in the spectrum seen from Earth. This is none other than a real life occurrence of atomic spectra and a consequence of the quantum nature of atoms.

British astronomer Joseph Lockyer discovered helium by studying the lines in sunlight. Helium has two electrons; its line spectrum is more complicated than that of hydrogen but much simpler than other heavier elements. However, even the simple helium lines have a strange property.

Up in the ultraviolet, invisible to the human eye, are helium lines where by rights there shouldn't be any. These lines are at such high frequency that the corresponding photons have more energy than is required to rip an electron clean out of the helium atom. The reason for this apparent paradox can be traced to helium having two electrons. Spectral lines in the visible range correspond to the ladder of energy levels of one or other of the electrons. The very high frequency lines are related to the two electrons behaving in concert. Energy levels corresponding to raising electrons to excited states *together* seem to be possible, but to create them means inserting more energy into the helium atom than is required to completely remove a single electron. A helium atom will try to absorb the larger amount of energy but is unstable, and instead of forming an excited helium atom, the state decays rapidly into a positively charged helium ion and an electron.

This is resonance. The light striking the helium atom matches the amount of energy required to create an excited state, at least in principle. In practice, the energy of the light is so great that this excited state cannot be stable, and it decays.

In contrast, the spectral lines at the lower frequencies correspond to helium atoms absorbing light as electrons rise to higher energy levels before dropping back down again, emitting light in the process. The crucial difference is that these are bound states, since although the electron is in an excited state it does not have enough energy to leave the confines of the atom altogether. In a resonant state, however, the excitation energy is enough to allow an electron to leave the atom, so a resonant state is not a bound state.

The characteristic peak of the resonance scattering cross-section is due to absorption of light of frequencies both at and close to the resonance frequency. This means the peak shows a spread in frequency, a much larger spread than that shown by "normal" spectral lines. This spread is a feature of the unstable nature of the resonance, and is caused by a smearing out of the energy level by the continuous levels all around it. As a result, energy of a limited range of frequencies can be absorbed. Via the uncertainty principle this spread, or uncertainty, in the energy translates into the lifetime of the resonance.

Resonance occurs when the scattering energy matches that of some excited state that is unstable. The result is a short-lived resonance state. Resonances are seen in atomic, nuclear, and particle physics.

FIGURE 4.8 The Italian navigator. Fermi, whose name lives on in "fermion," "Fermilab," and the element "fermium," won the Nobel prize in 1938 for his work using neutrons to induce nuclear reactions and create new isotopes. Fermi demonstrated the first-ever controlled self-sustaining nuclear reaction on 2 December 1942. The news was relayed to the Manhattan Project managing committee in a phone call from Arthur Compton, then working with Fermi in Chicago: "The Italian navigator has just landed in the New World." Credit: The University of Chicago.

Piano strings and organ pipes inspire the nomenclature and the idea, but resonance here is really about energy levels, and in this respect is a quantum phenomenon.

Fermi (Figure 4.8) was well placed to realize that the pion–proton cross-section peak was a likely resonance effect. In the mid 1930s Fermi, then at the University of Rome, had led the way in the study of neutron scattering, especially the scattering of low-energy neutrons. Scientists knew from the day neutrons were discovered that they would offer the ideal tool to probe the nuclear core since they have no electric charge, and so can easily reach the nucleus without the problems of electrostatic repulsion faced by protons. Fermi and others studied numerous interactions with many elements, prefacing the 1938 discovery by Otto Hahn and Fritz Strassmann in Berlin of barium in the remnants of neutron-irradiated uranium. Their long-time collaborator and by then refugee, Lise Meitner, along with her nephew Otto Frisch, realized that this signaled a new energy-generating process they christened nuclear fission. Hahn got the Nobel prize (in chemistry); Meitner, a victim of politics and personalities, sank into relative obscurity, and the world got both nuclear power and atomic bombs. Those developments were in the future, though, when Fermi and his colleagues recorded many instances where neutrons were remarkably strongly absorbed, and speculated that some kind of resonance effect was at work. Neils Bohr noted that the time taken for neutrons to pass by the nucleus should be about 10^{-21} s, whereas Fermi's experiments suggested that neutrons dallied in the vicinity of the nucleus far longer than was necessary. This inspired Bohr's compound nucleus mechanism, proposed in 1936, in which the incoming neutron enters the nucleus and shares its energy amongst the nuclear constituents. Eventually the unstable excited nucleus decays, perhaps by emitting a neutron or a high-energy photon, a gamma ray.

Bohr's model remains a valid picture of nuclear resonances. A direct nuclear reaction takes about the same time as a particle transiting the nucleus. An unbound resonant state, on the other hand, lives for about 10^{-18} s, about a thousand times longer. For comparison an excited nuclear state, a bound state, may well decay by emitting a gamma ray photon and have a lifetime of maybe 10^{-12} s. Weak radioactive decays are slower still.

Slow neutron experiments were on the minds of many at that time. In the same year that Bohr outlined his model of compound reactions, in effect a picture of resonances, Breit and Wigner, at the Institute of Advanced Study in Princeton, New Jersey, published one of the most important results of scattering theory ever. In their paper entitled simply "Capture of Slow Neutrons," they presented what is now known as the Breit–Wigner formula, which describes the characteristic shape of resonance curves in scattering cross-sections. The Breit–Wigner approach compares the phase of an incoming matter wave with that of the post-scattering wave, whose phase has been shifted by the scattering center. The resonance peak corresponds to the two waves being completely out of step. The Breit–Wigner result, derived decades ago in the context of neutron scattering, actually applies to a host of resonance scattering processes.

The particle explosion

In the four years following Fermi's pion–proton scattering result, the peak was confirmed as a resonance. Experimenters were able to show that it followed the Breit–Wigner form, and that it had a lifetime of about 10^{-23} s: Fermi had discovered the first particle resonance.

Unknown to Fermi, his Δ particle resonance, as it's now known, was the first rumble of an explosion, the particle explosion. Though by 1950 several "fundamental" particles had been discovered, including the electron and positron, the proton, neutron, neutrino, muon, pion, and in 1944 the first of the K mesons, nobody was prepared for what was to happen. A flurry of discoveries from cosmic ray studies introduced the neutral Λ of mass 1115.7 MeV in 1951, the Σ^+ of mass 1189.4 MeV, and the Ξ^- of mass 1321.3 MeV, both in 1953.

That was also the year accelerators wrested the torch from cosmic ray experiments. The world's first proton synchrotron, the Cosmotron, came into operation the year before, and in 1953 confirmed the existence of the Σ^+, then went on to discover its sister, the Σ^- particle. High-energy pions produced with the Cosmotron's 3 GeV proton beam helped to confirmed Fermi's $\Delta(1232)$ resonance, and showed it was accompanied by others, including the $\Delta(1920)$ and the $N(1680)$, where the numbers in brackets are the masses of the resonances, the total combined energy of the pion–proton system at the resonance peak.

Using the same letter to denote different resonances, as for the $\Delta(1232)$ and the $\Delta(1920)$, is madness with a method. Recent data tables list 22 distinct Δ resonances including these two, and 21 N resonances of which $N(1680)$ is the sixth in order of increasing mass. This is a particle explosion indeed. The method in the madness is that resonances sharing the same identifying letter are related in some way, even though their masses are different. These family relationships turn out to be central to making sense of the zoo of "elementary" particles.

Most resonances are not as simple to spot as the $\Delta(1232)$, which is relatively easy because it has the lowest mass resonance in its family. Higher resonances overlap with others from their own family and from other families, producing complicated bumps in cross-section plots that are tricky to analyze.

The $\Delta(1232)$ resonance, like the rest of its family, is also an example of a resonance formation, meaning that colliding particles combine into the resonance, which then decays, and nothing else. In resonance production, on the other hand, the resonance is produced along with one or more other final state particles. In these cases, a specific group of particles appears with a combined mass equal to the mass of the resonance. Experimenters have to monitor all particles produced in the process, and plot all possible combinations, to find evidence of the resonance, making these experiments much harder.

They can be very productive, however, and many important resonances have been found this way. For example, the ρ meson, mass 770 MeV, was found in 1961 at the Cosmotron by studying π^-

collisions with protons. The collision produced pairs of pions and either protons or neutrons, and revealed itself as a peak in the mass distribution of pion pairs. That same year one of the big names among resonance hunters, the American Luis Alvarez, and his colleagues used the Bevatron to find another important meson, the ω of mass 781.9 MeV. They studied groupings of three pions in proton–antiproton collisions that produced five pions.

New resonances came thick and fast: in 1960 experimenters found the $\Sigma(1385)$ and the $K^*(892)$. Other discoveries of 1961 were the $\Lambda(1405)$ and the η meson of mass 547.3 MeV. In 1962 came another heavy resonance, the $\Sigma(1530)$ and two more mesons, the ϕ of mass 1019.4 MeV and the f_2 of mass 1275.0 MeV.

Resonances have a typical lifetime of about 10^{-23} s, making them the most transient phenomena ever studied. So are they real particles? The answer is yes, they're as real as protons and neutrons, but there is one crucial difference. Protons and neutrons are at the end of the line; there is nothing else for them to decay into via strong interactions and still satisfy various conservation laws. But otherwise resonances are fundamentally the same as protons and neutrons and all are "particles."

And that hints at the short lifetime. Resonances in particle physics are particles that are created by and decay via the strong interaction, and this decay is very fast. When decay via the strong force is inhibited, for example due to some conservation principle, then the particle may be stable or may decay relatively slowly via some other interaction that doesn't respect the conservation law. For example, neutrons and protons are both stable against decay via the strong force. But free neutrons decay with a mean lifetime of nearly fifteen minutes into protons (and electrons plus antineutrinos) – a vast time compared to a resonance lifetime. The neutron decay is so slow because it takes place through the weak interaction, which shows scant respect for the conservation rules that prohibit such a decay via the strong force. Protons, the lightest of the baryons, are stable because there is no lighter baryon into which they can decay. Stable means they have a mean lifetime of at least 1.6×10^{25} years; people check these things using big detectors at the bottom of mines since "grand unification" theories that combine strong, weak, and electromagnetic forces indicate that protons do in fact decay. So far nobody has seen it happen. Some decays are electromagnetic, and these have lifetimes longer than resonances but shorter than weak decays, reflecting the relative strengths of the three forces.

For example, the neutral pion decays electromagnetically into two photons with a mean life of about 8×10^{-17} s, much faster than the 2.6×10^{-8} s needed for the weak decay of charged pions.

Resonances could not have been discovered and explored without particle accelerators – they are so transient they could not be seen any other way. Understanding resonances played an important part in the development of QCD. Amongst other things, TRIUMF and the other meson factories continue to measure properties such as mass and decay modes of resonances and continue to explore the way resonances feature in low-energy processes.

The new resonances were only part of the particle explosion. Dirac, through his relativistic quantum mechanics, had predicted antiparticles – in particular the existence of the positron, and this had already been found. However, doubts remained that the equivalent prediction would apply to heavy particles such as protons, partly because this in turn would imply antimatter and hence antigalaxies, things that nobody had ever seen.

The issue of the existence or otherwise of an antiparticle partner to the proton was so important that the Bevatron energy was selected with the hunt for the antiproton in mind. The investment was rewarded. In 1955, a team led by two American physicists, Owen Chamberlain and Emilio Segrè, discovered the antiproton by examining the debris from collisions of protons with a copper target. The discovery of the antineutron followed the next year, also at the Bevatron.

Other antibaryons were discovered shortly after, including the $\bar{\Lambda}$ in 1958, the $\bar{\Sigma}$ in 1960, and $\bar{\Xi}$ in 1962. By 1964, the particle count exceeded 80, including particles and antiparticles. The sheer number left little doubt that few, if any, could be truly fundamental particles, an idea supported by emerging patterns. Physicists knew there had to be a way of organizing and understanding the confusing multitude of particles. Indeed the revolution that would ultimately see strongly interacting particles organized into neat families, based on fundamental new building blocks, was already unfolding. Those building blocks have a name – quarks.

THE EIGHTFOLD WAY

There's an inherent problem with text books and popular books on physics. That there are Great Ideas in physics cannot be questioned; the problem is, in books they often appear to have been the result of perfect reasoning, to have burst from thin air, to have been instantly recognized as Great Ideas, and to have been accepted as such by all. The impression can come across that the pressing problem of the day was totally resolved, and that researchers were then ready and waiting for the next step in a logical and tidy sequence.

The story of QCD illustrates many times over that this is frequently not the case. But both technical and popular accounts of physics must cut through the uncertainties, the misunderstandings, the false starts, the unwieldy notation, and the multitude of external inputs that underlie all Great Ideas, so as to avoid swamping and confusing both the student and casual reader alike. Which is a pity really, since it creates a sanitized version of how physics unfolds, at odds with the far messier reality.

Quarks are a Great Idea, one of the greatest. They are also a good example of how Great Ideas have a history, how they usually fail to solve the whole problem – at least at first – that they are designed to fix, and how they can be seen as crazy or worse by other researchers. This is how it is, in varying degrees, for many developments in QCD, and in physics as a whole. The quark story is a nice place to see real physics in action.

A cosmic puzzle

Perhaps the story of quarks really begins with a couple of murky gray images created by George Rochester and Clifford Butler of the University of Manchester, UK. In 5000 cosmic ray cloud chamber photographs, they found two showing "forked tracks of a very striking character," as they described them in their 1947 report in *Nature*.

Rochester and Butler believed they had found evidence for new elementary particles. According to their analysis, one picture showed the decay of a neutral particle, of mass somewhere between 340 MeV and 800 MeV, into two charged particles. The second picture showed the decay of a charged object, though not a proton, of mass between 490 MeV and 940 MeV, into another charged particle plus a neutral particle. With only nucleons, muons of mass 105 MeV, and pions of mass around 140 MeV to pick from, it's easy to see how they reached the conclusion that they had something new on their hands.

Striking the tracks may have been, and exciting their implications, but for a while nobody paid a great deal of attention. Then, two years later, came confirmation of the Manchester results, when Carl

Table 4.1 *The known particles, circa 1952.*

Particle	Mass (MeV)	
γ	0	photon
e^-	0.51	electron
e^+	0.51	positron
μ^-	105.7	muon
μ^+	105.7	antimuon
K^0	497.7	strange mesons
K^+	493.7	
K^-	493.7	
π^0	135.0	pions
π^+	139.6	
π^-	139.6	
p	938.3	proton
n	939.6	neutron
Λ	1115.7	strange baryon
Δ^{++}	1231.0	resonance

Anderson and his team at the California Institute of Technology (Caltech) picked out similar tracks in their cloud chamber pictures. Further experiments by the Manchester group, and by several other cosmic ray groups around the world, ultimately revealed two new neutral particles: today these are known as the Λ, of mass 1115.7 MeV, and the K^0, mass 497.7 MeV. Early experiments also showed that the Λ decayed into a proton and some other unidentified object, known today to be a negative pion. The K^0 turned out to have charged partners, now called the K^+ and K^-. In fact, the first observation of the K^+ is sometimes put at 1944, a single track found by Louis Leprince-Ringuet and Michel L'héritier working in the French Alps, though their claim is now disputed. The K^+ itself was eventually seen to decay into a μ^+, along with some other unidentified object, now known to be a neutrino: this is most likely the charged particle decay seen by Rochester and Butler in one of their two original pictures. Their other decay is thought to have been $K^0 \rightarrow \pi^+ + \pi^-$.

Table 4.1 lists the meager population of the particle zoo of that era, inflated somewhat by the new particles. Little was known about the new particles: the data were fragmentary, and the statistics dreadful: handfuls of cloud chamber tracks are not much to go on when it comes to measuring masses and estimating cross-sections. The results were confusing, too, since given two different decays it was hard to tell if they arose from different particles, or from two different decay modes of the one particle type. Not a single decay event had had both decay products identified.

But one thing *was* clear: the new particles carried with them a puzzle. The tracks were long enough to show that the new particles decayed into others in about 10^{-10} s. This is the time frame of a weak decay, far longer than the time frame of a strong decay, so hinted at decay via the weak interaction. Yet enough data existed to suggest that these particles were generated with a frequency comparable to that of pions, indicating that they were created by the strong interaction. But if that were the case, why did they not *decay* via the strong interaction? Surely something created via the strong force, like Fermi's newly revealed pion–proton resonance, should decay the same way to give an intermediate state with a lifetime of around 10^{-23} s, not 10^{-10} s? The basic problem, then, was to understand how the lifetimes of the new particles could be so long, given that they seemed to be created in a strong interaction. It was because of this weird behavior that these particles were collectively dubbed "strange."

Armed with limited data, and without a sound theory of how any nuclear particle interactions really worked, in 1952 Abraham Pais at the Institute for Advanced Study in Princeton was nevertheless able to make some progress in untangling the confusion. Taking seriously the idea of a strong creation and weak decay mechanism, Pais devised a simple number system with one rule for strong (and electromagnetic) interactions and a different rule for weak interactions. His rule blocked strong decays, as required, and predicted that strange particles should be created in pairs.

It may not seem much. In fact, Pais only billed his idea as an "attempt to codify the present information": the context was that of a search for ordering principles. His rule was an ad hoc device to try and make some sense of mysterious and sparse experimental results, and this it did. It even made a prediction, which turned out to be at least partially true. His paper contains some false notions, and some deep insights that were ahead of their time. What it does not contain, however, is a watertight, comprehensive theory. Nor was his idea unique, since it turned out that Yoichiro Nambu, Kazuhiko Nishijima, Sadao Oneda, and others followed a similar line of reasoning. But his idea helped to point the way forward. In short, Pais' paper was a piece of real science.

What about that prediction? At the Cosmotron, experimentalists witnessed the double production of strange particles, as predicted by Pais. To do this, they fired a 1.5 GeV pion beam into a cloud chamber filled with hydrogen, giving the reaction $\pi^+ + p \rightarrow \Lambda + K^0$: with an accelerator, creating strange particles became easy. In fact, both the Cosmotron and the Bevatron were able to mass-produce strange particles, in a controlled way, to the tune of thousands or even millions a day. There was no way that cosmic ray researchers, picking out the occasional strange particle from thousands of photographs taken over weeks or months, could ever compete. Accelerators heralded the demise of the classic era of cosmic ray research, at the same time providing the means of exploring the new strange particles. They soon added to the list: the Σ^+ and Σ^- are both heavy, strange particles. With the advent of accelerators, things moved very quickly.

Even before the discovery of the next heavy strange particle, the Ξ^-, was published, a half-page letter in *Physical Review*, published in 1953, set the stage for an improved rule system governing strange particles. The new scheme was to prove crucial. The paper's author was a junior faculty member at the University of Chicago: his name was Murray Gell-Mann (Figure 4.9).

FIGURE 4.9 Murray Gell-Mann. No other name appears so often in the story of QCD. Credit: Louis Fabian Bachrach.

Strange symmetries

If there's one name in particular that's associated with both quarks and QCD, it is Murray Gell-Mann. But both quarks and QCD were well down the track when Gell-Mann grappled with the problem of strongly created, weakly decaying strange particles.

Gell-Mann took inspiration both from Pais and from Columbia University physicist David Peaslee. The latter had suggested that the principle of charge independence should hold for the new particles, an idea that Gell-Mann took on board and explored using isospin. He supposed that the Σ^+, Σ^0, and Σ^- particles form an isospin family, and that as family members they all have the same value of isospin, in this case isospin 1. Further, he figured the K^+ and K^0 should form a second family of isospin 1/2, with the $\bar{K}^0$ and K^- forming another isospin 1/2 family.

All up, his short paper contains some pretty wild ideas. His family groupings included the Σ^0, which had not even been discovered at the time and would not be found until 1955. Added to that, he omitted the then-known Λ from his scheme. Added to that, his allocations for the K mesons required that the antiparticle partner to the K^0 was not simply the K^0 again, but was the K^-, and that the antiparticle partner to the K^+ was the $\bar{K}^0$. This was unpleasant and surprising to some, who felt that since the photon is its own antiparticle, $\bar{\gamma}$ equals γ, and the π^0 meson is likewise its own antiparticle, $\bar{\pi}^0$ equals π^0, then the K^0 should follow suit. But there is nothing to say it must, and Gell-Mann broke with popular expectation. His efforts met with some opposition, but in due course his scheme, arrived at independently the same year by Osaka City University's Nishijima and Tadao Nakano, would triumph.

Over the next couple of years, Gell-Mann and Nishijima separately refined the family grouping idea, deducing equivalent versions based on a whole new property of matter, "strangeness." Strangeness is a simply additive, conserved quantum number, with those particles in a given isospin family all sharing the same strangeness value. According to this rule the three Σ particles are all assigned a strangeness value of $S = -1$. The K^+ and K^0 are $S = 1$, and the $\bar{K}^0$ and K^- have strangeness $S = -1$; the proton and neutron do not posses the strangeness attribute, so for them $S = 0$. The idea is that in a strong interaction process, the strangeness values on the two sides of the reaction must add up to give the same value, but for a weak interaction process they don't. For example, with the Λ particle assigned $S = -1$, the strangeness scheme forbids the neutron–neutron process $n + n \rightarrow \Lambda + \Lambda$, a reaction allowed by Pais' numbering system, since the left-hand side has strangeness of zero, but the right-hand side has strangeness of -2. The reaction $\pi^+ + p \rightarrow \Sigma^0 + K^+$ is a viable strong process since the strangeness allocations run $0 + 0 = -1 + 1$. However, the decay $\Lambda \rightarrow p + \pi^-$ cannot be a strong decay since the strangeness allocations go as $-1 = 0 + 0$, which is false; this decay is controlled by the weak interaction.

Introducing a whole new property of matter, with a new conserved quantum number, is not something physicists do every day. But conservation laws for electric charge and for isospin had already set the precedent. Added to which there was another simple, additive, conserved property of matter linked to the very stability of the material Universe itself.

Aging aside, so far as anybody knows we are not slowly disintegrating. Nor is the material world we inhabit. This is a sign that protons do not decay. Neutrons, on the other hand, are happy to decay in just a few minutes into a proton and some other objects. Yet even neutrons are discerning about their possible decays, refusing, for example, to decay into just pions. All of which is very suggestive of a nucleon conservation number of some kind. This number is called the baryon number, denoted B.

Protons and neutrons have baryon number $+1$; antiprotons and antineutrons have baryon number -1. Baryon number is conserved by *all* interactions, so far as anybody has ever seen. For example, the neutron weak decay $n \rightarrow p + e^- + \bar{\nu}$ is allowed under this scheme since the baryon number assignments go $+1 = +1 + 0 + 0$. There's no corresponding decay for protons, since protons are slightly lighter than neutrons so, in terms of weak decays into baryons, they are the end of the line. Decays such as $n \rightarrow \pi^+ + e^-$ and $p \rightarrow \pi^0 + e^+$ are not allowed, since for both $+1 = 0 + 0$, which is false. Baryons – strongly interacting matter particles at least as heavy as protons – are $B = 1$. Antibaryons have baryon number $B = -1$.

Baryon number conservation, on one hand an empirical rule, is also inspired by quantum field theory, in which the nucleon can be viewed as a fundamental fermion in the same vein as the electron. Then, just as the electron has an antiparticle partner, so too will the nucleon – hence antiprotons and antineutrons. But, in addition, quantum field theory requires that the particle number, the number of particles minus the number of antiparticles, be conserved. For QED, with electrons minus positrons, this "conserved charge" is identified with electrical charge. For nucleons, the number of nucleons minus the number of antinucleons is likewise conserved, and the corresponding conserved charged, once called "heavy charge," is the baryon number, B. Conservation of B equates to a global $U(1)$ symmetry of the master equation for nucleons, just as a global $U(1)$ symmetry equates to a global charge conservation for electrons and positrons. But in the case of electric charge, the $U(1)$ symmetry and charge conservation become local, and out pops the corresponding force transmitter, the photon. There is no matching step for the baryon number $U(1)$ group, in the sense that it does not lead to a viable theory of particle forces.

There's a lepton number, too, and its story parallels that of the baryon number. The creation of an electron is accompanied by a matching positron, but creation of a μ^+ rather than a positron would suffice if electric charge conservation was all that mattered. Put another way, a process such as $e^+ + e^- \rightarrow \mu^+ + e^-$ is not allowed in nature, even though it conserves electric charge, though $e^+ + e^- \rightarrow \mu^+ + \mu^-$ is fine. As an empirical rule, electrons and the neutrino associated with electrons are assigned a lepton number of $L_e = +1$, and their antiparticles are assigned $L_e = -1$. The muon, and the neutrino associated with it, which is distinct from the electron's neutrino, are assigned a lepton number $L_\mu = +1$. Here, the μ and e subscripts highlight the distinction between the muon and electron neutrinos. The corresponding antiparticles are ascribed $L_\mu = -1$. So according to this scheme, the early cosmic ray K^+ decay goes as $K^+ \rightarrow \mu^+ + \nu_\mu$ with the muon lepton number assignments $0 = -1 + 1$, the ν_μ representing a muon–neutrino having a lepton number of $+1$. A third lepton number, L_τ, labels another lepton called the tau. Martin Perl and his team at SLAC found the heavy tau lepton in 1975. Perl's discovery earned him the 1995 Nobel prize, which he shared with Reines for the latter's detection of the electron neutrino. The tau also has a matching neutrino, only seen directly for the first time at Fermilab as recently as 1998, and both have a tau lepton number L_τ of $+1$. The separate lepton numbers are always individually conserved by all interactions, like the baryon number, and like the baryon number they also relate to a global $U(1)$ symmetry of the corresponding fundamental field.

Which these days is all a bit of a mystery. Pure, global symmetries are seen as accidental, apparently having little role in gauge theories based on local symmetries. Could it be that they are not so pure after all? That is, the corresponding "conserved quantities" may not be truly conserved. In 1974 Howard Georgi and Sheldon Glashow, both at Harvard University, introduced what remains the simplest attempt at a unification scheme bringing strong, weak, and electromagnetic interactions all under one theoretical umbrella. Their strategy was a logical extension of the unifying theory of Glashow, Weinberg, and Salam for the weak and electromagnetic forces, and was based on the larger symmetry group $SU(5)$. The most dramatic prediction of their grand unification theory is that B and L are not individually conserved, though the combination $B-L$ should be. If they are right, then matter is actually unstable, and we are all slowly decaying, as is everything around us. A number of proton decay experiments, typically watching for occasional decays in a large quantity of substance at the bottom of a mine shaft, where cosmic rays are less of a nuisance, have – as yet – found no evidence of proton decay. So if the Universe is decaying, it's doing it very slowly. According to the simplest implementation of Georgi and Glashow's theory, the proton should prefer to decay according to p $\rightarrow$ π^0 + e^+ – clearly violating simple baryon and lepton number conservation – with a mean life of no more than 2×10^{31} years. Since recent experiments give a value exceeding 9×10^{32} years, protons are more stable in terms of this decay channel, or route option, than "simple" theory suggests. Thus the basic version of Georgi and Glashow's theory has to be ruled out.

Such matters were still two decades away, however, when Gell-Mann and others were struggling to understand the ground rules of strong interactions. Baryon number and strangeness turned out to be key labels. So too did isospin.

Like spin, like isospin

Isospin is modeled on spin. One way of thinking of spin in "everyday life" is in terms of the rotations of a rigid rod about some fixed point, to which the rod is attached at one end. These rotations form a group, $SO(3)$. For spin applied to particles such as electrons, the appropriate symmetry description is the group $SU(2)$, which is only subtly different from $SO(3)$. And mimicking the fact that spinning tops go round either clockwise or anticlockwise, spinning electrons have restrictions on the allowed values of their spin, revealed in the splitting of some atomic spectral lines.

Isospin mirrors spin, with two states of the nucleon, one the proton and the other the neutron, in place of two states of spin, up and down. The "rotations" of isospin apply to some abstract "isospin space" rather than any kind of real space: isospin is an internal affair, not a space-time transformation. But all of the mathematics carries over in a direct way, illustrated schematically in Figure 4.10.

The proton and neutron form an "isospin doublet," where doublet simply means a pair having some shared property, in this instance that of being nucleons. The crucial point is that the strong interaction does not distinguish between the two possible isospin states: all it knows about is the nucleon. Those two nucleon states are labeled $I_3 = +1/2$ for the proton, and $I_3 = -1/2$ for the neutron. Notice that "3" subscript. The labels 1, 2, 3 for components of isospin, I, rather than the perhaps more conventional x, y, and z, often used when discussing spin and rotation, emphasize that isospin space is an *internal* space, not a physical space in which particles reside.

Other particles besides protons and neutrons can be grouped into isospin families. The three pions – the π^+ (139.6 MeV), the π^0 (135.0 MeV), and the π^- (139.6 MeV) – all look the same so far as the strong force is concerned, though their electromagnetic properties expose them as three separate particles. They form an isospin triplet, or family of three – a single pion with three possible isospin states. If "pion" is the family name, isospin $I = 1$, then the three "forenames" are pi-plus ($I_3 = +1$), pi-zero ($I_3 = 0$) and pi-minus ($I_3 = -1$). A four-member family, a quadruplet, is the $\Delta(1232)$ resonance, with $I = 3/2$, and I_3 values of $+3/2$ (Δ^{++}), $+1/2$ (Δ^+), $-1/2$ (Δ^0), and $-3/2$ (Δ^-).

Using isospin, physicists can predict some relative strong interaction cross-sections. To do this means thinking about *combinations* of particles and their isospin labels. Once again the prototype is spin.

For spin, a vector, not only does its magnitude matter – the length of the "spin arrow" – but so too does its orientation. The rules for combining vectors must respect both orientation and length information, and for "quantum vectors," such as the spin of an electron, they must also accommodate the restrictions on the possible values of the vector. For a spin 1/2 electron, the spin arrow has length 1/2 and points either up or down. Two such particles can form a combination having a total spin 1 or 0. But because each has two possible states, up or down, there are *four* possible combinations. If both spins are oriented in the up direction, then the z-component of the total spin will have a value

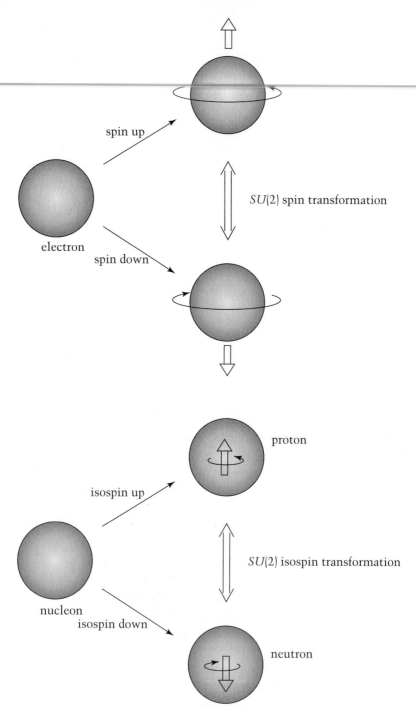

FIGURE 4.10 The mathematics of spin carries over to isospin, where the "rotations" are in an internal space.

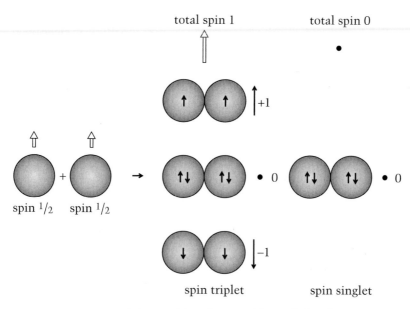

FIGURE 4.11 Spins – and isospins – combine such that the two spin 1/2 states, having either spin up or spin down, give three possible states of total spin 1. These three states have a third spin component of +1, 0, or −1. In addition there is a single spin 0 state.

of +1. If both are down, then the z-component of the spin will have a value of −1, Figure 4.11. These are the two obvious z-component possibilities for a combination having total spin of 1. Presumably, the two particles may have opposing spins too, so that the z-component is zero.

There's something missing, though. That's three cases, but where's the fourth? Not only that, the spins-oriented-opposite option, in terms of z-components, could be spin up and spin down, or spin down and spin up, implying that two possibilities are involved. Thirdly, a combination having a total spin of 1 should behave, in terms of its spin possibilities, like an "elementary" spin 1 object. Such a spin 1 object should, according to the rules, have three allowed z-components: +1, −1, and 0.

This last possibility seems to be missing: it provides the clue of how to fix up the rules for combining spins. Combining the spins of two particles A and B as {A(up) B(down) + B(up) A(down)} gives a combination having a z-component of spin of 0. So too does the combination

{A(up) B(down) – B(up) A(down)}. These two combinations seem very similar, but are in fact quite distinct. One way to see this is to swap over the particles A and B. In other words, swap the particle labels, but leave the spin assignments alone, so that A(up) → B(up), B(down) → A(down), and so forth. Then the first combination becomes {B(up) A(down) + A(up) B(down)}, which is the same as the initial version. However the second combination goes to {B(up) A(down) – A(up) B(down)}, in other words *minus* {A(up) B(down) – B(up) A(down)}, the *negative* of the starting combination.

Tracking what happens to combinations of two particles when they are interchanged has a familiar tang to it – recall that a combination of two electrons is antisymmetric under a swap. For spin, the combination {A(up) B(down) + B(up) A(down)} is *symmetric* under the interchange of particle labels A and B, whereas the combination {A(up) B(down) – B(up) A(down)} is *antisymmetric*, on account of the appearance of an overall minus sign when A and B are swapped. This option is the odd one out relative to the other three possibilities, since both A(up) B(up) and A(down) B(down), the z-component $+1$ and -1 cases, are also symmetric under the interchange of A and B. So, it's the lone antisymmetric "odd man out" combination that's the zero spin case, for which the *total* spin is 0 and z-component likewise zero. The mixed combination {A(up) B(down) + B(up) A(down)} is the spin 1 case with a z-component of 0. In this way, all three spin 1 states share the same symmetry. That symmetry is exactly opposite to that of the spin 0 combination.

Since isospin exactly mimics the spin prototype, two isospin $1/2$ particles in combination give total isospin of one, with the third component taking values $+1$, 0, and -1, and a single isospin zero case, with a third component of zero.

Combining isospin $1/2$ particles might apply to scattering of protons and neutrons, for example. Historically, however, it was the scattering of pions and nucleons that was important to isospin's reputation. With isospin $1/2$ particles combined with isospin 1 particles, such as nucleons with pions, the total isospin possibilities become $I = 3/2$ and $I = 1/2$. For example, negative pions colliding with protons can create the resonance named N(1520), having isospin of $I = 1/2$, Figure 4.12. Only the I_3 values trivially add to give the same number on both sides of the reaction: the I values satisfy the more subtle vector combination rules. Another example is the collision of positive pions with protons

	π^+	p	$\longrightarrow$	Δ^{++} (1232)	
I	+1	$+1/2$		$+3/2$	
I_3	+1	$+1/2$		$+3/2$	
	π^-	p	$\longrightarrow$	N^0 (1520)	Δ^0 (1232)
I	+1	$+1/2$		$1/2$	$3/2$
				or	
I_3	-1	$+1/2$		$-1/2$	$-1/2$

FIGURE 4.12 Rules for combining isospin are the rules of vector addition. Only the third component, I_3, adds up in a simple way. Isospin predicts different isospin channels, with different resonances resulting, for reactions such as $\pi^- p$.

to create the Δ^{++} resonance, which has $I = 3/2$ and $I_3 = 3/2$. Strong interactions do not depend on I_3 values, so for example the cross-section for $\pi^+ + p \rightarrow \pi^+ + p$ is the same as that for $\pi^- + n \rightarrow \pi^- + n$, processes for which only the $I = 3/2$ option is available. However, strong interaction processes do depend on total isospin I, so that the cross-section $\pi^+ + p \rightarrow \Delta^{++}$, for which $I = 3/2$, differs from the cross-section $\pi^- + p \rightarrow N$, which has $I = 1/2$. Fermi's pion–nucleon scattering experiments agreed with the isospin-based expectation that cross-sections for the three processes $\pi^+ + p \rightarrow \pi^+ + p$, $\pi^- + p \rightarrow \pi^0 + n$, and $\pi^- + p \rightarrow \pi^- + p$ are in the ratios 9 to 2 to 1.

The power of isospin thus established, it's not surprising that researchers looked to $SU(2)$ symmetry as a candidate for the fundamental symmetry of the strong interaction.

It didn't work out. But in trying to build a theory based on $SU(2)$ something of tremendous value emerged. In 1954 Chen Ning Yang and Robert Mills (Figure 4.13) at Brookhaven National Laboratory reasoned

FIGURE 4.13 Chen Ning Yang and Robert Mills in 1999 at Stony Brook, New York, during a symposium marking Yang's retirement. The theoretical structure that Yang and Mills explored in 1954 forms the basis of both electroweak theory and quantum chromodynamics. Credit: Courtesy C. N. Yang.

that internal $SU(2)$ symmetry should hold at each point of space and time separately, thereby making it a local symmetry. They wrote:

Once one chooses what to call a proton, what a neutron, at one space-time point, one is then not free to make any choices at other space-time points. It seems that this is not consistent with the localized field concept that underlies the usual physical theories.

Their prototype was QED, where a global $U(1)$ symmetry, forced to become local, sees the introduction of a gauge field mediated interaction. In the case of QED, that gauge field is the photon. Yang and Mills hoped to do the same based on $SU(2)$. In terms of building a working theory of the strong force, they failed. But what they did do was figure out how a gauge theory should work for a non-Abelian rather than an Abelian group, thereby creating what's now called Yang–Mills theory. The difference sounds like a minor technical step but the consequences are enormous. In fact, both the electroweak theory and QCD are

Yang–Mills theories. Yang and Mills didn't invent either, mainly because they hadn't got the right symmetry inputs.

Their efforts, generalized in 1956 by the Japanese theorist Ryoyu Utiyama at the Institute for Advanced Study in Princeton, from today's perspective seem a logical and almost obvious step along the road to modern gauge theory. Yet at the time the work was tinged with controversy: a nasty problem relating to the mass of the gauge bosons could have seen the whole idea shot down. Pauli criticized Yang and Mills on this point, having himself written down, but not published, a version of the Yang–Mills result. Another unpublished version came from Cambridge graduate student Ronald Shaw. But it was Yang and Mills who published, and it's their names that are attached to the theoretical framework underpinning QCD.

The families grow

The nucleons, pions, and the Δ resonances are all isospin multiplets, families of particles linked via symmetry – in this instance, isospin symmetry. These families differ in size: the nucleons form a doublet, the pions a triplet, and the Δ resonances a quadruplet. Correspondingly, the appropriate representations of isospin $SU(2)$ symmetry linking particles within a family must also be of different sizes, corresponding to the numbers of family members.

This idea harks back to rotations on a tabletop, represented most easily using two-by-two matrices, or via three-by-three matrices if the rotations are referred to the walls of the room containing the table, rather than the edges of the table itself. For protons and neutrons, $SU(2)$ symmetry works through two-by-two matrices, the group's smallest representation. The three-by-three representation of $SU(2)$ links the three possible pion states to one another. The four-by-four representation embraces all four Δ states.

A possibly subtle point is that a symmetry applicable to different sized families of particles corresponds to different sized representations of the *one* group, in this case $SU(2)$. The groups $SU(3)$ and $SU(4)$ do not feature here, although their smallest representations are indeed three-by-three and four-by-four matrices, respectively: they are *different* groups describing *different* symmetries and conservation laws.

The smart way of figuring out these larger representations is to first work out how large the particle families are. The essential idea here is to form larger symmetry multiplets from combinations of smaller ones. The rules physicists and mathematicians have devised for doing this codify the example above of a combination of two isospin doublets, say two nucleons. Recall that the result was a three-strong isospin one family, and a single isospin zero "singlet," or family having just one member. This is written as $2 \otimes 2 = 1 \oplus 3$. The numbers show how many family members inhabit each $SU(2)$ multiplet. They are shown bold as a reminder that it's multiplets that are being added and multiplied. And the special add and multiply symbols, with circles round them, are shown thus for the same reason.

Actually, just what the representation matrices themselves look like is very much secondary. It's the multiplets that contain the physics: since the rules show how to build larger multiplets from smaller ones, physicists know right away they have families of symmetry-linked particle states. And that is what's interesting as a way of bring order to the particle zoo.

With the inclusion of strangeness those family groupings were set to grow. The strangeness-based prediction of the Σ^0 and Ξ^0 particles, discovered in 1955 and 1959, respectively, confirmed both baryon number and strangeness as reliable labels of the strong interaction. In fact, the sum of the two is a *second* conserved label of the strong force, in addition to isospin, and has its own name: baryon number plus strangeness is called hypercharge.

Two labels rather than one implies a larger symmetry group, larger than the single-label group $SU(2)$. The group to do the job is the two-label group from the same stable, $SU(3)$.

Inspiration for turning to $SU(3)$ came, in part, from a doomed attempt to understand some observed particles as composites of more fundamental ones. In 1956, the Japanese physicist Shoichi Sakata proposed that the proton, neutron, and lambda were the fundamental building blocks of the baryons and mesons, the lambda's presence being necessary to get strangeness in on the act. His motivation was the search for a meaning behind the strangeness rule, his inspiration the way the neutron had resolved many puzzles of the atomic nucleus twenty-five years earlier: he reasoned that the lambda might be the missing mysterious ingredient of compound particles, in much the same way that the neutron had been the missing ingredient of the atom's nucleus.

The idea that pions might be a nucleon–antinucleon combination, for example, seems odd by today's standards. But Sakata, and others like him, who sought to devise composites from more elementary objects to make sense of the mushrooming number of "elementary" particles, was pursuing a line of attack that would eventually prove very fruitful.

But not right away. Sakata's model initially stalled, largely because there was no hint of how the three particles might stick together. Progress came in 1959 from three fellow Japanese physicists, Mineo Ikeda and Shuzo Ogawa at Hiroshima, and Yoshio Ohnuki who, like Sakata, was at Nagoya. They took seriously the idea that, in the same way the proton and neutron are two states of a single underlying entity, the triplet of proton, neutron, and lambda were three states of a single object, in other words three members of a triplet. This being the case, the three particles would be equivalent in some sense, at least as far as strong interactions are concerned. So, in parallel with the way the isospin group $SU(2)$ relates the two states of the nucleon, they introduced the $SU(3)$ group to express the larger symmetry of the p, n, and Λ triplet. These three particles, and their antiparticles, were then to be the building blocks from which particles such as mesons were built.

Making the apparently simple enlargement from $SU(2)$ to $SU(3)$ introduces a couple of unexpected twists. For the smaller group, there is a single choice for the smallest representation. However, for $SU(3)$ mathematicians know there are *two* possibilities, both in terms of three-by-three matrices and both equally valid. Yet the two possibilities in effect reside in parallel mathematical worlds. An analogy is a photograph, which is a representation of some scene. Then the photographic negative is an alternative representation of that same scene, linked to the first in a way that isn't totally trivial. Since there are two smallest representations of $SU(3)$, there are two multiplets to match, both triplets since they correspond to the three-by-three representations. One of these $SU(3)$ triplets is written "**3**," where the number says there are three particles inhabiting the $SU(3)$ multiplet. The other triplet is distinguished by an asterisk, "**3***." Ikeda and his colleagues turned the existence of these *two* triplets to their advantage. Since particles and antiparticles are also similarly distinct, living in a world and an antiworld, they identified one multiplet, the "**3**," with particles, and the other, the "**3***," with antiparticles.

There is another twist. Not all groups have representations of every conceivable size. For $SU(2)$, the starting point is the two-by-two matrix: all other sizes, or dimensions, of representations are allowed. Correspondingly, in principle there's an $SU(2)$ multiplet for any number of symmetry-linked particles. But the same is *not* true of $SU(3)$. Here, the available multiplets are **3, 6, 8, 10, 15**, and so on. Having gaps in the sequence is actually more typical – it's $SU(2)$ that is the special case. What the gaps mean is that, for example, there's no appropriate four-strong family of particles linked through $SU(3)$ symmetry. Nor is there a five-strong family.

As before, the multiplet combination rules reveal the larger multiplets. One of the most interesting combinations is that of a "**3**" with a "**3***" to give a singlet and an eight-strong multiplet, an octet. In multiplet-speak, this is written $\mathbf{3} \otimes \mathbf{3}^* = \mathbf{1} \oplus \mathbf{8}$. This was exploited by Ikeda and his co-workers as a way of combining triplets in the Sakata model, and led them to assign eight mesons to the "**8**" of $SU(3)$, one of several of their ideas which would endure even though their model would not.

Assigning strongly interacting particles to various multiplets of $SU(3)$ was a productive line of reasoning. That, at least, was what a not-so-young graduate student at London's Imperial College, Yuval Ne'eman, was thinking. Ne'eman's late arrival in physics was attributable to, amongst other things, his having served with the Jewish underground resistance movement in Tel Aviv. And later he fought with the Israeli army in the 1948 War of Independence. In due course, he would go on to become an Israeli cabinet minister, too, but for the time being he was juggling his duties as Israeli military attaché to London, and physics research under the wing of Abdus Salam. His goal was to organize the inhabitants of the particle zoo according to the patterns suggested by group theory. It was a strategy that was to lead Ne'eman, and independently Gell-Mann, by now at Caltech, to the "Eightfold Way."

Familial enlightenment

To the expectant eye, a chart setting out particles against both their isospin and hypercharge labels (Figure 4.14) shows hints of patterns.

It was the pattern of eight baryons, and a second pattern comprising pions plus K mesons, which Gell-Mann explored in his 1961 paper "CTSL-20," an internal Caltech report (the CTSL stands for *Caltech Synchrotron Lab*) that somehow was never published in the scientific journals. It was here that Gell-Mann introduced the expression "Eightfold Way" to the world.

Gell-Mann and Ne'eman treated the eight baryons as though they all have roughly the same mass, and assumed that they are all

(a)

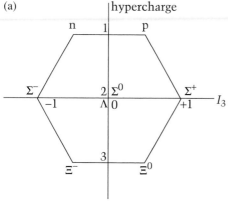

(b)

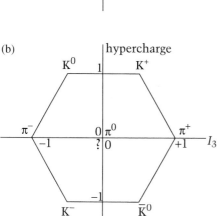

FIGURE 4.14 Isospin and hypercharge assignments for the known baryons (a) and mesons (b), circa 1961. The numbers on the vertical axes are the hypercharge assignments: in the meson diagram, hypercharge is equal to strangeness since the baryon number is zero. The horizontal axis is the third component of isospin. The "?" in the meson diagram was predicted to be the η meson via the Eightfold Way. It was discovered that same year.

members of a single eight-strong family, an octet. Likewise the pions and K mesons inhabit a second octet. The symmetry they chose to relate members of a family to one another was $SU(3)$, on the grounds that it was the "simplest generalization" of the symmetry group $SU(2)$, and they proceeded to follow the $SU(2)$ prototype as closely as possible. The central role of the octet is reflected in the "eight" of the Eightfold Way, an allusion to Buddhists' Eightfold Way to enlightenment.

Since it's possible to build the octet from smaller multiplets, according to $3 \otimes 3^* = 1 \oplus 8$, Gell-Mann thought that the n, p, Σ, and Ξ baryons must contain antiparticles, suggested by the presence of the 3^*. Though this would turn out to be incorrect, the successes of the Eightfold Way were manifest. In particular, the octet inhabited by pions and K mesons led Gell-Mann and Ne'eman to postulate the existence of the η meson, the first evidence for which was published later that same year. They also predicted an entire new meson family, subsequently discovered, and that the particles within octet families should all share certain common properties, another feature found to be true.

Both Gell-Mann and Ne'eman thought in terms of a Yang–Mills theory based on their Eightfold Way, and in this both were ahead of themselves: the Eightfold Way did not provide a complete theory of the strong interaction. In fact, for a while it was simply one of several options. But its stature as a way of making sense of the particle zoo was secured with the 1962 prediction, via the Eightfold Way, of the existence of a new particle called the Ω^-. By this time, Gell-Mann and Ne'eman had figured that the $\Delta(1232)$, $\Sigma(1385)$, and $\Xi(1530)$ resonances should belong to a ten-strong family which, plotted with isospin across the page and hypercharge down the page, gives a triangle with a missing tip, Figure 4.15. Note that the $\Sigma(1385)$ and $\Xi(1530)$ resonances are heavier relations of the Σ and Ξ appearing in the baryon octet, Figure 4.14. The Ω^- was to be the missing tip, so would therefore have labels $I = 0$

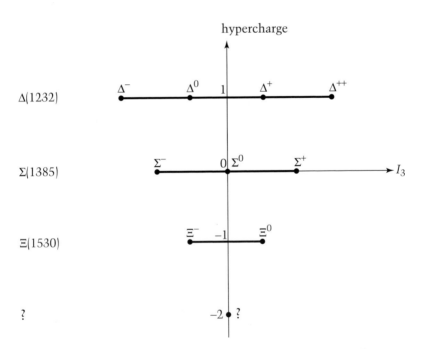

FIGURE 4.15 The baryon decuplet with a missing particle at the apex. The vacancy was filled by the Ω^-. The $\Sigma(1385)$ and $\Xi(1530)$ are resonances related to the lighter mass Σ and Ξ appearing in the baryon octet.

FIGURE 4.16 Yuval Ne'eman (left) and Murray Gell-Mann at Caltech in 1964, with the famous first picture revealing the Ω^- particle. The discovery of the Ω^- provided convincing support for the Eightfold Way. Photo Courtesy of the Archives, California Institute of Technology.

and $S = -3$ (hypercharge is the sum of B and S, so hypercharge of -2 for a baryon, $B = 1$, means strangeness of -3). They also predicted its spin to be $3/2$, and that it should decay by weak interaction, making it much longer lived than others in its family. And by exploiting the fact that the masses of particles in a multiplet are not really identical, but show systematic differences, Gell-Mann and Ne'eman were even able to estimate that the Ω^- mass would be about 1680 MeV. The Ω^- particle was found two years later at Brookhaven, complete with the expected weak decay sequence: its mass, according to more recent experiments, is 1672.4 MeV (Figure 4.16).

Muster Mark's quarks

"The mathematics of the unitary group $[SU(3)]$ is described by considering three fictitious 'leptons' . . . which may or may not have something to do with real leptons," wrote Gell-Mann in his 1961 Eightfold Way report. Gell-Mann felt it was useful to discuss $SU(3)$ symmetry and its workings in terms of three fictitious particles "to help fix the physical idea."

His three fictitious particles he called the electron, muon, and neutrino, all set to zero mass in the interests of simplicity, and possibly having some relation to the real particles bearing those names, and possibly not. These three objects then became the inhabitants of an $SU(3)$ triplet. Subsequently endowing his fictitious fermions with some mass allowed Gell-Mann to understand the mass difference patterns of the baryons and mesons.

The notion of three fictitious fermions, having an $SU(3)$ symmetry, which somehow could be used to understand eight baryon states also possessing $SU(3)$ symmetry, was very significant. The reason is that the Eightfold Way harbors a puzzle. The important multiplets seem to be the octet and decuplet – how come the basic triplet itself seems to be ignored in nature? After all, compared with the isospin prototype, though the larger multiplets are relevant, the isospin doublet at the root of it all is occupied by the neutron and proton. In short, what fulfils the role in $SU(3)$ of that played by the proton and neutron in $SU(2)$?

A clue lay in a simple mathematical result. *Three* copies of the $SU(3)$ triplet strung together make an octet, and more besides. In fact the full haul is $3 \otimes 3 \otimes 3 = 1 \oplus 8 \oplus 8 \oplus 10$. In 1964 Gell-Mann proposed three fundamental entities, updates of his pedagogic fictitious leptons, which were to inhabit the **3** appearing in this product. Thus, since there are three copies of this triplet, three of these entities would be needed to build a member of the baryon octet. He originally termed these entities "kworks," later becoming "quarks," from "three quarks for Muster Mark" in James Joyce's *Finnegan's Wake*. Gell-Mann's paper contains a reference to Joyce's book, a rare, if not unique, attribution to classic literature in a physics journal. Actually, quark is also a type of German soft cheese.

His three types, or flavors, of quarks Gell-Mann christened up, down, and strange, or u, d, and s, names that have endured to this day. In this picture, baryons are composed of three quarks, qqq. Just as electrons and other fermions must, according to field theory, be accompanied by corresponding antiparticles, the quarks must be partnered by antiquarks, written $\bar{q}$, or in terms of their individual flavors $\bar{u}$, $\bar{d}$, and $\bar{s}$. According to this scheme, mesons are a quark–antiquark pair, $q\bar{q}$. Associating the three quarks with the "**3**" of $SU(3)$, and the antiquarks with the "**3***," the corresponding combinations of multiplets are **3** $\otimes$ **3** $\otimes$ **3** and **3** $\otimes$ **3*** for baryons and mesons, respectively. In this scheme, the proton is composed of three quarks, two up quarks and one down quark, written uud: a neutron is udd, and the Σ^+ is made from suu quarks. Figure 4.17 shows the quark assignments for the baryon and

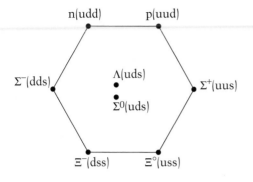

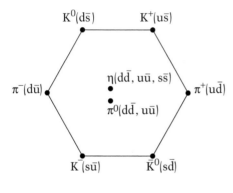

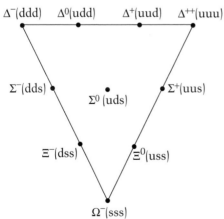

FIGURE 4.17 Quark model assignments for the baryon octet, the light meson octet, and the baryon decuplet.

Table 4.2 *Properties of quarks and antiquarks.*

Quark/antiquark	Electric charge	Isospin, I_3	Strangeness	Baryon number	Spin
u	+2/3	+1/2	0	+1/3	1/2
d	−1/3	−1/2	0	+1/3	1/2
s	−1/3	0	−1	+1/3	1/2
ū	−2/3	−1/2	0	−1/3	1/2
d̄	+1/3	+1/2	0	−1/3	1/2
s̄	+1/3	0	+1	−1/3	1/2

meson octets of the Eightfold Way, along with the baryon ten-strong family, or decuplet. The similarities between this "quark model" of particles, and the old Sakata model are striking: Sakata just used the wrong three building blocks and tried to combine them in the wrong way.

The quark model comes at a price, however; an unpalatable feature that Gell-Mann himself at one time perceived as an obstacle. For the scheme to work, quarks must have fractional electric charges. The u quark has charge 2/3, and both the s and d quarks have charge −1/3, the charges being fractions of the electron charge. Such fractional charges challenge the received wisdom that all electric charges come in whole number multiples of the basic electron charge. Charges of one, two, three, or more times the electron's charge would be fine, but charges of 1/3 and 2/3 mean dumping a fundamental principle. And as if that wasn't enough, since baryons are made of three quarks, then all quarks have a baryon number of 1/3. Small wonder, then, that the quark model met with opposition.

Other properties of quarks are, however, a little more palatable. Quarks are spin 1/2 fermions, and in this sense they match the three fictitious fermions of the Eightfold Way. The u and d quarks form an isospin doublet, with $I_3 = +1/2$ and $I_3 = -1/2$, respectively, which accounts for the isospin properties of the neutron and proton. The s quark has no isospin, but a strangeness of −1 (Table 4.2).

The notion that the Eightfold Way could be explained by envisaging the proton and its family as built from three component parts was exercising the minds of others, too. As early as 1962 Ne'eman and his collaborator Haim Goldberg realized that an octet linked to three triplets suggested three constituents for the proton. But realizing that this implied fractional quantum numbers, an idea that would be seen as ludicrous, they rested with the notion that these constituents would be best considered a calculational device.

Hiding under the bush of theoretical abstraction was not the way of ex-Caltech student George Zweig, however. As a graduate student, Zweig had heard about the Eightfold Way from Gell-Mann. Later, at CERN, Zweig independently invented his version of quarks, which he termed aces, that he envisaged as very real constituents of the proton. Yet such was the antipathy towards the quark model that Zweig's version was never even published in a journal, emerging only as a pair of CERN reports early in 1964. Rather belatedly his work appeared as part of a compilation volume on quarks many years later.

Gell-Mann himself argued that quarks, with their controversial properties, couldn't be seen individually, but only as composites having the better-behaved properties of particles such as protons and neutrons. In this way, fractional electric charges and baryon numbers were not "real" as such. But where does that leave the poor quark? Is it a particle that experimenters can look for, or is it some mythical mathematical device useful for explaining a bit of group theory?

In 1964 Gell-Mann wrote:

It is fun to speculate about the way quarks would behave if they were physical particles of finite mass (instead of purely mathematical entities as they would be in the limit of infinite mass).

On the basis that electric charge and baryon number really are conserved, Gell-Mann deduced that at least one species of quark must be stable. There might be a small stable quark contamination of matter at the Earth's surface arising from cosmic rays, he suggested. In the finish he wrote: "A search for stable quarks...at the highest energy accelerators would help to reassure us of the non-existence of real quarks." According to Gell-Mann, "mathematical" quarks are ones that would not emerge singly, and could never be seen alone; "real" quarks could emerge and could be seen in isolation. It's a subtlety that has dogged him ever since. In 1992 he wrote:

What I meant by "mathematical" for quarks is what is now generally thought to be both true and predicted by QCD. Yet, up to the present, numerous authors keep stating or implying that when I wrote that quarks were likely to be "mathematical" and unlikely to be "real," I meant that they somehow weren't there. Of course I meant nothing of the kind.

Semantics aside, the quarks are there: it's just a question of whether or not we get to see them in splendid isolation. If there are free quarks, the challenge facing experimentalists is to find them.

Quarks revealed

There are several pieces of evidence supporting the idea that quarks *are* real, fundamental components of particles such as protons and neutrons, despite the thorny issue of actually seeing quarks "in the flesh." One pretty piece of experimental work that supports the idea of three-quark baryons arises from quarks' capacity to behave like tiny magnets.

In 1995, after years of effort experimenters at Fermilab finally completed their program of precision measurements of the magnetic moments of all the baryons made from u, d, and s quarks, Figure 4.18. Included in their table of values are magnetic moment measurements for the neutron, the Λ, and the Ξ^0. If these three *neutral* particles were elementary point-like entities, then their magnetic moments would be *zero*, since elementary particle magnetic moments are associated

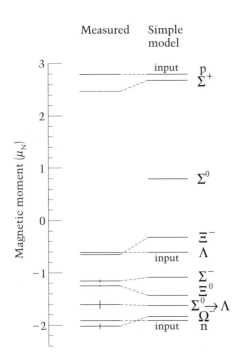

FIGURE 4.18 Measured baryon magnetic moments compared to simple quark model predictions. Credit: Particle Data Group.

with spinning charge. But the Fermilab measurements reveal values for these three particles, respectively, of −1.913, −0.613, and −1.253 (in units called nuclear magnetons, the magnetic moment of a point-like proton according to relativistic quantum mechanics). Not only are these values very different from zero, they are also different from one another. The implication is that these particles are made up from charged constituents, whose electrical charges combine to zero, but whose magnetic moments combine in a different way to give a value different from zero.

In fact, the proton and neutron magnetic moments turn out to be a triumph of the quark model. Simply assume that each quark is a point-like, spinning, charged object, and add the magnetic moments of the component quarks respecting the standard rules for combining spin 1/2 entities. Then the calculated ratio of the proton magnetic moment to that of the neutron comes out at −2/3, in good agreement with the experimental value of −0.685. For the Λ, composed of uds quarks, the quark model directs that the magnetic moment is due solely to that of the strange quark, whilst the moment of the Ω^- is simply three times that of the strange quark moment. So, the quark model predicts the ratio of magnetic moments of Λ to Ω^- to be +0.33, and experiment gives +0.304, using the precision value for the Ω^- reported in 1995. The agreement looks good. Using the measured moments for p, n, and Λ, it's easy to extract moments for each of the three quarks, which come out as +1.852, −0.972, and −0.613 for the u, d, and s quarks, respectively. It's equally simple to then use the quark model to calculate the moments of other baryons. Generally, the calculated values agree with the measured ones to within around 10%, except for the Ξ^-, for which the match is a more modest 30%. For such a simple approach, this level of agreement is impressive.

Besides some constants, the magnetic moment for a point-like spinning particle depends only on its charge and mass. This means that, if quarks inside baryons really are like spinning point-like particles, and the magnetic moments listed for each quark are to be taken seriously, then experimenters can extract values for the mass of each of the three quarks. In MeV, these are 338 for the u quark, 322 for the d quark, and 510 for the s quark. These masses are "effective masses," reflecting the idea that they are the masses which the quarks *appear* to have when nestled inside composite baryons, as opposed to the masses of free quarks, whatever they might be.

All of which makes baryon magnetic moment experiments good value. The fact that neutral particles such as n, Λ, and Ξ^0 have a finite magnetic moment attests to charged components. Simple quark model calculations based on spinning charged quarks and neat symmetry principles, without any complicated theory of how the whole system sticks together, give an impressive match with experiment. Not only do magnetic moment studies suggest that the quark model is on firm ground, a handy by-product is an estimate of the effective masses of the quarks themselves.

COLORFUL QUARKS AND WHITE PROTONS

Quarks as a substructure for protons, neutrons, and their companions look very promising, but in the mid 1960s theorists still had a lively problem on their hands. There remained a slightly technical, but nevertheless very significant, objection to the quark model. It was a problem Gell-Mann knew about, and worried about. The problem was this: the particle statistics didn't make sense.

Quarks must be fermions. As fermions, they are governed by the exclusion principle, and only for fermions is it meaningful to talk in terms of a definite number of quarks comprising some particle such as a neutron or proton. Were quarks to be bosons, the exclusion principle would not apply, and any number of quarks could be involved in building a composite particle. Just as laser light photons cease to have separate identities on account of being bosons, boson quarks could not be distinguished one from another, and the quark model would not work.

In addition, quarks need to be spin 1/2 if the spins of the objects built from them are to come out right. The baryon octet particles, including the proton and neutron, all have spin 1/2, while the decuplet baryons have spin 3/2, and the mesons have spin 0 or spin 1, yet all are supposed to be built from the same spin 1/2 building blocks, the quarks. In fact, quark spin is an essential ingredient of the quark model. It also features in the particle statistics problem.

This problem is really in evidence in the baryons, and nowhere more clearly than in the spin 3/2 Ω^- particle, one of the decuplet baryons. The Ω^- is made up of three strange quarks, sss, a symmetric configuration since all three quarks are identical, and swapping any pair of them changes nothing. To give a total spin of 3/2, all three quarks are spin-aligned (Figure 4.19) and this is likewise a symmetric configuration. That being the case, the combination of quark flavor

FIGURE 4.19 The three spin-aligned strange quarks of the Ω^- are indistinguishable from one another.

with quark spin is symmetric. But how can this be? The Ω^- itself is a fermion – it has spin 3/2 – and so it must be overall *antisymmetric*.

The Ω^- is a special case only in that it makes the problem plain to see. The same problem exists for the spin 3/2 Δ^{++}, made up of three up quarks, uuu, and the Δ^- comprising three down quarks, ddd. Both these particles have obviously symmetric quark and spin configurations and, like the Ω^-, give the wrong overall symmetry. In fact, all ten of the quark combinations in the baryon decuplet are symmetric under interchange of any two quarks, even for those particles made up from mixtures of up, down, and strange quarks. For example, the Σ^0 is made up from uds quarks, Figure 4.17, which is actually short for (uds + dsu + sud + sdu + dus + usd) multiplied by some numerical constant. Since all ten family members are symmetric spin combinations, on account of being spin 3/2, the entire family has the same overall "wrong" symmetry.

Ignoring spin (and quarks whirling round one another), the baryon octet is neither totally symmetric nor totally antisymmetric. Remarkably, when spin enters the picture the "mixed" symmetry of the octet interlocks exactly with a similarly mixed symmetry of the three quark spins. The result is an overall symmetric combination: all eight family members of the baryon octet, including the neutron and proton, are symmetric in terms of the quark flavor *and* quark spin. This is quite some coincidence, since each quark having one of three possible flavors, and either spin up or down, implies a total of 91 "simple" baryons, some with spin 1/2 and others with spin 3/2. Yet only those particles apparently having exactly the *wrong* overall symmetry are singled out to appear in nature.

Colorful solutions

Oscar Greenberg, of the Institute for Advanced Study in Princeton, dreamt up a fix. His idea was to insist that quarks in sets of three behave

FIGURE 4.20 Oscar "Wally" Greenberg, at home in Maryland, 2003. Greenberg introduced color as a way of fixing problems with the statistics of quark states. The color attribute would turn out to be as central to QCD as electrical charge is to QED. But the path of a good idea is not always smooth. Greenberg recalls showing his paper to J. Robert Oppenheimer, then Director of the Institute of Advanced Study in Princeton. "Your paper is beautiful," Oppenheimer told Greenberg, "but I don't believe a word of it." Credit: Courtesy Oscar Greenberg.

like a fermion with respect to another set of three quarks. In other words quarks, according to Greenberg, should obey a variant of fermion statistics, whereby sets of three quarks obey the usual fermion statistics with respect to other sets, even though within each set of three the quarks have a symmetric configuration. In this way, his scheme automatically guaranteed the correct overall statistics for baryons. The price was the need for a new label on each quark that could take one of three possible values (Figure 4.20).

A new, three-valued label likewise featured in a fix proposed by Moo-Young Han, at the University of Syracuse in New York, and Nambu, at the University of Chicago, who like Greenberg published their idea in 1964. Han and Nambu attached a three-valued label to each quark and proposed that the "common" baryons – the baryon octet and decuplet – were to be overall antisymmetric due to their antisymmetry with respect to this new label. And since the label could take three possible values, they further insisted that, like the three quark flavors u, d and, s, it too should exhibit $SU(3)$ symmetry: they introduced a second $SU(3)$ symmetry into the quark picture.

Han and Nambu disliked the idea that quarks should have fractional electric charge, and set up their model to try and insure whole-number quark charges. Adjust Han and Nambu's model to match today's prevailing view, that quark electrical charges really are fractional, and their scheme becomes equivalent to Greenberg's. The new label both attached to quarks now has a name: it is called color.

Color is a new quantum number having three possible values. So what *do* you call a new quantum number? The choices are to invent some totally new word, or extract some nonsense word from classic literature (as in the case of "quark" for example), or use an existing word such as "color." The problem with color is that to some people it might imply quarks have a real pigmentation property, which is *not* the case. Color is simply the name of a quantum property of quarks, although the choice of the name "color" actually turns out to be quite handy. An individual quark can be red or green or blue, corresponding to the three possible values of the color label: the actual color names used are entirely a matter of convention. The color label, being different for each of the three quarks comprising a baryon, means that the three quarks are now no longer identical, and this is the key to how color fixes the baryon statistics problem. So in Figure 4.21, for example, one of the three strange quarks of the Ω^- is red, one is green, and the third is blue.

What about that second copy of the group $SU(3)$? Is there really anything magic about *two* copies of $SU(3)$ entering the frame? The first of them comes about because, in the common baryons that inspired the quark model, there are three different quark types on offer. Two flavors are needed to distinguish between, say, neutrons and protons, and a third to accommodate the new strangeness property. The copy of $SU(3)$ linked to the quark flavors is sometimes labeled $SU(3)_f$, or $SU(3)$-flavor, to emphasize its role. But despite the "f'" or "flavor" label, mathematically it is still simply *the $SU(3)$* group. If there were more

FIGURE 4.21 The three spin-aligned strange quarks of the Ω^- can now be distinguished from one another via their colors, red or green or blue, represented here by three different shadings.

than three types of quark – now there's a story – then the flavor $SU(3)_f$ group would be replaced by some larger SU group. The three of the color $SU(3)$ group, on the other hand, usually denoted $SU(3)_c$ to distinguish it from the flavor $SU(3)_f$, comes about because there are three (possibly identical) quarks in a baryon, and so three different labels are needed to distinguish between them.

Yet even with the introduction of color, there was still something wrong. Greenberg's scheme seemed to complicate matters, allowing for the existence of a huge number of unseen particles – he listed symmetry groupings for 2600 baryons in his 1964 paper. Han and Nambu likewise allowed for many unseen particles, though their paper contained the germ of what was to follow. This was the realization that an antisymmetric color singlet – more on that in a moment – was the way to fix the symmetry problem for the common baryons. What they did not do was anything to suppress all the other options permitting baryons having color configurations that were not singlets. They, like Greenberg, allowed for the possibility of real observable quarks and, implicitly, the ability of experiment to see and measure the color attribute.

Through numerous experiments color's seat at the table is secure, yet experimenters never get to see color directly. For nature decrees that *observable* particles should have no *overall* color. According to this key principle, quarks themselves are not observable since they carry color. Baryons, however, are made from three quarks, each having a different color. A real, observable baryon, such as a proton, is a quantum mix of colored quarks yielding overall a "white" composite particle that therefore has no overall color.

The idea works for mesons too. Observable mesons are a quantum mix of quarks and antiquarks. Mimicking the fact that an electron has a charge of -1 and its antiparticle, the positron, has an "anticharge" of $+1$, antiquarks carry "anticolor," denoted $\bar{r}$, $\bar{g}$, and $\bar{b}$. The color and anticolor of the quarks and antiquarks in a meson cancel one another to give "colorless" observable particles. Color is indeed a useful and graphic name for this quantum number.

This is the elevation to fundamental tenet of the color singlet idea mooted by Han and Nambu: real, observable particles are "color singlets." The prototype antisymmetric singlet has already debuted earlier as the combination of two spin (or isospin) 1/2 particles according to {A(up) B(down) – B(up) A(down)}. In a corresponding way, a specific combination of colors yields a similar singlet, a "white" state. Just as the three flavors of quark make up a triplet, or **3**, of $SU(3)_f$, so three colors, labeled r, g, and b, make up a triplet, or **3**, of $SU(3)_c$. Since a baryon comprises *three* quarks, *three* copies of the color **3** will feature in the baryon's total color description. These three copies combine as $\mathbf{3} \otimes \mathbf{3} \otimes \mathbf{3} = \mathbf{1} \oplus \mathbf{8} \oplus \mathbf{8} \oplus \mathbf{10}$. The **1** is the all-important color antisymmetric singlet corresponding to real particles. Explicitly, the color antisymmetric singlet for baryons reads {rbg − brg + bgr − gbr + grb − rgb}, just the right mix of colors to guarantee "white" particles. The other color combinations do not appear in nature.

Since mesons are quark–antiquark combinations, things there look a little different. The three anticolors, r̄, ḡ, and b̄, inhabit the **3*** triplet of $SU(3)_c$. These will combine with the color **3** triplet according to $\mathbf{3} \otimes \mathbf{3}^* = \mathbf{1} \oplus \mathbf{8}$. The **1** is the color singlet state, explicitly {rr̄ + gḡ + bb̄}, the colors of the quarks canceling the anticolors of the antiquarks. This singlet combination is *symmetric* in color labels. The **8**, on the other hand, has an overall net color. It has a special role to play, as will become clear in due course.

Color $SU(3)_c$ arises because it takes three quarks to build a baryon. Flavor $SU(3)_f$ arises because there are just three types of quark to pick from when doing that building, at least for the lightest baryons. Flavor $SU(3)_f$ symmetry means that the strong interaction, which binds protons to one another or to neutrons, is insensitive to blending of the three quark flavors, u ↔ d ↔ s. This is why three quarks, having any mix of these three flavors and with spins suitably aligned, give baryons in the same multiplets, the octet or decuplet. Actually the symmetry is really only approximate. There is a pattern of mass differences within a given multiplet, implying some "breaking" of an otherwise "exact" symmetry. Color $SU(3)_c$ symmetry, combined with the requirement that observable particles are color singlets, means that, for any baryon, the force binding the quarks together is unchanged by blending r ↔ g ↔ b of the quark colors. Quark flavor and quark color are distinct: each flavor of quark can have any one of three colors, as listed in Table 4.3.

Color, and the rule that observable particles are color singlets, not only fixes the particle statistics problem, it explains much more besides. For example, combinations such as qq, qqq̄, and qqqq are forbidden on color considerations, and indeed no such particles seem to exist. So not all of the available multiplets of $SU(3)_f$ are populated by particles. For instance, the flavor **6** multiplet is uninhabited. It arises from flavor $\mathbf{3} \otimes \mathbf{3}$, corresponding to a forbidden qq state.

Table 4.3 *Each of the three quark flavors, u, d, and s, can have any one of the three possible colors, r, g, and b.*

	Color		
	u_{red}	u_{green}	u_{blue}
Flavor	d_{red}	d_{green}	d_{blue}
	s_{red}	s_{green}	s_{blue}

Color as charge

Fixing up the statistics of the quark model of baryons is just the half of it: the best bit is yet to come. For the perfect color symmetry amongst quarks offers the basis for a whole new gauge theory based on *local* color symmetry, with a set of colored gauge bosons mediating interactions between colored quarks. This idea, which took root during an intense period of activity in the early 1970s, is the foundation of a new theory, a theory called quantum chromodynamics. The last of nature's four fundamental forces had succumbed.

The basic idea is that each quark in a baryon has one of three possible colors, and the strong force between two quarks originates in their colors, in much the same way that the force between two electrons originates in their electrical charges. The electrical force is carried by photons. A new type of force carrier, the gluons, carry the analogous color force. The "chromo" of quantum chromodynamics – the name was originally coined by Gell-Mann – is in recognition of the fundamental role played by color as the origin of the inter-quark force.

The idea that some kind of "gluon" might be involved in the description of the strong force had already been around for several years: as early as 1962 Gell-Mann made reference to a "strong gluon coupling" between the constituents of hadrons as proposed in a model predating even that of Sakata. But according to QCD a single gluon is not enough: what's needed is a set of eight electrically neutral *color-carrying* gluons, a gluon octet. The fact that the gluons themselves carry color is hugely significant. To start with, it means that naked gluons are just as impossible to see as raw quarks. That is, since both quarks *and* gluons carry color, under the principle that real observable particles should have no net color, direct observation of both quarks *and* gluons is equally impossible.

It also means that one gluon can sense the presence of another, so gluons interact directly with each other. This is in stark contrast to photons, which carry no electrical charge; one photon can have no direct electrical interaction with another photon. At the simplest level, photons do not interact with one another.

That gluons carry color, and so may interact directly with one another, can be traced all the way back to the simple fact that $SU(3)$ is a non-Abelian group. Photons do not interact with one another directly because, from a theorist's point of view, the symmetry group of QED is Abelian. Thus what seem to be details of the mathematics of groups turn out to be reflections of fundamental features of nature's forces.

Other basic properties of the $SU(3)$ group translate directly into other significant features of QCD. In particular, the function of the $SU(3)$ group relies on eight independent parameters. This translates into *eight* conserved charges, one per group parameter.

Obliging the electron–positron field to satisfy a local $U(1)$ invariance means that the photon has spin 1. The fermion field, as it migrates from point to point, must be continually modified in order that the symmetry is preserved. This modification is realized through the photon. Applying the gauge principle to the quark color symmetry, making the symmetry local, is to do the same thing but with a more complicated symmetry group. The gauge principle yields a gauge interaction based on *eight* gluons, one per conserved charge, which are all massless spin 1 bosons. In effect, the eight gluons are all like colored photons, but they couple to the color on a quark rather than its electrical charge.

Color mix and match

A picturesque way of visualizing the interaction between two electrons is in terms of a photon thrown off by one, and absorbed by the other, illustrated in Figure 4.22(a). With the electron having a single possible electric charge, its antiparticle the corresponding anticharge, and the photon no charge at all, the possibilities of QED are relatively limited.

Quantum chromodynamics has a much richer structure, with three colors and three anticolors, plus a colored intermediary. Mimicking the simple QED case, a red quark, for example, can throw off a gluon and stay a red quark, Figure 4.22(b), where gluons are represented as whirling spirals rather than the wavy line of a photon. But there are

(a)

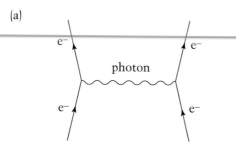

(b)

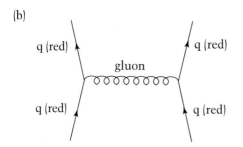

(c)

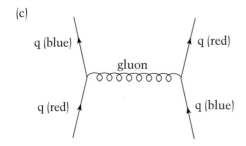

FIGURE 4.22 (a) Photon exchange mediating a simple QED interaction. Gluon exchange mediates the equivalent process in QCD (b) and processes involving quark color changes (c).

novel alternatives. In particular the quark could flip color to become a blue or green quark, the "color difference" passing on by means of the gluon to flip the color of the gluon-absorbing quark, Figure 4.22(c)

To ensure color conservation, the possible gluon color configurations match the quark color changes resulting from interaction with the gluon. Of the nine options available to an interacting quark's color, there are six in which the color changes, $r \to g$, $r \to b$ or one of four others, and three in which the color does not change, $r \to r$, $g \to g$, and $b \to b$. Of this constant color trio, only two – any two – are necessary. The third would allow for the exchange of a color singlet gluon. That's forbidden under the tenets of QCD since a color singlet object

Table 4.4 *The eight gluon color configurations.*

$r\bar{g}$
$b\bar{g}$
$g\bar{b}$
$r\bar{b}$
$b\bar{r}$
$g\bar{r}$
$\frac{1}{\sqrt{6}}(r\bar{r} + g\bar{g} - 2b\bar{b})$
$\frac{1}{\sqrt{2}}(r\bar{r} - g\bar{g})$

cannot feel the color force. All up, this leaves nine minus one, or eight gluon color options. So gluons demand a color octet. There is one, the octet revealed in the combination $\mathbf{3} \otimes \mathbf{3^*} = \mathbf{1} \oplus \mathbf{8}$. This is the color-carrying octet flagged up earlier as playing a special role, and it does: it accommodates gluons.

The complete set of gluon colors is shown in Table 4.4. The first six are easy to understand, and their origin is suggested in Figure 4.23(a) for quark–quark interactions, where colors running "against the flow" are depicted as the corresponding anticolors. Figure 4.23(b) illustrates color flows for quark–antiquark interactions. The origin of the final two gluon color combinations in Table 4.4 is less physically intuitive, though, and despite appearances they do carry color.

Not all conceivable quark and gluon color combinations are allowed. Figure 4.24 shows two impossible cases. In the first, a quark would need to change to an antiquark, impossible for a host of reasons, and not least because electric charge isn't conserved, since there's no electric charge carried away by the gluon. The second is impossible since overall color is not conserved – there's r and b in, g and b out.

In QED, like charges repel, opposites attract. In QCD, colors will do something similar. Two quarks having the same color repel, but quarks of different colors attract, unless their colors are interchanged, in which case they repel. A quark and antiquark with corresponding color and anticolor attract, but repel if the anticolor of the antiquark does not correspond to the color on the quark, as illustrated in Figure 4.25.

The basic strength of the interaction between any quark and gluon is always the same. This is not an assumption fed into QCD, but

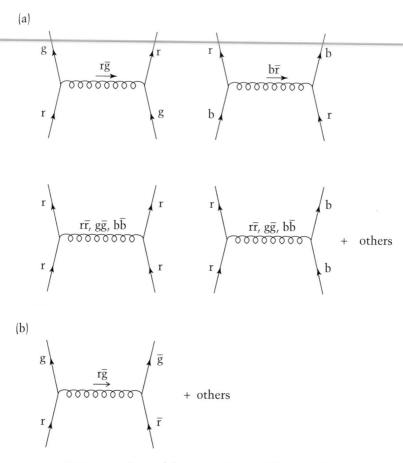

FIGURE 4.23 Some of the gluon color possibilities in (a) quark–quark interactions and (b) quark–antiquark interactions.

a feature of the theory called universality: the basic coupling strength of a quark to a gluon does not change with different quark and gluon colors.

To illustrate the significance of this, it's worth comparing QCD with QED once again. In QED, the strength with which a charged particle interacts with a photon is given simply by its electrical charge, which might be 1, 2, or more units. There is nothing in QED to constrain the charge, and nothing within QED to say why it must come in multiples of some basic unit, as already discussed in Chapter 3. In QCD, however, the corresponding strength of the basic coupling between any

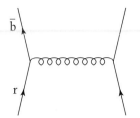

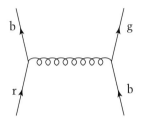

FIGURE 4.24 The disallowed: the first case is not possible since it requires a quark–antiquark conversion. The second does not conserve overall color.

gluon and any quark has the same (unspecified) value for all. Thus a red quark couples to a gluon with exactly the same strength as would a green or blue quark. In addition, the color carried by each quark can only ever be one unit: a quark carrying two units of red instead of one, for example, would couple twice as strongly to a gluon, and this is forbidden. If it were not this way, then gauge invariance would be destroyed and QCD would fall apart.

This difference between QED and QCD can also be traced back to the non-Abelian nature of QCD. In terms of basic operations, non-Abelian means that the difference between a *pair* of group operations performed in one order, and the same operations in reverse order, is given in terms of a *single* operation, multiplied by some constant. The important point is that this establishes a relationship between *single* operation and *pairs* of operations, which is therefore very restrictive. Universality is a direct consequence of those restrictions. It might seem a simplifying feature, but in fact linking transformations of different powers, in this case powers of one and two, means that Yang–Mills theories are what mathematicians term non-linear. Children are non-linear: having two children in the family is not simply twice the pain and pleasure of having one, since the two of them interact, creating a vastly more complex situation. And, as with children, non-linear mathematical problems are virtually impossible to deal with, and QCD turns

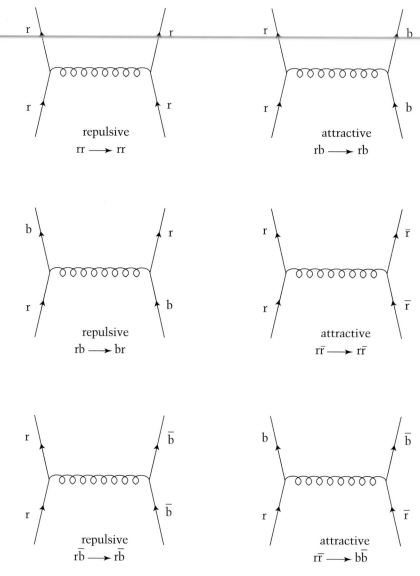

FIGURE 4.25 Quark–quark and quark–antiquark attractions and repulsions.

out to be no exception. Most of the more tractable problems in physics are, by contrast, of a class termed linear – they involve only single powers of transformations or variables. Figure 4.26 illustrates a trivial example of a linear system that can readily be made a non-linear one.

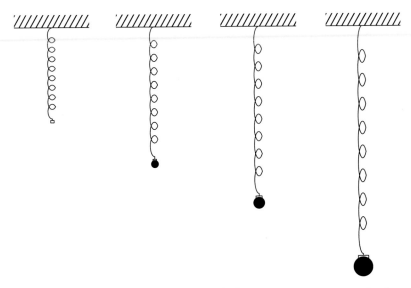

FIGURE 4.26 A small mass attached to a spring gives an extension from the unloaded position, far left, to the position shown next door to the right. Double the mass gives double the extension, next door again. This is a linear relation – double the mass gives double the extension. But far too much mass, far right, and the simple linear rule breaks down: the system is then non-linear.

The non-linearity of QCD stemming from the non-Abelian nature of $SU(3)$ may make it difficult to deal with mathematically, but it's what makes QCD so exciting. It is manifest in the fact that gluons carry color themselves, and hence interact directly with one another. Indeed when the master equation of QCD is run through the Feynman path integral formulation and perturbation theory to generate the Feynman diagrams of QCD, two totally unfamiliar diagrams appear, diagrams having no parallel in QED. They are the three-gluon and four-gluon diagrams, depicted in Figure 4.27, showing explicitly gluons interacting with one another. As a consequence of universality, the strength of the coupling at the three-gluon vertex is the same as for a quark and a gluon: at the four-gluon vertex the coupling is the square of the basic quark–gluon value.

Colors, charges, and currents
The origin of the force between two electrons is their electrical charge. Electric charge is readily identified with the single conserved charge

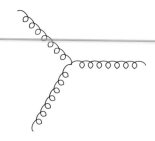

FIGURE 4.27 The three- and four-gluon vertices. These are the only gluon self-interaction diagrams.

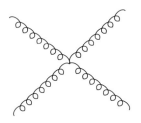

given by Noether's theorem, and which clearly resides on the matter particles. In QCD, the idea is that color is similarly the origin of the color force, but there are fresh angles to consider – the force-carrying particles themselves carry color, and there is a mismatch between the numbers of quantum labels and conserved charges. How does this relate to the simple QED case of conserved charge carried on fermions?

QCD is based on three colors, in place of the single electrical charge of QED. There are three related anticolors, just as there's positive and negative electrical charge. In place of the single conserved charge of QED, however, QCD offers something more luxuriant: the mathematics of groups says that, for the $SU(3)$ group, there must be eight conserved charges in all. There's one conserved charge per gluon.

There may be eight charges, but the rules of quantum mechanics do not permit specifying all eight simultaneously for a single particle, just as simultaneously labeling position and momentum of a quantum system is banned. In fact, the rules of group theory show that the maximum number of simultaneous labels for QCD is just two. Therefore the eight conserved charges of QCD cannot be trivially matched one for one with the two individually conserved quantum numbers, since there's not enough quantum numbers to share round. This is in contrast to QED, where things are simpler – the one quantum number equating to the one conserved charge.

None of which should come as a surprise. After all, the color group $SU(3)_c$ is just another copy of the $SU(3)$ group, and we have already seen how this equates to two labels. Recall that the $SU(3)_f$ quark model is based on two quantum labels, hypercharge and the third component of isospin, which tag the particles in a given family. All the baryons of the baryon octet and all the eight mesons of the meson octets, for example, can be set out on a chart of hypercharge versus third component of isospin, Figure 4.14.

Aping the quark model, the two available labels for the $SU(3)_c$ case could be called *color* hypercharge and third component of *color* isospin. This is not common practice since observable particles have no net color, so there's little point in giving names to quantum labels that can never be seen. However, it makes the parallel clearer, since the color hypercharge and third component of color isospin may then be used to label the eight gluons, exactly mimicking the meson octet labeled using regular hypercharge and third component of isospin. Mesons inhabit an **8** multiplet, from a flavor **3** ⊗ **3***, whereas the gluons inhabit an **8** from **3** ⊗ **3*** of color, and the two groupings can be labeled according to their respective quantum numbers, Figure 4.28. The two particles appearing in the middle of the meson octet, the η and the π⁰, both having zero hypercharge and zero third component of isospin, parallel the seventh and eighth gluon color combinations, respectively, listed in Table 4.4. The first six on the list form the perimeter of the gluon octet of Figure 4.28.

In classical electromagnetism, a point charge is a source of electric field. A moving charge becomes a current, which is also conserved. Electric current is a source of both electric and magnetic fields, which in quantum theory becomes the idea that a current is the source of photons. The same idea applies to QCD, and this is perhaps a more fruitful approach than looking for eight conserved charges: look instead for the eight corresponding conserved currents.

Again in classical electromagnetism, when a charge changes direction, it emits electromagnetic radiation, the principle by which radio transmitters work. In quantum-speak, the emission of a photon is associated with a change in motion of the charged electron, represented diagrammatically in Figure 4.29. The electron's four-momentum will be changed as four-momentum is carried off by the photon, indicated by the four-momentum labels k and h in the figure. The electron's charge is unchanged, since photons are unable to carry charge. So the small

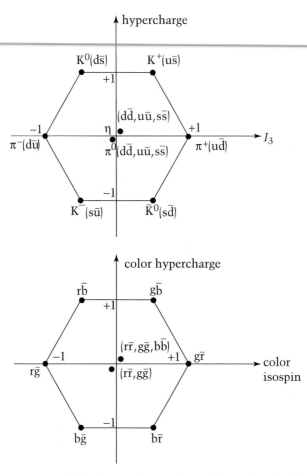

FIGURE 4.28 The gluon *color* octet (lower diagram) can be labeled in a way that mimics the meson *flavor* octet (upper diagram).

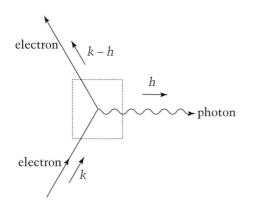

FIGURE 4.29 An electron of momentum k emits a photon of momentum h, thus becoming an electron with momentum k–h.

dashed box represents a conservation zone, where four-momentum in equals four-momentum out and electric current in equals electric current out.

A colored quark could emit a gluon in an analogous way, represented schematically in Figure 4.30(a), the dotted line suggesting color flow: no color flow line can start or stop inside such a diagram. Just as the electron retains its electric charge after emitting a photon, the quark color – shown as red, though green and blue are equally likely – is unchanged. But quark colors can also flip, for example Figure 4.30(b). Figure 4.30(c) shows an even more complicated example, including a three-gluon vertex. In all three cases, a box delineates a region where *color* current in equals *color* current out. There are eight conserved color currents, effectively labeled by the eight color combinations ascribed to the gluons, which in turn correspond to the eight ways in which a quark's color evolves following interaction via a gluon.

In QED there is a single conserved electric current that is a source of photons, whereas in QCD there are eight conserved color currents that are the source of gluons. In all gauge theories, *currents* associated with symmetry are the sources of the force-carrying fields.

Signs of color
The simple blunt truth is that without color there would be no QCD since, according to QCD, color is the particle attribute through which those particles interact. That means experimenters need to come up with some pretty rock-solid evidence for color. And not just that: they must be able to show there are three colors, no more and no less.

Experimenters have risen to the challenge. They have produced support for the existence of color, or at least something that behaves in just the way folk expect color to behave. They have also found evidence that the number of colors really is three. They have even gone a step further, homing in on the precise form of the symmetry group itself.

Some experiments are more direct than others. Arguably the most direct of all involves annihilating electrons and positrons together in head-on collisions, and looking at the number of muons produced relative to the number of hadrons.

Wait a moment. This is an electron–positron collision, so where do the hadrons come from? Electrons and positrons are two particles that are still, even today, understood to be as elementary as any particle can be. So they are not composite objects, and they do not feel the

(a)

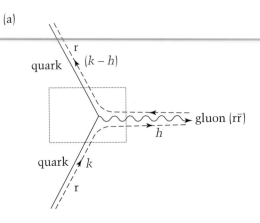

(b)

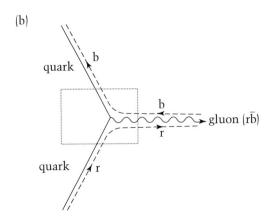

(c)

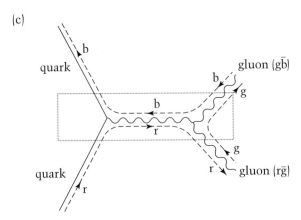

FIGURE 4.30 Color flows in quark–gluon and gluon–gluon interactions.

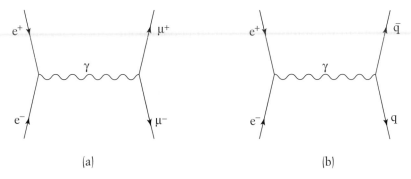

strong force. The fact that hadrons appear is testament to the validity
of both special relativity and quantum field theory. In one sense, that of
special relativity, the collision yields a bundle of energy resulting from
the combined masses plus motion energies of the colliding particles,
which is redistributed amongst the particles in the shower of debris
that results. In another sense, that of quantum field theory, particles
can be created where none existed before. Some of these will be strongly
interacting, so long as essential conservation principles are respected.

In terms of a simple Feynman-type diagram, Figure 4.31, a high-
energy electron–positron pair creates a very energetic virtual photon,
which is free to create a charged fermion–antifermion pair in what is,
effectively, the reverse of the original collision. Indeed, an electron–
positron pair could be produced, but this would be hard to see amongst
all the other electrons and positrons in the experiment, including those
that interacted without actually annihilating. A muon–antimuon pair,
on the other hand, could only be created via annihilation, so would be
a signature that an electron and a positron had destroyed one another.
And in a similar way, a quark–antiquark pair might instead result from
this annihilation. But unlike the muon and antimuon, those quarks and
antiquarks would not be directly visible in the experimenters' detectors. Instead, each quark and antiquark would spawn a shower of particles flashing through the detectors.

In fact, given three quark flavors, three different quark–antiquark
combinations are possible: u plus ū, d plus d̄, and s plus s̄. The charges
on these three quark flavors ($+2/3$, $-1/3$, and $-1/3$ for u, d, and

s quarks, respectively) dictate the strength of their electromagnetic coupling to the intermediate virtual photon from which they spring as quark–antiquark pairs. Thus, because muons and antimuons have charges of −1 and +1, relative to producing muon–antimuon pairs the probabilities of producing quark–antiquark pairs go as $(2/3)^2$, $(−1/3)^2$, and $(−1/3)^2$, or 4/9, 1/9, and 1/9. That, in turn, means that the ratio of quark–antiquark pairs produced relative to muon–antimuon pairs should be $4/9 + 1/9 + 1/9 = 6/9$.

It isn't. That's because there's something missing from this calculation: color. With three possible colors for each of the three quark flavors, each of the quark–antiquark combinations is three times as likely to be produced, so the overall answer must be multiplied by three to give 18/9, or 2.

It would be neat and tidy to say that this is what experimenters see, but that's not true either. At electron–positron collision energies below about 3 GeV, the ratio *is* about 2, though it's not so easy to see this or to believe it. Then, if the ratio is plotted against increasing energy, there's a step increase, followed by some wobbles, and finally above about 10 GeV a flat, straight line. The value of the ratio is remarkably constant all the way up to about 45 GeV, after which it gently sweeps upwards. Its value over the long flat portion is 11/3, not 2, as shown in Figure 4.32. So, the good news is that there is indeed a plateau in a plot of the ratio, but the bad news is that it has the wrong value. The apparent discrepancy is due to the best possible cause: there are two more quarks contributing, one of charge +2/3, called the charmed quark, discovered in 1974, and the other of charge −1/3, the bottom quark, first seen in 1977. Using these *five* quarks, and applying the same arithmetic idea as before, that is squaring then adding the charges, then multiplying the result with the color factor 3, gives 11/3. This agrees with the experimental value for the plateau above 10 GeV: experiment supports three colors and five quarks.

An accelerator interlude

A huge amount of experimental effort and hardware lies behind the ratio plot shown in Figure 4.32. Much of the stack of data comes from a host of experimental groups using the 1970s and 1980s generation of electron–positron colliders. The first of these was the Stanford Positron Electron Asymmetric Ring, or SPEAR, at SLAC. With a circumference of 234 m, SPEAR was eventually able to collide 4 GeV electrons with

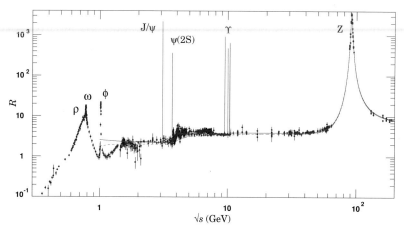

FIGURE 4.32 The ratio R of the cross-section for electrons colliding with positrons to yield hadrons, over electrons and positrons yielding muon-antimuon pairs. The J/ψ, Υ and other states are shown. Prior to the J/ψ, the ratio is 2, not 6/9. The factor of 3 difference is evidence for color, first pointed out by Bardeen, Fritzsch, and Gell-Mann at a conference presentation in 1972 in which they also discussed color in neutral pion decay. Above the Y resonance, R jumps to 11/3: without color this figure would be 11/9. Credit: Particle Data Group.

4 GeV positrons, giving a total collision energy of 8 GeV. SPEAR produced many discoveries that feature in the story of QCD.

One of SPEAR's earliest tasks was to measure the key ratio of hadron production versus muon production, but it could not reach very far up the energy scale. To move to still higher energies, the baton was passed to the larger SLAC collider, dubbed the Positron Electron Project collider, or PEP. This machine, which started up in 1980, had a circumference of 2.2 km and collided electrons and positrons at a total energy of 30 GeV. After a decade of running, PEP was closed down to make way for a 177 million US dollar upgrade called PEPII or, more imaginatively, the Asymmetric B-Factory, which started up in 1999. This machine studies a heavy breed of mesons called B mesons.

Between SPEAR and PEP, both in energy and timing, came CESR, the Cornell Electron–positron Storage Ring, at Ithaca, New York. A more modest 768 m circumference, CESR collides its particularly bright 8 GeV beams inside a single, massive particle detector with the delightful name of CLEO, invented early on as a suitable partner to the name CESR, as in Caesar and Cleopatra.

The United States has by no means cornered the market in electron–positron colliders. Operating in Europe are DORIS and PETRA, both at DESY, the German accelerator center. DORIS, or DOppel-RIng-Speicher (Double Storage Ring), the earliest and smallest of the pair, started up in 1974, and reached an energy of around 11 GeV as DORIS II. PETRA, the Positron Electron Tandem Ring Accelerator, which came on line in 1978, is much larger. This machine, 2.3 km in circumference, collided 23.4 GeV electrons with positrons of the same energy, giving 46.8 GeV in all. PETRA will feature again in the story on QCD, not least because the accelerator is still working in particle physics, though not in its original form.

Particle accelerators do not tend to die, they usually become injectors or radiation sources. An injector is a "small" accelerator that supplies a larger sibling with a beam of pre-accelerated particles: PETRA now performs this task for the large HERA electron–proton collider, which began running at DESY in 1992. DORIS, meantime, after more upgrading, went on to become a powerful and well-controlled source of X-rays for use in medical, materials, and chemistry research. The European Molecular Biology Laboratory has a station at DESY to exploit this facility, a perhaps surprising spin-off of particle physics research.

Gathered together, the data from experiments using this generation of machines give a remarkably consistent value of the magic ratio of muon–antimuon pair production to hadron production in electron–positron collisions. This ratio tells physicists that, with five quarks, the number of colors is three. But a new breed of electron–positron colliders, with CERN's LEP in the vanguard, offered even more energy, and the prospect of a variant on this classic result.

LEP – the world's largest accelerator

Nobody makes giant particle accelerators as a stock item. Each is customized and hand-built, and each is a triumph and a marvel of modern engineering. What makes particle accelerators so remarkable is that they embody the highest precision on the largest scale.

Take LEP, for instance, CERN's Large Electron Positron collider, Figure 4.33. Housed in a 26.67 km circular tunnel beneath the Franco-Swiss border near the Jura mountains, until its closure at the end of 2000 LEP was the world's largest particle accelerator. Completed in 1989 at a cost of over a billion US dollars, for half of the 1980s LEP was

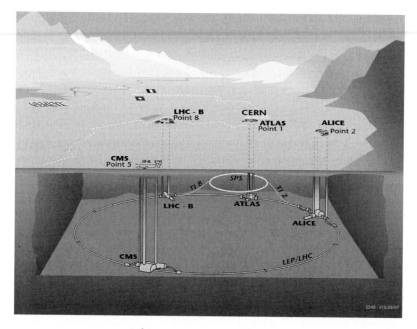

FIGURE 4.33 Subterranean monster. The LEP tunnel, now inhabited by
the Large Hadron Collider (LHC), beneath the Franco-Swiss border. The
"notches" on the ring are the experimental access areas. LEP
experiments took up four of these: L3 (Point 2); ALEPH (4); OPAL (6);
and DELPHI (8). Newer LHC experiments, whose names are marked on
the graphic, are: ATLAS (1); ALICE (2); CMS (5); and LHC-B (8). Credit:
CERN.

the single largest civil engineering project in the whole of Europe. This
really is big science.

For a hole in the ground, even LEP's tunnel was a piece of precision
work. The plane of the tunnel was tilted at 1.4° to ensure that as much
of the tunnel as possible passed through solid rather than loose rock.
By using laser interferometry, engineers insured the tunnel circumference was accurate to less than a centimeter over the entire 27 km. It
was also a big hole. To dig the tunnel, which was to be about 4 m in
diameter and lay between 50 and 100 m below ground, together with
the experimental caverns and other shafts, 1.4 million cubic meters of
earth and rock had to be excavated. Above ground were over 70 surface
buildings, some of which housed the 160 or so control computers and
microprocessors.

Inside the tunnel, over 4000 magnets manipulated the beam, with a total of 22 km of the beam vacuum chamber inside a magnet of some kind. Some relatively low-power magnets bent the beam into a roughly circular orbit. Others applied alternating gradient focusing to the beam by first focusing then defocusing it. More than 750 dc power supplies, ranging in power from 1 to 7 kW and accurate to twenty parts per million, were at the disposal of operators who used the magnets to control the beams, whose positions were monitored by 500 sensors.

The alternating voltage required to accelerate particles was applied as a radio-frequency electromagnetic wave. Because electrons and positrons have opposite electrical charge, the same electric fields accelerate them in opposite directions around the accelerator ring. The accelerating radio-frequency field was applied at specific points round that ring, and between those points particles were in free flight, apart from forces due to the magnets. In the original incarnation of LEP, radio-frequency energy entered the beam via 128 radio-frequency resonant cavities. These 2 m long copper chambers encircled the beam, channeling into it energy from sixteen 1 MW klystrons (generators of radio-frequency electromagnetic waves).

LEP was an electron–positron synchrotron, relying on the use of phase stability using changing magnetic fields and changing accelerating voltage frequency. One consequence of this was that particles could not be accelerated from cold, but had to be pre-accelerated and then injected into LEP, a common practice in modern accelerators.

At LEP, the starting electron beam came from a heated filament, just as in a television tube. The positron beam came from collisions between 200 MeV electrons and a suitable target. Each beam then passed through a 600 MeV linear accelerator, or "linac," and on to a 600 MeV accumulator to build up a stash of particles. From here, particles were transferred to the CERN Proton Synchrotron (PS) running at 3.5 GeV, and then to another accelerator, the CERN Super Proton Synchrotron (SPS), before both electrons and positrons were injected into LEP itself with energies of 20 GeV. Both the PS and SPS were modified to allow them to accelerate electrons and positrons together. In the case of the SPS, the electron–positron acceleration cycles were slotted in between proton acceleration cycles, so that the SPS could still function as a fixed target proton accelerator despite being an injector for LEP. Particle acceleration to high energies entails rather more than

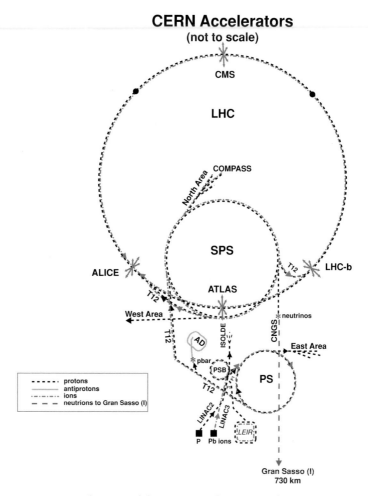

FIGURE 4.34 Schematic of the CERN accelerator complex. LHC, Large Hadron Collider; SPS, Super Proton Synchrotron; AD, Antiproton Decelerator; ISOLDE, Isotope Separator OnLine Device; PSB, Proton Synchrotron Booster; PS, Proton Synchrotron; LINAC, LINear ACcelerator; LEIR, Low Energy Ion Ring; CNGS, CERN Neutrinos to Gran Sasso. Credit: CERN

simply switching on an accelerator each morning and waiting a little while for it to warm up, Figure 4.34.

The need to inject pre-accelerated particles is just one of the compromises necessitated at LEP and elsewhere, for example at the Tevatron. Another is that particles arrive, not in a constant stream, but in bunches. At LEP, each bunch was just 1.8 cm long, and there were just a few bunches each of electrons circulating one way, and an exactly equal number of bunches of positrons circulating the other way. Arguably one of the most amazing statistics of LEP was that bunches of electrons and positrons approach each other at close to the velocity of light, around $300\,000$ km s^{-1}, yet only met one another exactly at the centers of the four detectors located around the ring. Collisions between electron and positron bunches occurred at the centers of the detectors every few microseconds.

The brighter the beams, the more collisions experimenters see in their detectors since there's more chance of particles colliding. The effective brightness of an accelerator is called its luminosity. Luminosity is a quantity that accelerator engineers often tinker with as one of the most significant ways of improving the physics output of an accelerator. At LEP, special magnets at the collision points squeezed the beams to about 8 microns high and 200 microns wide, the smaller the better since more particles are crammed into a smaller area. This meant that, at a collision point, each bunch had a cross-sectional area comparable to that of a pin head, yet the bunches still manage to hit one another. Another way to improve luminosity is to increase the number of particles per bunch. With 41.6×10^{10} electrons or positrons per bunch, for example, LEP's luminosity clocked in at 11×10^{30} particles per square centimeter and per second, a figure that continued to rise over the years. Experimentalists often gather their data over weeks or months of run-time, and give an accumulated luminosity called the integrated luminosity, the brightness integrated over time. A typical integrated luminosity figure for a LEP experiment might be 35 inverse picobarns, for instance, where a picobarn is 10^{-12} barns or 10^{-40} square meters. That's equivalent to about 37 days of uninterrupted collisions between beams at the luminosity given above.

Electrons and positrons are supposed to collide with each other and not with anything else. To ensure that this is the case, the beams must find themselves traveling in as good a vacuum as possible, so there's little chance they will collide with air or other gas atoms at some point in their long journey round and round the ring. Maintaining a high-level vacuum on a large scale is tough. Pumps do most of the work of sucking out the air from inside the vacuum chamber through which the particles fly. At LEP a zirconium–aluminum strip 3 cm wide and 22 km long soaked up remaining individual atoms of contaminating gas, much as a fly strip catches flies. The strip was "cleaned" periodically by heating it to 400 °C, so that the atoms of contaminant diffused into the bulk of the strip to become irreversibly buried within.

With junk atoms out of the way, electrons and positrons were able to collide with each other. And this they did, almost as soon as LEP began operation on 14 July 1989, one day ahead of schedule. Within four months, LEP experimenters had already bagged over 100 000 Z particles. This established LEP as a "Z factory," capable both of producing copious numbers of Z particles, and of exploring them in detail, thereby fulfilling one of the main goals of LEP's master plan.

Another item on the original agenda was to eventually upgrade LEP to higher energies, so that it could create pairs of W particles. The Z particles can be produced individually since, being neutral, there's no objection on charge conservation grounds to producing a single Z from an electron and positron collision. At an energy of 50 GeV per beam, or 100 GeV total collision energy, LEP in its original configuration was well-able to create Z particles, whose mass is 91.2 GeV. On the other hand W particles, being charged, must be produced in pairs as W^+ and W^- so, with a pair of W particles having a mass of 160.6 GeV, LEP would have to exceed this energy to see W pair creation.

This was a main objective of the LEP upgrade to a much higher energy machine called LEP2. To reach these higher energies, engineers replaced the old copper accelerating cavities with new superconducting ones. By July 1996, 144 superconducting cavities had been installed, and LEP2's total collision energy reached 161 GeV. Just two days after restarting the machine at this new, higher energy, LEP experimenters witnessed their first W pair production.

This increase in energy, and the use of superconducting cavities, brought new challenges. Electrons and positrons now circled the 27 km accelerator at around 11 200 times per second, and on each revolution lost 2% of their energy, all of which had to be replaced to prevent the particles from slowing down. The energy is lost because any accelerated charged particle throws off electromagnetic radiation. In the case of particles constrained to move in a circle, they are constantly accelerated towards the center of the circle, so constantly emit electromagnetic radiation. At the energies of a particle accelerator, this so-called synchrotron radiation takes the form of X-rays. There's so much synchrotron radiation produced that accelerators are sometimes built underground, to exploit the ground itself as a radiation shield, as is the case with LEP. In addition, the vacuum chamber through which the beams travel must be specially shielded, so that the radiation does not damage cables and other components sharing the tunnel. Even then, people were not allowed in the tunnel while LEP was running, due to the radiation hazard. With every cloud having a silver lining, however, synchrotron radiation can turn an accelerator into a useful X-ray source for medical and materials research, for example DORIS at DESY, but in general it's undesirable. In fact, the large circumference of the LEP ring was selected to minimize synchrotron radiation, since the radiation is greater for a more tightly curved path. But it's energy that hurts most: doubling the accelerated particle energy increases the synchrotron radiation loss by a factor of sixteen. Doubling the particle energy is exactly what happened at LEP when its upgrade was completed in 1998, the total collision energy leaping to 192 GeV. It's the energy loss from synchrotron radiation that effectively limits the overall power of electron and positron accelerators such as LEP.

The superconducting cavities used to get to the higher energy of LEP2 also brought new challenges. Oscillating electromagnetic fields lose a lot of energy in conventional copper cavities through electrical resistive heating in the cavity walls. In superconducting cavities, however, the electrical resistance of the superconducting niobium coating of the cavity is almost non-existent. This makes superconducting cavities much more efficient and able to work at far higher electric fields, but here too there is a trade-off: they must be cooled to liquid helium temperatures, around −269 °C. To achieve this, CERN had to produce and store 60 000 liters of liquid helium, which meant building the world's largest liquid helium plant.

Forward planning for the LEP project included plans for LEP's demise: in 2000 LEP closed down to make way for the Large Hadron Collider, or LHC, which will take over LEP's tunnel. The LHC, scheduled for completion in 2005, will collide protons together, rather than the more usual proton–antiproton combination, and will be the largest and the most energetic accelerator on the planet.

Across the Atlantic was LEP's prime competitor, the other Z factory of the 1990s: the Stanford Linear Collider, or SLC. Firing up in 1989, the same year as LEP, and closing down two years earlier, on the one hand the 100 GeV SLC was an upgrade of the original SLAC two-mile linear accelerator, and on the other hand it was a totally unique machine.

The basic idea of the SLC was to use the one linear accelerator to accelerate both electrons and positrons. The positrons were created in the positron source using electrons siphoned from the linac, and were then transferred back to the start of the accelerator. Prior to their main acceleration run, both electrons and positrons passed through small storage rings, called damping rings, whose function was to reduce each beam's spread. In the accelerator itself, each bunch of positrons was pursued by a bunch of electrons just 60 ns (nanoseconds) behind. A second trailing bunch of electrons was redirected out of the linac and used to generate positrons. When the two sets of accelerated particles left the end of the linac, the 50 GeV electrons were swept into one arc of bending magnets, the 50 GeV positrons into an opposing arc, and the two beams were brought together into a head-on collision. Starting in 1991 an elaborate detector, the SLD or SLC Large Detector, watched and recorded these collisions.

The SLC was also the test bed for a new generation of electron–positron colliders. Higher energy machines that are simply scaled-up versions of LEP become either unacceptably large, or lose too much energy to radiation. The answer could be to accelerate electrons and positrons in a straight line, and then bend them so that they collide head-on, as in the SLC. A successor machine to the SLC, tentatively dubbed the Next Linear Collider (NLC), is already under active study by SLAC scientists and engineers. As things stand, LEP2 seems likely to remain in the history books as the biggest and most powerful circulating electron–positron collider ever built.

The color of OPAL

Relative to the earlier generation of electron–positron colliders, the higher energies of LEP and the SLC opened up a whole new realm of possibilities.

At lower energies, electrons and positrons can annihilate to a photon, and this photon can, in turn, be a source of a daughter electron–positron pair, or a muon–antimuon pair, or even a quark–antiquark pair. The latter two were exploited to look for evidence of color in PETRA, PEP, and in other colliders.

At LEP and SLC energies, however, the electron and positron can, in addition, annihilate to give a neutral Z particle, and this too can decay to give a fermion–antifermion pair, for example an electron–positron pair or a quark–antiquark pair. So the higher energy opens up a whole new channel for electron–positron annihilation via a Z boson. Not only that but, focusing on the quark–antiquark possibility, each quark or antiquark leaving the collision point has so much energy that there's a realistic chance that they, in turn, throw off an energetic gluon, a form of bremsstrahlung.

If each quark and antiquark can emit a gluon, then an annihilation should sometimes produce four collision products, a quark, an antiquark, and two gluons. But there are three other ways in which a total of four quarks, antiquarks, or gluons can be created. Either the quark or the antiquark may throw off two gluons on its own, for instance. Or the gluon thrown off by either may in turn spawn two gluons. Or the gluon thrown off by either may spawn a new quark–antiquark pair. These four distinct processes, all roughly equally likely, are represented by the Feynman diagrams of Figure 4.35, which shows just a few of the full set of relevant diagrams.

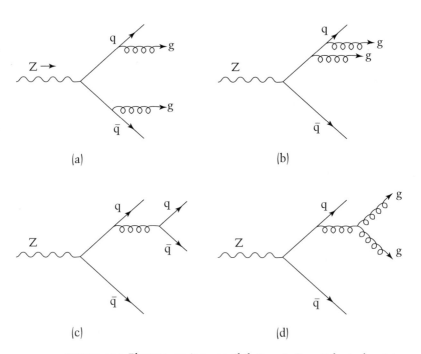

FIGURE 4.35 Electron–positron annihilation via Z particles to four jets: (a) and (b) are double-gluon bremsstrahlung, yielding $e^+e^- \rightarrow q\bar{q}gg$. Diagram (c) contains pair production to give $e^+e^- \rightarrow q\bar{q}q\bar{q}$. Diagram (d) contains a three-gluon vertex, $e^+e^- \rightarrow q\bar{q}gg$.

Table 4.5 *Color factors from theory and experiment. OPAL results support SU(3) as the color group.*

| | Theory | | | Experiment |
	$U(1)$	$SO(3)$	$SU(3)$	(OPAL)
Ratio 1	0	1	$9/4 = 2.25$	2.11 ± 0.32
Ratio 2	3	1	$3/8 = 0.375$	0.4 ± 0.17

The experimenters at the SLD or at the LEP detectors do not see quarks, antiquarks, or gluons as the products of these collisions, however. Instead, each of these collision products spawns its own complement of strongly interacting particles, and it's these showers, or jets, of particles that the experimenters pick up. The signature of the group of processes depicted in Figure 4.35 is four distinct jets of strongly interacting particles, called hadronic jets, which coincide with a Z particle production peak in the electron–positron cross-section.

Production of four such jets via Z bosons provides an important way of checking out the color group. When theorists calculate these cross-sections, they must include averages and sums over all possible color contributions. This leads to the emergence of three distinct color factors, numbers representing the relative strengths of the three processes $q \rightarrow qg$, $g \rightarrow gg$, and $g \rightarrow q\bar{q}$. The actual values of these factors track back directly to the properties of the $SU(3)$ group. Should the underlying symmetry group of the strong force between quarks *not* be $SU(3)_c$, but for example $SO(3)$ or $U(1)$ (actually, the product of three copies of $U(1)$), then this fact would be revealed in the measurable values of the color factors.

As it happens, it's not even necessary to determine the color factors directly to test which group they match: their ratios are enough, which is very convenient for experimenters. Three color factors give just two independent ratios. Labeling the two options as ratio 1 and ratio 2, if the underlying symmetry is $SU(3)_c$ then ratio 1 should have a value of $9/4$, and ratio 2 a value of $3/8$. If instead the strong interaction obeyed an $SO(3)$ symmetry, then both ratios would have the value 1. A $U(1)$-based strong interaction yields ratio 1 as zero, and ratio 2 as 3, Table 4.5. For this final possibility, the differences stem principally from the fact that $U(1)$ is an Abelian group, and so prohibits three-gluon interactions of the kind that feature in Figure 4.35(d). For an $SU(3)$-based theory, however, three-gluon interactions should be plentiful, and consequently around 95% of all the four-jet events should be due to qq̄gg, and only 5% due to qq̄qq̄.

There were four experimental detectors on the LEP accelerator, each the size of a house, and each run by scientific collaborations hundreds strong from universities and institutes from around the world. The first alphabetically was ALEPH, which stands for "Apparatus for LEP Physics," possibly a crude attempt at ensuring pole position in any alphabetic listing of LEP's detectors. One graduate student working on the machine suggested "A Large Expensive Piece of Hardware" as an alternative. Next in the list was DELPHI, or Detector with Lepton, Photon, and Hadron Identification. Providing a break from the somewhat tortured and occasionally nested acronyms was L3, and last but not least came OPAL, the "Omni Purpose Apparatus for LEP," Figure 4.36. All four used similar technologies to track and identify particles, all four were different, and each had its own particular features.

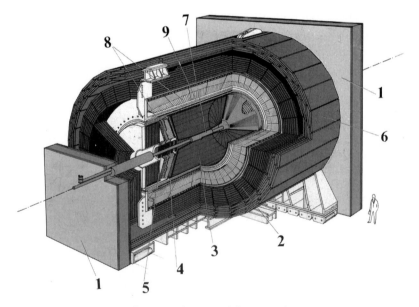

FIGURE 4.36 Cut-away diagram of the OPAL detector at CERN: 1, muon end cap; 2, hadron calorimeter; 3, jet chamber; 4, magnet coil; 5, forward detector; 6, muon barrel; 7, vertex detector; 8, lead glass; 9, presampler and time-of-flight detector. Credit: CERN

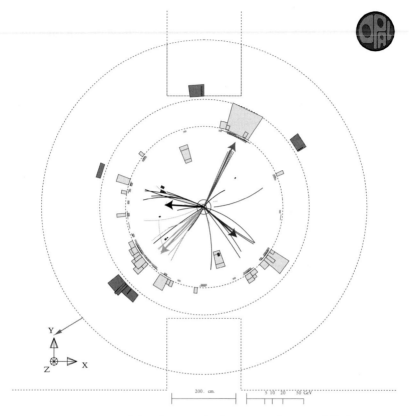

FIGURE 4.37 A four-jet event in electron–positron annihilation captured in the OPAL detector, viewed end-on. Arrows indicate the four jets. Figure prepared by Stefan Kluth. Courtesy: CERN and OPAL collaboration.

All have examined Z to four-jet events in some way or other as a means of exploring the color group.

One of these studies, involving over a million Z decays into hadrons, was published in 1995 by the OPAL collaboration. From their million-plus events, the OPAL team identified just 21 732 four-jet events, which they analyzed one by one in terms of the angles between the emerging jets, Figure 4.37. Using this angular information they extracted values for the color factor ratios, obtaining 2.11 ± 0.32 for ratio 1 and 0.4 ± 0.17 for ratio 2; the respective values from QCD theory based on three colors and $SU(3)_c$ are 2.25 and 0.375.

Theorists have calculated color factor ratios for a long list of available groups. Many of these are ruled out straight away by the OPAL result, and by measurements by the other LEP collaborations. But not all other candidates can be so readily wiped from the slate. For instance, the errors on OPAL's result are large enough to allow four colors with the color group $SU(4)$, at least in principle. However, if the number of colors is taken to be three, as demonstrated by, amongst other things, the electron–positron to hadrons versus muons ratio discussed above, then there are just three possible color group candidates: $SU(3)$, $SO(3)$, and $U(1)$. The OPAL result counts against both $SO(3)$ and $U(1)$, and offers powerful backing for the choice of $SU(3)$ as the symmetry group of color.

A THEORY FIT FOR QUARKS

Though in the 1950s Yang and Mills, and later Utiyama, set out the framework for what are now called non-Abelian gauge theories, of which QCD is one, there was plenty left for theorists to do. This framework directly relates underlying symmetries, for example the $SU(3)$ group that OPAL and other experimenters see at work, to physical interactions. But before non-Abelian gauge theory can be used as a basis of a physically meaningful description of the world, two key questions need to be answered. Can quantum rules be incorporated properly? Is the theory renormalizable? The answer to both questions needs to be yes.

Obeying the quantum rules . . .

Quantization, as the process of enforcing the quantum rules is called, means, at one level, feeding the master equation of the theory into the Feynman path integral program and extracting the Feynman rules. But here, a problem rears its ugly head. The very feature that makes gauge theories so remarkable, the property of gauge invariance, becomes a source of trouble. It causes an over-counting that seems to make the final answers to some calculations shoot off to infinity.

The over-counting problem is a little like conducting a census in a town full of traditional families, each comprising mom, dad, and a few kids, and mistakenly equating the number of people in the town to the number of families. Since each family contains several members, a simple head count would over-estimate the number of families, a trivial example of an over-counting problem.

The path integral recipe says to include all possible contributions from all possible configurations of the fields in the theory. The trouble is, for a gauge theory, some of these field configurations are related to one another via gauge transformations. So working with field configurations gives an over-counting in an analogous way to a head count that over-estimates the town's families. Not only that, each field configuration can be linked by a gauge transformation to infinitely many others. As a consequence, naively applying the Feynman path integral algorithm fails, with yet another infinity appearing in the final answer.

Compared to the census example, it's as though there are a limited number of families in the town, but each family has an infinite number of children. So counting people will give an infinite answer, but counting just families will give a reasonable answer. Similarly, to quantize a gauge theory means finding some way of including only a single representative field from each family into which the complete population of fields naturally divides: a gauge transformation links fields within a family, but not those in different families.

Picking out a single member of each family of fields is termed gauge fixing, and deciding just how to do this from the many available options is called choosing the gauge. In terms of the census example, it's rather like deciding to count only adult males, rather than all people, as a means of counting families.

With this new restriction imposed, just one field from each family of gauge-related fields remains, and the infinity appearing in the path integral takes up residence in a harmless overall constant, having no role in deriving the all-important Feynman diagrams. As a consequence of limiting the path integral in this way, however, something very odd happens. When the gauge-fixed master formula is put through the Feynman path integral mill, the new constraint gives rise to extra Feynman diagrams, in addition to the expected ones. These new diagrams reflect the presence of a new and mysterious "particle" in the problem. These bizarre newcomers are called ghosts.

Feynman was the first to appreciate the need for ghosts, way back in the early 1960s, but it was two Russian researchers based in Leningrad, Ludwig Faddeev (Figure 4.38) and his student Victor Popov, who sorted out how ghosts fix the gauge over-counting problem. For this reason, and in part as a reflection of their clear presentation of their results, ghosts are often called Faddeev–Popov ghosts. In Texas, Bryce deWitt, interested in quantum gravity, also figured out how ghosts work but his account, likewise published in 1967, was relatively impenetrable though prescient in scope.

FIGURE 4.38   Ghostbuster. Ludwig Faddeev, pictured here in 1986 in Alushta, at a conference in the Dubna "holiday house" overlooking the Crimean coast. Credit: Courtesy Ludwig Faddeev/ Yuri Tumanov.

So what exactly are ghosts? They are unphysical entities, and this really means unphysical, not simply unobservable in the sense of quarks. Ghosts are essentially artifacts of the calculational recipe: nobody has ever mounted an experiment to look for them, and nobody in their right mind ever would. They have no spin, yet behave as fermions, with antighosts distinct from ghosts, so their spin-statistics are unphysical – another hint that there's no point looking for them.

Ghosts only appear in calculations involving loops of gauge bosons. Simple tree Feynman diagrams are spared the additional complication of corresponding ghost diagrams. For every gauge boson loop, there will be a matching ghost loop diagram, Figure 4.39. The details of the coupling between ghosts and gauge bosons depends on the choice of gauge, but nothing can make ghosts couple to fermions such as electrons or quarks. Ghosts are a feature only of the gauge boson part of the theory, and they would be necessary even if there were no fermions in the theory, just gauge bosons such as gluons.

It might seem that introducing an unphysical particle to spirit away an unwanted infinity is a bit of a fudge, a way of extracting a

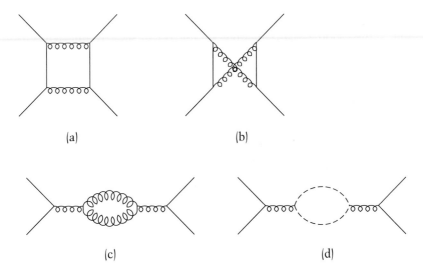

FIGURE 4.39 The four Feynman diagrams contributing to q̄q → q̄q at the
one-loop level. Diagram (d) shows a ghost loop which parallels the gluon
loop of diagram (c). Without diagram (d) the three other diagrams would
yield an infinite result.

meaningful answer from a theory that seems to be trying to do otherwise. That would be a bit unfair. Ghosts come about as a result of the
demands of special relativity, in the form of the requirement of covariance. Enforcing this means that unphysical gauge boson contributions
get incorporated by accident. These must be removed, but there is no
way of doing so without destroying covariance. Ghosts fix the problem by canceling out the extraneous portions, but without sacrificing
covariance.

There's another way of viewing the need for ghosts. The orientation of a particle's spin relative to its direction of motion is called
its polarization. A photon with its spin pointing along its direction of
motion, a spin of +1, is said to be right-hand polarized. A left-hand polarized photon has its spin pointing back from whence it came, and has a
spin of −1. Plane polarized light, manifest as the sheen from the surface
of a lake, is a mix of the two. But what's missing in all this is the spin 0
option. This would ordinarily feature in a discussion of spin 1 particles
whose measurable spin, according to quantum theory, comes out to be
+1, 0, or −1. But photons are no ordinary spin 1 particles, because they
have no mass, and for this reason the spin 0 case is missing.

Physical gluons, likewise massless, also have just two possible spin states, though experimenters do not get to see left and right circular polarized gluons in the raw since nature has opted to hide them from view. The trouble is that, in QCD, Feynman diagrams with internal closed gluon loops *include* contributions from spin 0 gluons, a consequence of Lorentz covariance. All contributions to the final answer due to these unphysical contributions must be removed. This is done by the ghosts, which cancel out portions due to unphysical gluon spins. In so doing they restore unitarity: without ghost diagrams, unitarity is violated. In their key 1971 work, 't Hooft and Veltman demonstrated that ghost diagrams do indeed restore unitarity for non-Abelian gauge theories. The result of the Faddeev–Popov trick is a consistent set of Feynman diagrams, both tree diagrams and those with loops. In other words, non-Abelian gauge theories can indeed be quantized.

How is it, then, that ghosts didn't pop up in QED? The gauge fixing issue arises in the same way, and for the same reasons, but there is one crucial difference. Because QED is an Abelian theory, ghosts there do not couple with photons, and so they couple with nothing at all. For this reason, the ghost contribution can be separated out and safely ignored, though it is possible to make the ghost parts appear explicitly if one so desires. Ghosts are not an issue in QED, though with hindsight theorists now see how they fit in. It's probably very fortunate that QED was discovered before QCD or the electroweak theory, and is devoid of complications such as ghosts. Discovering QCD first, with all its additional complications, would have been much harder without the guidance and inspiring successes of QED!

... and beating the infinities

There are many roads to immortality. One of the more efficient is to write a brief letter to a scientific journal containing an absolutely key idea or equation. Because this result will feature in so much that is to follow, it begs for a name, and its name is almost certainly going to be that of its originator. And so it was for John Ward.

In 1949, Ward, at the University of Oxford, published a half-page paper in *Physics Letters* that put his name in chapter subheadings in field theory texts ever since. Not that Ward knowingly set out to achieve immortality in this way, of course. The focus of his attention was Dyson's speculation that two of the three infinite constants featuring in the renormalization of QED should actually be the same. Ward's note was a proof that this is indeed true. In getting there, however, he wrote down a version of the equations that now bear his name: the Ward identities. For proving the renormalization of QCD and the electroweak theory, the Ward identities are indispensable.

Ward's result was generalized by Yasushi Takahashi at the State University of Iowa in 1957. It was extended to non-Abelian theories in 1971 by Andrei Slavnov, a colleague of Faddeev and Popov at the Steklov Mathematical Institute in Moscow, and Oxford University's John Taylor. Bracketing together as Ward identities both Ward's original version and these extensions, Ward identities are a consequence of the symmetry of a gauge theory. They are actually relations between Green's functions, the amplitudes for "something happening" in a particle collision process, for example. Thus the Ward identities inforce the cancellations that are so crucial to the renormalization program.

That trusty template gauge theory, QED, is a good place to look to see what this means. The basic interaction of QED involves an electron, a positron, and a photon, Figure 4.40. The object on the left-hand side of 4.40(a) is supposed to represent the sum of all the Feynman diagrams of all possible complexities based on the electron–positron–photon vertex diagram. The first couple of entries in this list appear in Figure 4.40(b). The right-hand side of the "diagram equation" in Figure 4.40(a) is about propagators, which describe a particle's propagation from point to point. The difference between two fermion propagators, differing in momentum by an amount equal to the momentum of the photon, divided by the product of the same two propagators, is equal to the vertex diagram of the left-hand side. The propagator pieces themselves are each sums over all the contributing Feynman diagrams, represented in Figure 4.40(c). Equations of diagrams may be unfamiliar, but that doesn't much matter. What does matter is that there's a relationship between the vertex diagram – a three-point Green's function – and propagators, that is, two-point Green's functions. So the Ward identities relate different Green's functions. The diagram equation is a reminder that entire armies of Feynman diagrams are included: the Ward identities do not merely link individual Feynman diagrams of similar complexities, but hold to all orders of perturbation theory.

Ward identities are another spin-off of gauge invariance. Recall that to beat the gauge field over-counting problem inherent in the path integral as outlined above, at some stage a particular gauge has to be chosen. Once this is done, by definition gauge invariance is sacrificed.

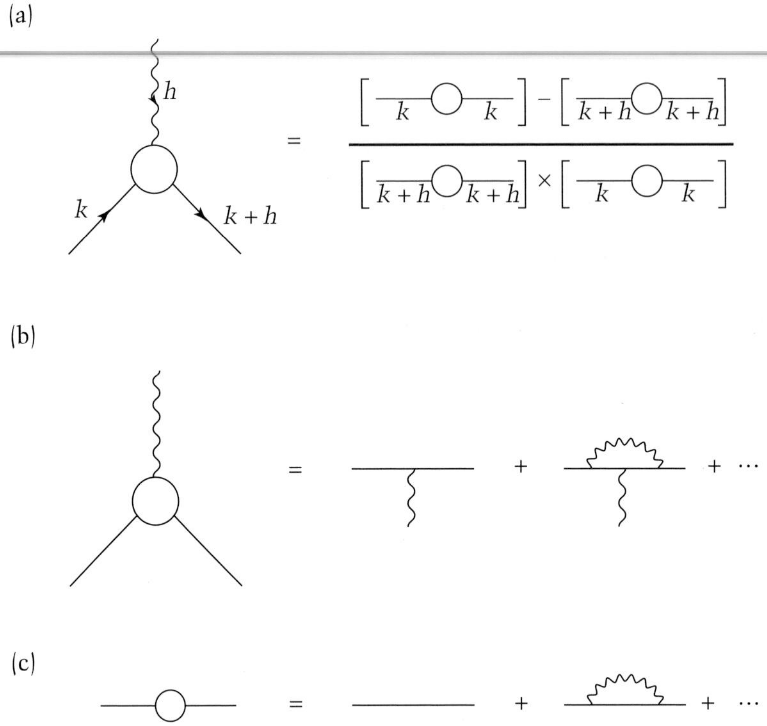

(a)

(b)

(c)

FIGURE 4.40 The Ward identity in QED (a), relating the vertex to propagators. The vertex on the left-hand side in (a) represents an infinite series of Feynman diagrams, depicted in (b). Each of the propagators on the right-hand side of (a) also represents an infinite series of diagrams, indicated in (c). Symbols k and h both represent momentum.

Yet the calculated probability of some physical process occurring cannot care about the choice of gauge, a mere convenience, and has to remain gauge invariant. Reconciling these two apparently conflicting demands leads to relationships between different Green's functions – the Ward identities.

So what? Ward's original paper gives a hint of how this is all going to help with the problem at hand, the renormalization of gauge theories. By equating two of the three infinite renormalization constants of QED, which relate bare quantities to the corresponding observable ones, Ward showed that only two constants are needed where originally it looked like three were necessary. That's nice, but the physical meaning actually runs much deeper. Quantum electrodynamics applies to electromagnetic interactions between other charged species, such as muons for example, just as much as it applies to the interactions between electrons and positrons. That two of the constants turn out to be equal means that, if the unrenormalized electric charge is the same for both electrons and muons (and other species), then the renormalized charges are also the same: the renormalization is universal, guaranteed by the Ward identities. That different particle species such as electrons and muons have identical charges, and couple in exactly the same way in QED, only really makes sense if the Ward identities hold good. And, in a similar vein, in QCD Ward identities guarantee that the gauge-enforced property that all quarks, no matter their color, couple in the same basic way to gluons, survives renormalization.

In terms of the unfolding of QCD, however, the Ward identities are most significant as an aid to proving the renormalizability of the theory. Proving renormalization relies on tracking complicated cancelations between Feynman diagrams. For instance, in QCD there are unphysical contributions from gluons that must cancel out with ghost contributions in order that the theory gives sensible predictions. The Ward identities inforce these cancellations. It was by exploiting Ward identities that 't Hooft proved the renormalizability of a whole Yang–Mills family of theories, and therefore QCD, earning himself a Nobel prize in the process. Not bad for a graduate student.

THE BIRTH OF QCD

Exactly when should scientists celebrate QCD's birthday? And exactly how old is QCD? Quantum chromodynamics is a Yang–Mills theory in which the basic particles are quarks, the gauge group is color $SU(3)$, and the force mediators (gluons) are an eight-strong team that each carry color. Han and Nambu said as much as long ago as 1965, though in a model in which quarks had integer electrical charge. They also noted that observable hadrons would be overall color-neutral, in other words color singlets. Was this QCD's birth? No. That is, their work is not generally recognized as marking the beginning of QCD. It's almost as though they were ahead of their time: the threads that were ultimately to be woven together in the making of QCD had not been spun.

One of those (not entirely independent) threads concerned Yang–Mills theories themselves. The basic idea of Yang–Mills theory appealed to many, and around 1967 Glashow, Weinberg, and Salam had set out a Yang–Mills theory of the weak force. But nobody knew if this or any

FIGURE 4.41 Harald Fritzsch (left) and Murray Gell-Mann (right) smile for the camera at a conference in Berlin. Credit: Courtesy Harald Fritzsch.

other Yang–Mills theory was renormalizable. Come 1971 and 't Hooft and Veltman showed they were. In the run up to 1971, Yang–Mills theory was a dusty corner: suddenly in 1971 it was interesting again. The electroweak theory of Glashow, Weinberg, and Salam was resuscitated, and it became natural to think in terms of Yang–Mills theory for the strong interaction too.

A second thread was the irresistible march of the new color quantum number. Beginning at the end of 1971, Gell-Mann spent some time working at CERN with German theorist Harald Fritzsch and William Bardeen, from Princeton (Figures 4.41 and 4.42). Bardeen in particular was well-versed in a niggling problem concerning the decay of the neutral pion into two photons: for pions built of fractionally charged quarks, the calculated decay rate came out at just one ninth of the measured rate.

In a paper that first appeared as a CERN preprint in 1972 Bardeen, Fritzsch, and Gell-Mann showed that the decay rate comes out just right if each of the quarks is assigned three colors. In this same paper they also showed how the factor of three due to color fixes up the ratio of hadrons to muons in electron–positron collisions mentioned earlier. These two

FIGURE 4.42 Bill Bardeen in his office at Fermilab, 1996. Bardeen's father John Bardeen is the only person to have won the Nobel prize in physics twice over: once for the discovery of the transistor effect in semiconductors and later for the Bardeen–Cooper–Schrieffer theory of superconductivity. Credit: Fermilab.

findings remain two of the most oft-cited pieces of evidence for color, but there were more. Also in 1972, three University of Paris theorists – Claude Bouchiat, John Iliopoulos, and Philippe Meyer – showed that the color attribute was essential to preserving renormalization in electroweak theory embracing both quarks and leptons. Because of the subtle cancelations involved, this result (which was also derived by David Gross and Roman Jackiw) has particular appeal for theorists.

Again in 1972, Gell-Mann presented a paper on behalf of himself and Fritzsch at the XVI International Conference on High Energy Physics, at the National Accelerator Laboratory, as Fermilab used to be called. Here they grappled with the problem of a model of strong interactions involving colored quarks and, for the most part, a *single* gluon devoid of color. So this is not QCD as we know it. But there were some important elements in their paper. There was the statement "real particles are required to be singlets with respect to the $SU(3)$ of color," echoing a more tentative version in the 1972 preprint and, of course, the work of Han and Nambu. Fritzsch and Gell-Mann also referred to the idea that not only quarks but gluons too should be colored. "For example, they could form a color octet of neutral vector fields obeying the Yang–Mills equations," they said. This sounds like QCD. Yet after a couple of brief observations of the potential benefits of such an

FIGURE 4.43 David Gross at the conference podium, 2003. Credit: Courtesy David Gross.

approach, they chase after a doomed single-gluon approach. So does *this* paper represent the birth of QCD? Some theorists think so.

A third thread offers up another candidate. In 1969 experimenters at SLAC showed that protons seemed to contain point-like scattering centers. This finding, couched in terms of a phenomenon called "Bjorken scaling," to be discussed shortly, represented a serious challenge to theorists. It was the inspiration for David Gross and his student Frank Wilczek at Princeton and graduate student David Politzer at Harvard University, to study the high-energy behavior of the newly fashionable Yang–Mills theory (Figures 4.43–4.45). Both groups showed that the interaction strength for Yang–Mills theories falls away to nothing as the energy increases. Gross and Wilczek in particular noted that, with Yang–Mills theories, "one can construct many interesting models of the strong interaction." Aware of the emerging evidence for color, for example citing the 1972 work of Bardeen, Fritzsch, and Gell-Mann on pion decay and electron–positron annihilation, Gross and Wilczek describe "one particularly appealing model" – that sounds pretty tentative still – in which $SU(3)$ color symmetry accommodates the strong interaction. Quarks having the same flavor but different colors are mixed by virtue of neutral "vector mesons" (i.e., gluons). Now, this

FIGURE 4.44 Frank Wilczek at
the American Physical Society
meeting in Philadelphia, 2003.
Credit: Courtesy Betsy
Devine.

FIGURE 4.45 David Politzer in
1979. Politzer discovered the
key property of asymptotic
freedom independently of
Gross and Wilczek. Credit:
California Institute of
Technology.

sounds like QCD. Having written down the master equation of Yang–
Mills theory, and identified the symmetry aspect as the color group of
QCD in their "appealing model," Gross claims that he and Wilczek
were the first to publish the master equation of QCD. Then again some

might argue that once the framework is given as Yang–Mills theory, and the group is specified as $SU(3)$, actually writing down the master equation is straightforward. That aside, what Gross and Wilczek, and independently Politzer, discovered, in the modern language of quarks and gluons, was that the gluon-mediated force between quarks is weaker when quarks are closer together. At high energies, *quarks behave increasingly like free particles* as the strong force loses its grip. This property is termed asymptotic freedom, and without doubt rates as a cornerstone of QCD. Gross and Wilczek soon submitted two large papers fleshing out asymptotic freedom and its consequences, presenting the general case though with the $SU(3)$ color group clearly occupying the front seat.

Some text books point to yet another paper, by Fritzsch, Gell-Mann, and Heinrich Leutwyler (Figure 4.46), as their generic reference for QCD. This 1973 work, entitled "Advantages of the Color Octet Gluon Picture," emerged after the papers of Gross and Wilczek and Politzer, citing their work as one of the advantages implied by the title. In this paper, Gell-Mann and his collaborators finally toss out

FIGURE 4.46 Heinrich Leutwyler in Bern, 2003; Fritzsch, Gell-Mann, and Leutwyler discussed the structure of QCD in their paper of 1973. Credit: Courtesy Heinrich Leutwyler.

the single-gluon attempt, and focus instead on what is unmistakably QCD. Now all the key elements are here: the master equation; the asymptotic freedom attributed to Gross, Wilczek, and Politzer; the idea that observable particles are color singlets; Yang–Mills theory; and the correct color symmetry, $SU(3)$. Yet even here there's no sense of some eureka moment, more the feeling of a fancy new automobile being compared to a well-loved but failing wreck finally consigned to the scrap heap. To be fair, the single-gluon model was essentially a device Gell-Mann used to extract results in the study of quarks: writing twenty years later he described it as a "throw away" model.

Insofar as there is consensus, it is that by the end of 1973 QCD *was* born, even though it as yet lacked a name and was hardly an overnight sensation: acceptance was to gather momentum after the theory had marinated for a year or so.

"Asymptotic" is a word mathematicians use to describe a quantity approaching, but never quite reaching, a specific value, often zero or infinity. The "freedom" in this context is freedom from interaction. Asymptotic freedom, the drop in the strength of the force as energy rises, is central to QCD, to calculations using QCD, and to understanding why we don't see free quarks and gluons. In fact as Gross, Wilczek, and Politzer showed, it's a general property of non-Abelian gauge theories. And, as if to ram the point home, Gross and Sidney Coleman, Politzer's thesis supervisor, also showed that no renormalizable field theory possesses this property of asymptotic freedom unless it is non-Abelian. Since scaling demands asymptotic freedom, a theory capable of explaining the 1969 SLAC results *must* be a non-Abelian gauge theory.

The key piece of mathematical technology underlying the discovery of asymptotic freedom was the renormalization group. This translates the essential arbitrariness of the line between "short" and "long" distances in a theory into an energy-dependence of key parameters such as mass and interaction strength, a point to be revisited in Chapter 5. Swiss physicists Ernst Stückelberg and André Peterman first discovered the renormalization group in 1953. However Gell-Mann, working with University of Illinois' Francis Low, invented it independently and applied it to the study of QED. In modern parlance, they showed that QED does not possess asymptotic freedom. In fact, QED shows quite the reverse – the strength of interaction increases at higher energy.

FIGURE 4.47 Asymptotic freedom dawns: Curtis Callan, left, with David Gross, 1972. Credit: Courtesy David Gross.

Following a lead provided by Italian theorist Giorgio Parisi, who suggested that point-like scattering behavior requires a strong interaction that effectively switches off as energy rises, Gross, Wilczek, and Politzer examined Yang–Mills theory under a renormalization group spotlight. They found that at high energies the coupling parameter locks in to a certain fixed value, and that value is zero – asymptotic freedom. Ironically, due to a minus sign error in an early version of their calculation, Gross and Wilczek originally thought they had proved the exact opposite.

The intertwined paths of the renormalization group and asymptotic freedom encompass landmarks due to Curtis Callan (Figure 4.47), Kurt Symanzik, Kenneth Wilson, Weinberg, 't Hooft, and many others. In 1970 Callan, at Caltech, and Symanzik, at DESY, independently derived a version of the renormalization group equation that could accommodate quarks having mass. In turn, they had rediscovered a result due to Lev Ovsiannikov dating from 1956, a period when several Russians, including Dmitrij Shirkov and Nicolai Bogoliubov, had done much to extend the renormalization group idea. The Callan–Symanzik equation, as it came to be known, became the launch pad for the discovery of asymptotic freedom. Wilson, who started out as a student of Gell-Mann's studying strong interactions, applied renormalization group ideas to condensed matter physics or, more properly, to so-called "critical phenomena." This work, in the early 1970s, not only earned him the 1982 Nobel prize in physics, it established the renormalization group as a powerful tool in a wide range of physical problems. But in fact hints of asymptotic freedom actually date back to 1965, when two Russians, Vladimir Vanyashin and Mikhail Terentev, saw in their calculations what they viewed as an undesirable *decrease* in interaction strength in a toy Yang–Mills theory, as distinct from the *increase* in strength at short distances for electrodynamics. This was the signal of asymptotic freedom. Another Russian, Iosif Khriplovich, derived a comparable result four years later. So did 't Hooft, who mentioned his work at a discussion session following a lecture by Symanzik at a 1972 conference in Marseilles. But 't Hooft didn't see his result as especially novel or interesting, and never published. In 1973, Weinberg produced a paper in which he pulled together the strands of asymptotic freedom with the group theory elements of both weak and strong interactions, initiating what is now dubbed the Standard Model.

Weinberg also discussed one immediate problem facing the fledgling theory: what about free quarks and gluons? The answer, as theorists now understand it, was to be found in the first of Gross and Wilczek's larger asymptotic freedom papers, where they had considered the same question. They pointed out that if the strength of the interaction falls as the energy rises, it correspondingly increases at low energies. The idea is that at low energies the coupling between quarks becomes so great that free quarks are never seen: they are confined. This is the notion of confinement, the idea that quarks – and gluons – are always *confined* within composite particles. Experimenters only see color-neutral particles in their detectors. Confinement, a pretty reasonable idea given asymptotic freedom but one which has never been rigorously proven, is to be explored further in Chapter 7. This low-energy or, equivalently, long-distance hike in interaction strength in addition offers a natural explanation of why gluon-mediated processes are short range. Elsewhere, massless force carriers, such as photons and gravitons, are associated with interactions infinite in extent.

Asymptotic freedom is such an important feature of QCD it's fair to say that, without its discovery, QCD is incomplete. And its discovery was ultimately inspired by the search for an explanation of an amazing experimental result.

GLIMPSING QUARKS

Red illuminated "Pizza Parlor" signs are harder to read than green ones, which in turn are harder to read than blue ones. At least that's how things are in theory. The reality is a little more complicated since, for starters, the eye is not a perfect optical instrument. But the fact remains that there's a link between wavelength and the ability of an instrument, be it a telescope or the human eye, to create a sharp image.

All things being equal, the golden rule is that shorter wavelengths make for crisper images, since shorter wavelengths are less prone to the intrinsic wave spreading through diffraction that ultimately limits resolution. Microscopists know all about this. The ability of their conventional optical microscopes to pick out detail, in other words their resolution, has a theoretical limit set by the wavelength of light, around a few hundred nanometers. To see finer detail, microscopists need shorter wavelengths. That's where the electron microscope comes in. It's a direct application of the wave nature of "particle" matter: electrons in an electron microscope, accelerated through a voltage of between 10 and 200 kV, will have wavelengths down to tens of nanometers, fine enough to see large protein and polymer molecules.

The trick, then, to seeing fine detail, is to use very short wavelengths. A proton is around a fermi (10^{-15} m) in diameter, and to see something that small, or even to see something even smaller within, requires a really big electron microscope. The north German city of Hamburg is home to what is, in effect, the world's biggest electron microscope.

HERA – the world's biggest electron microscope
In 1995 the H1 collaboration at HERA (the Hadron–Electron Ring Accelerator at DESY, Germany) was able to show that light quarks, meaning the u, d, and s quarks, have a radius less than 2.6×10^{-18} m. That still doesn't mean that quarks have no size, though that remains a possibility. It just means that quarks are very, very small. HERA, the "electron microscope" that made this measurement possible, probes down to distances about one hundredth millionth the size of an atom.

When HERA, Figure 4.48, came into operation in 1992, it ushered in a new era of accelerators. It is a collider, not of two identical particles, for instance protons with protons as at the LHC, or of particle–antiparticle pairs, as at LEP and the Tevatron, but of *electrons* and *protons*, two totally dissimilar beasts. HERA comprises a circular,

FIGURE 4.48 A view down the 6.3 km HERA tunnel. Credit: DESY, Hamburg, Germany.

6.3 km tunnel containing two independent accelerator rings. The upper ring uses powerful, superconducting bending magnets to guide protons as they travel in one direction round the ring. Immediately beneath is the second ring, relying on lower power conventional magnets to guide electrons (or positrons) traveling in the opposite sense. At two places the rings intersect, and the beams collide with one another.

Two detectors, H1 and ZEUS (in Greek mythology Hera was the wife of Zeus) sit astride these intersections, watching for electron–proton collisions. These detectors face a special challenge, since although HERA is a collider, the protons carry much more momentum than do the electrons. This tends to blast the products of collisions into a narrow cone around the proton direction. The fragments of smashed protons tend to continue along the beam pipe unseen, Figure 4.49.

By colliding 920 GeV protons with 27 GeV electrons to give a total available collision energy of 318 GeV, HERA offers a view of the nucleon interior a hundred times sharper than that offered by any other machine. At a cost of about 650 million US dollars for a resolution roughly one thousandth of a proton diameter, HERA offers ten orders of magnitude improvement in resolution over a commercial electron microscope for only a few thousand times the price. That's value for money!

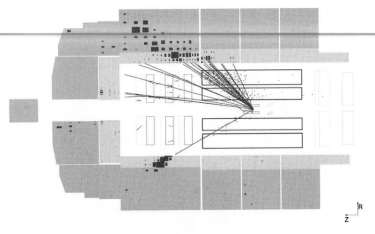

FIGURE 4.49 An electron–proton collision captured in the H1 detector. Credit: DESY, Hamburg, Germany.

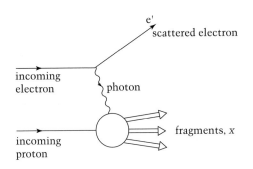

FIGURE 4.50 Deep inelastic electron–proton scattering. The incoming electron throws off a photon which smashes a proton into fragments, x.

So what exactly does HERA see? An incoming electron bounces off a charged constituent of the proton, and into a detector that records its energy and scattering angle. This is an electromagnetic interaction, so is mediated by photons. The process is well-represented by the simplest Feynman diagram, Figure 4.50, in which the incoming electron throws off a single photon, and it's this photon that probes the proton. In effect, the target is being viewed by the photon thrown off by the scattered electron. Photons carrying the largest momentum, and imparting the biggest kick, have the shortest wavelength and therefore give the sharpest "pictures" of the nucleon's contents. This is deep inelastic scattering: the electron loses energy in smashing up the proton, hence the inelastic, and the intermediate photon has a short enough wavelength to probe the depths of the proton, hence the deep. Deep inelastic scattering of electrons on protons, for example, is denoted e + p → e' + X, where the X stands for the stuff resulting from blasting a proton to pieces, and e' the scattered electron. Deep inelastic scattering is a major tool for studying nucleon structure.

Particle physicists use two key numbers to describe this kind of scattering. One is the energy lost from the electron, the difference between the incoming and outgoing electron energies, which is therefore called the energy transfer. The second is the (negative square of) the four-momentum transferred to the nucleon by the photon, already discussed in Chapter 3. This "size of kick" is tuned via both the incident and outgoing electron energy, and by the angle at which the electron scatters. Were the photon to be an ordinary real one, then there would be no need for this second label, since for a real photon the four-momentum squared is always zero since real photons are massless. But the photon probing the proton's heart is a virtual photon, its effective mass given by this momentum transfer. The larger the momentum transfer, the shorter the wavelength, and the finer the detail resolved in the experiment.

Researchers at HERA, and at other deep inelastic scattering experiments, measure the inclusive cross-section in an electron–proton collision, for example. That is, they look for and measure the scattered electron, and ignore the wreckage coming from the break-up of the struck proton. So far as the incoming electron is concerned, it sees, via the photon, another charged entity rather like itself, only with a range of energies and momenta. That's because the charged scatterer lives inside the proton and, although the proton's energy and momentum might be well-known, the energy and momentum of the charged scatterer are less certain. In fact, the apportioning of scattering center energies and momenta relative to those of the host proton is given by what particle physicists call structure functions. These they can reveal through cross-section measurements. Deep inelastic scattering exposes the distribution of point-like charged scattering centers – quarks – inside the nucleon, an ongoing program that experimentalists, with the start-up of HERA, are taking to new heights.

The trouble with cyclotrons

There are lots of cyclotrons scattered round the world, some used for research, others in medicine – neutrons are better at destroying cancer tumors than are X-rays, and protons are better than X-rays at delivering a tumor-annihilating radiation dose without destroying nearby tissue. Cyclotrons are also used for making medical radioisotopes and for fabricating advanced microchips.

As a tool for creating high-energy particle beams for particle physics research, however, cyclotrons have a problem: as the particles approach relativistic speeds, the timing condition so crucial to the machine's function falls apart. The fix, discussed earlier, is phase stability, resulting in the synchrocyclotron and synchrotron.

But there is a third way of putting phase stability to work to build a particle accelerator, and that is in a linear accelerator, or linac, in which particles are accelerated in a straight line. Instead of a circulating beam, a straight beam can pass along the axis of a line of charged cylinders. Acceleration takes place in the gaps between cylinders; while traveling the length of an individual cylinder a particle's velocity doesn't change. Counting along the line from one end, the odd-numbered cylinders are wired together, and so too are the even-numbered cylinders, Figure 4.51. So long as the transit time of a particle in a cylinder matches the time taken to switch over the voltage, a particle will emerge from a cylinder in time to see an accelerating voltage.

Phase stability is enforced by the cylinder lengths, which are greater at the higher energy end. A particle that arrives too late at a cylinder will receive a bigger kick and so will arrive at the next gap slightly ahead of time. A particle arriving too soon will receive a smaller kick and arrive slightly later than the norm. Though particles might

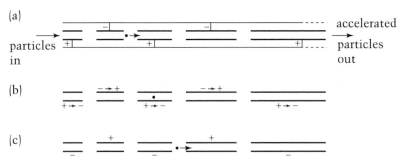

FIGURE 4.51 Schematic diagram of a linear accelerator. In (a), electrical couplings link cylinders. A negatively charged particle accelerates towards the third cylinder. When the particle is inside the cylinder (b) the cylinder voltages are switched over, indicated by + to − arrows, so that when the particle emerges it is accelerated towards the positively charged fourth cylinder (c).

be introduced continuously into the start of the accelerator, only those that catch the cycle just right will be accelerated – the output of this machine is a series of pulses of accelerated particles. Essentially, an electromagnetic wave travels the length of the machine, and electrons are carried along with it, "surfing" on the wave.

Linear accelerators offer some significant advantages over circular machines. Since the beam line is straight, there's no need for expensive guiding magnets. The accelerating voltage frequency is constant, and the beam tends to be well-focused compared with the beam of a circulating accelerator. And there is the advantage that particles traveling in straight lines do not throw off synchrotron radiation, an issue that is especially relevant for lightweight particles such as electrons.

One of the earliest large electron linacs was the 1 GeV Mark III at Stanford. It was using this machine that, in 1956, Robert Hofstadter's team bounced electrons elastically off protons, work that earned him the 1961 Nobel prize in physics (Figure 4.52). Hofstadter and his colleagues measured the way in which the cross-section depended on the

FIGURE 4.52 Robert Hofstadter, who bounced electrons off protons and showed that the proton is not a point particle but has a finite size. Hofstadter won the Nobel prize for physics in 1961. Credit: Stanford University, courtesy AIP Emilio Segrè Visual Archives, W. F. Meggers Gallery of Nobel Laureates.

electron scattering angle, showing conclusively that the proton is not a point-like entity, but that it has a finite size. The incoming electrons see the proton as a charged ball having a measurable radius, a radius Hofstadter and his colleagues determined to be 0.74×10^{-15} m, or 0.74 fermi. Hofstadter's results were the first direct evidence that nucleons have a size.

Partons in protons

The Mark III machine was, in turn, inspiration for SLAC's 3 km machine, sometimes called the two-mile linac. The accelerator came into operation in 1966, initially delivering 17 GeV electrons into a fixed target. It was in the late 1960s and early 1970s that, using this machine, Jerome Friedman and Henry Kendall, both from MIT, and Richard Taylor, from SLAC, and their colleagues first "glimpsed quarks." This sighting was to earn Friedman, Kendall, and Taylor the 1990 Nobel prize in physics (Figure 4.53).

FIGURE 4.53 Richard Taylor (left), Henry Kendall (center), and Jerome Friedman (right) join hands to celebrate winning the 1990 Nobel prize in physics. Guided by Bjorken's insight, their deep inelastic scattering of electrons from protons provided the first clear evidence of quarks. Credit: SLAC, courtesy AIP Emilio Segrè Visual Archives, *Physics Today* Collection.

Like Hofstadter, Friedman, Kendall, and Taylor fired electrons into a stationary hydrogen target and then monitored the electrons scattered at an angle relative to the incoming beam. But their electron beam was about a hundred times more powerful than Hofstadter's, and at such high energy most collisions are *inelastic*, the target proton being smashed up in the process. What they expected was something paralleling Hofstadter's results. What they got was something totally different.

They thought they would see the cross-section fall away as the wavelength of the probe photon dropped, reflecting the proton's size, as Hofstadter had seen for elastically scattered electrons. Instead, they were surprised to find that the cross-section was relatively insensitive to this changing wavelength, and so failed to register the proton's size. The implication was that the scattering source did not depend on the size of the proton.

Handy hints in that direction came from a young SLAC theorist, James Bjorken (Figure 4.54). Bjorken had spent a couple of years trying to understand what happens when electrons scatter off protons. At that time, the structure function simply embodied all that was unknown about the insides of a proton as seen by a scattered electron, essentially a description of how the proton departed from a simple point charge. The structure function should in general depend on both the wavelength of the probing photon, and the energy transferred via that photon from the incoming electron to the proton. Bjorken had figured out that the energy and photon wavelength were actually locked together in a combination that has no units of mass or energy, and in fact behaves like a simple number. While this might seem somewhat esoteric, the fact that the structure function depends only on some dimensionless quantity means that there is no mention of a length scale. The scattering is effectively elastic scattering off point-like entities that carry some fraction of the protons' momentum.

The kitchen tap gives some insight into Bjorken's suggestion. Turn the tap on slowly and the flow of water is smooth. Turn it on so the water flows faster, and the flow is turbulent. Engineers know all about the leap from smooth to turbulent flow. It's a change they characterize using a quantity called the Reynolds number, named after the British engineer and physicist Osbourne Reynolds. Below a Reynolds number of about 2000, flow is smooth; above about 3000, it's turbulent. In between it can be either.

FIGURE 4.54 Bjorken scaling: "bj" takes on Yosemite Park's Cathedral Peak, 1960, captured on film by Henry Kendall. Bjorken's revolutionary ideas, which underpinned a quark-based interpretation of deep inelastic electron–proton collisions, were a milestone in the unfolding of quantum chromodynamics. Credit: Henry Kendall, Courtesy James Bjorken.

Reynolds number is exactly that, a number. It's the product of a fluid's density, its velocity, and some length intrinsic to the problem, maybe a pipe diameter, divided by the fluid's viscosity, its resistance to movement. To prove it takes a few lines on the back of an envelope, a bit of thinking through the quantities relevant to fluid flow down a pipe, and making sure their dimensions – meters, seconds, and so forth – all hang together. There's no fancy theory behind it. The fact that the Reynolds number is just a dimensionless number means, for example, that if an engineer works out the Reynolds number using curious engineers' units for some pipe flow situation and a physicist uses SI units to do the same thing, their numbers will agree.

However, there's something much deeper here than merely allowing engineers and physicists to agree on the answer to a pipe flow problem. What's really happened in the Reynolds number is that a bunch of key quantities relevant to flow have been formed up into a dimensionless bundle. What does this achieve? Fix the type of fluid, say water, so that the density and viscosity are fixed. Then for a whole range of water velocities and pipe diameters it's possible to pick combinations so that the Reynolds numbers come out the same. And that means the flow characteristics – smooth or turbulent say – are the same for all those combinations. This makes it easy for chemical engineers, for instance, to scale up from a benchtop model to a full-scale plant.

Scale up the pipe diameter by a factor of ten, scale down the velocity by a factor of ten, nothing happens to the Reynolds number and the message is that the flow is the same. In fact, the Reynolds number is an example of a scaling relation, a dimensionless grouping of key quantities that shows how inflating one quantity is countered by deflating another. It's a common and powerful idea used across physics.

What Bjorken did was to devise a scaling relation for inelastic scattering of electrons off protons. His was a dimensionless grouping analogous to the Reynolds number and which, like the Reynolds number, contained no reference to a fundamental length scale. Bjorken scaling, as it's called, implies scattering from point-like scattering centers. Friedman, Kendall, and Taylor found the first evidence of Bjorken scaling and hard nuggets within the nucleon.

Richard Feynman had also started to think in terms of point-like proton constituents, using a breathtakingly simple model that has served particle physicists well ever since. He introduced what he termed partons, point-like, freely moving scattering centers that inhabit the nucleon, as a way of understanding the collisions between high-energy hadrons. Between them, for a nucleon (or some other hadron) having a large momentum, the partons share that momentum amongst themselves in a way that is a property of the nucleon, rather

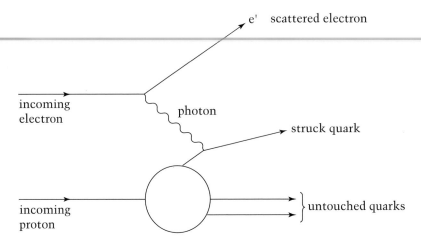

FIGURE 4.55 Deep inelastic electron–proton scattering in the parton model. The struck quark carries a fraction x of the incoming proton momentum.

than an artifact of the collision details. This fraction of the nucleon's momentum carried by a parton is universally denoted by x, and x therefore ranges between zero and one for any given parton. In the context of deep inelastic scattering, the momentum fraction x turns out to be simply the reciprocal of Bjorken's dimensionless bundle.

Feynman and Bjorken arrived at the same basic conclusions, of scaling and point-like scattering, but Feynman's parton model offered, in addition, a simple intuitive picture of what goes on when an electron smashes a proton: the incoming electron throws off a photon that strikes a point-like parton, Figure 4.55. The inelastic scattering seen at SLAC is just the sum of elastic scattering contributions from partons. In hadron–hadron collisions, for example in proton–proton collisions, a parton from one proton scatters elastically from a parton in the other, illustrated in terms of gluons in Figure 4.56.

A full description of electrons scattering inelastically off protons and neutrons requires not just one but two distinct structure functions, enough to accommodate all the photon polarization options. Hard on the heels of Bjorken's work, which suggested that *both* these structure functions depend only on the momentum fraction x, Callan and Gross linked the two by assuming that the hard scattering centers have a spin of one half. As early as 1969 Richard Taylor presented data at a conference in Liverpool, UK, supporting the Callan–Gross relation. More

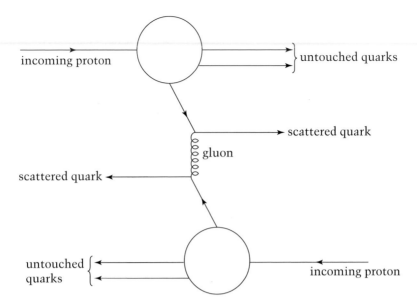

FIGURE 4.56 One of the ways in which two protons can collide, visualized in terms of the parton model.

SLAC data emerged, showing differences between electron scattering from protons compared with neutrons. And, in due course, experimenters at CERN scattered neutrinos off nucleons, a deep inelastic scattering process mediated by W bosons in place of photons, Figure 4.57. The combined data pointed to a conclusion that, from today's perspective, seems almost obvious: the point-like scattering centers are quarks.

Structure functions, then, are the momentum distributions of quarks inside the nucleon. To this day, they encode almost all particle physicists know about the structure of the proton and neutron. Measuring structure functions, and trying to explain those measurements in terms of QCD, is big business. HERA is so important because, as the electrons and protons collide, the mechanics of the collision mean that experimenters can see quarks having a smaller fraction of the parent nucleon momentum than at any other accelerator. In the jargon, they can explore the low-x region.

The CERN neutrino scattering results, using neutrinos produced via CERN's proton synchrotron and detected with the Gargamelle bubble chamber, were published in 1973, even as theorists were forging QCD. So soon, yet even then the Gargamelle team estimated that

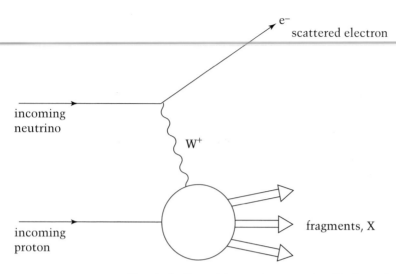

FIGURE 4.57 Deep inelastic neutrino–proton scattering. The incoming (electron) neutrino throws off a W^+ boson, which smashes the proton.

around 50% of the nucleon's momentum is carried by gluons, a conclusion that has been reaffirmed by numerous experimental groups since. It was also clear from the start that the three so-called valence quarks, those quarks corresponding to the quarks of the quark model of particle structure, were not the full story, since they could not reproduce the measured form of the structure functions. Instead, those three valence quarks interact with one another via gluons, now seen in experiment, and in turn those gluons can create quark–antiquark pairs. According to this picture, essentially the current view, valence quarks reside inside a nucleon awash with gluons and a sea of quarks and antiquarks. In fact, those early Gargamelle results were already showing evidence of *antiquarks* inside the nucleon.

The road to quark freedom
Even while experimenters were accumulating evidence for scaling, and Feynman and Bjorken were espousing a quark–parton view of the world, 't Hooft and Veltman were demonstrating that the entire magic family of non-Abelian gauge theories are indeed renormalizable. Glashow, Weinberg, and Salam's explanation of the weak force could now assume its place on the pedestal, and suddenly Yang–Mills theory seemed a very reasonable place to look for a way of understanding the strong force.

And scaling was a catalyst for looking at the high-energy behavior of that Yang–Mills theory.

Scaling certainly motivated Gross and Wilczek's study. To explore how Yang–Mills theories behave at high energy, both they and Politzer saw in Wilson's contemporary work on the renormalization group, and its recasting by Callan and Symanzik, the perfect tool: this was the renormalization group's coming of age. Of non-Abelian theories, Gross and Wilczek wrote "We have found that they possess the remarkable feature, perhaps unique among renormalizable theories, of asymptotically approaching free-field theory" in their 1973 letter on asymptotic freedom. For this is what asymptotic freedom means – that at high energy the interaction between participants drops away to nothing, and they behave as though they are freely interacting, in other words like "free fields." "Such asymptotically free theories will exhibit ... Bjorken scaling," they continued. "We therefore suggest that one should look to a non-Abelian gauge theory of the strong interaction to provide the explanation for Bjorken scaling, which has so far eluded field-theoretic understanding." The world listened, and did as they suggested.

Scaling, first seen in deep inelastic electron–proton scattering at SLAC, was the amazing experimental result so important to the emergence of asymptotic freedom and to the idea of scattering from point-like "nuggets" within the nucleon. Those hard lumps were identified with quarks, and from then on quarks became much more real, more than a mere device to explain patterns of hadron masses. Asymptotic freedom, a key property of the color-based theory describing interactions between quarks, says that under the right conditions quarks can behave as though they are free. There are hints and hypotheses, but no proof that, conversely, those quarks are never to be seen singly, in the raw, even though probing electrons can see them swarming inside the nucleon. But are there free quarks, outside of the hadronic cage that normally traps them, just waiting to be found by those who look in the right place? Particle physics must know.

THE HUNT FOR FREE QUARKS

Philosophers, presumably, have plenty to say about the logic of looking for something that you do not expect to find. After all, if you hope to find something and indeed you do find it, then you can throw a party to celebrate. But if you don't expect to find something, and in your experiments you fail to find it, your party might be clouded by the thought that perhaps you didn't do the right experiment. Perhaps the thing you didn't expect to find is really out there somewhere, but just hiding in a different place.

And so it is with quarks. These days, nobody seriously expects to find free, isolated quarks in the way that experimenters are familiar with free electrons or pions. That must make free quark searches pretty dull experiments to perform, since not only is there the logical problem of concluding there really are no free quarks simply because your experiment didn't find them, there's the added damper that you're expecting a null result from your equipment. It's hardly a very inspiring scenario.

And yet, the prize is potentially great, and the significance of finding free quarks increases with every passing year as the notion of confined quarks becomes more and more entrenched. Whoever finds free quarks could probably start work on their Nobel prize acceptance speech right away, and their discovery would galvanize the theoretical community. So, the stakes are high.

Equally, the experiments are varied and plentiful. Free quark searches date back as far as the idea of quarks themselves, the first search being made in 1964 in bubble chamber photographs taken from the CERN proton synchrotron and the Brookhaven National Laboratory's AGS proton accelerator at what were, in those days, high energies. After all, Gell-Mann himself had proposed the idea of looking in the "highest energy" accelerators to confirm the non-existence of "real quarks." And sure enough, the first searchers drew a blank.

But Gell-Mann had also mentioned the notion that cosmic rays might result in some stable quark contamination of matter at the Earth's surface. This inspired at least some of a whole range of other experiments that can at once be described as both non-accelerator experiments and, for particle physics, pretty offbeat.

Gell-Mann has recalled how University of Michigan optical sciences researcher Peter Franken used to call him at midnight to report on the progress of his hunt for quarks in oysters, which Franken believed would concentrate free quarks present in seawater. But the best-known free quark search in bulk matter is perhaps that of William Fairbank and his colleagues at Stanford University, who in 1977 reported finding fractionally charged particles, presumably quarks.

The Fairbank experiment was a re-run of the classic work by US physicist Robert Millikan, whose famous oil drop experiment earned him the 1923 Nobel prize. Millikan was out to measure the basic unit of electrical charge, which he managed to do by trapping tiny charged oil droplets and holding them motionless using the upwards pull of an electric field to counter the downwards pull of gravity. The oil drops acquired an electrostatic charge during their creation in an atomizer; examining thousands of oil drops over several years, Millikan and his co-workers at the University of Chicago found that the charge was always in some multiple of a basic charge unit, and this basic unit they were able to measure. Their experiments gave both a result for the electron charge, and also showed charge to be quantized. Fairbank and his colleagues replaced oil drops with tiny superconducting niobium balls, and used magnetic levitation to balance gravity. After applying oscillating electric fields, and monitoring very accurately how individual balls move, they claimed to have detected fractional charges of both $+1/3$ and $-1/3$. However, attempts to reproduce their results have failed. Closest in spirit to Fairbank's work was that of the University of Genoa's Giacomo Morpurgo and his colleagues who, in experiments starting in 1966 and spanning 15 years, levitated tiny steel balls and grains of graphite without finding even a whisper of fractional electrical charges. The widespread belief is that Fairbank's experiment was the victim of some extraneous influence, perhaps some magnetic background effect, and that he did not really see free quarks at all.

For Fairbank and his team, finding fractional charges was the *raison d'être* of their experiment. So it's fun to look back and take stock of a comment Millikan made in a 1910 report of his work in *Philosophical Magazine*. "I have discarded one uncertain and unduplicated observation apparently upon a single charged drop, which gave a value of the charge on the drop some 30 percent lower than e [the electronic charge]," wrote Millikan. A charge 30% lower than the electronic charge would, sign aside, be pretty much the charge on an up quark: a fluke, or had Millikan really found something? Unlikely: the drop in question was a *water* drop, which Millikan appreciated could evaporate during his experiment. In fact, he concluded that the errant drop was probably evaporating quickly because it was especially small. The following year, he published a new version of the experiment in *Physical Review*, using oil drops (and drops of other evaporation-resisting materials) and improved equipment. In every case he found that the charge on a drop was an exact multiple of some smallest quantity of electricity. So Millikan's work provides thin evidence indeed for the

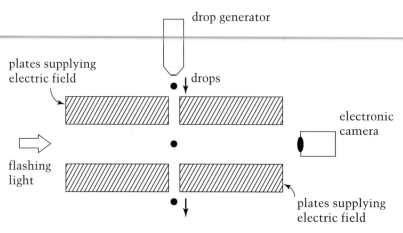

FIGURE 4.58 The SLAC re-run of the classic oil drop experiment, looking for fractional charges on oil drops.

existence of free quarks, though it does reveal him to be a superlative experimenter.

The spirit of Millikan's oil drops lives on. In 1996 a team based at SLAC, led by Martin Perl, reported a re-run of the oil drop experiment looking for fractional charges, the most distinctive feature of free quarks and the signature which all free quark searches rely on spotting. In the SLAC experiment, oil drops fell through a hole in an upper, horizontal plate, and then through about a centimeter of air before exiting via a second hole in a lower, parallel plate, Figure 4.58. The drops, each about 7 microns in diameter, were created by a custom-built dropper at the rate of about one every two seconds. An oscillating electric field applied between the plates altered the velocity of each drop, depending on the drop's charge, and the up or down orientation of the electric field. Using a stroboscope and a video camera, the group monitored nearly six million drops automatically, obtaining values for each drop's diameter, its terminal velocity in air, and its electric charge. In all, the experimenters searched through a total of 1.07 mg of oil. The experiment showed that, if there are isolated quarks with charges of $\pm 1/3$ or $\pm 2/3$, then, with a 95% confidence level, they are present in quantities fewer than one per 2.14×10^{20} nucleons.

Which is not the same as saying there are no free quarks. It's simply saying that, if they exist, then they are very rare. It's setting an upper bound on their occurrence, and this is the way experimenters typically express the result of something they are looking for but cannot find. It's certainly the normal way of expressing quark search results, the confidence level being an expression of the statistical uncertainty, and 95% confidence the accepted norm for most experiments in particle physics. To put the free quarks per nucleon numbers into perspective, it's roughly equivalent to being 95% sure there's fewer than one "rogue" grain of sand in a pile the size of the Great Pyramid of Cheops, which has a volume of over two million cubic meters.

There have been plenty of other searches for free quarks in bulk matter, for example in iron, mercury, and graphite, and several groups have tried niobium. Unil Perera and his colleagues at the University of Pittsburgh established a similar limit to the SLAC oil drop experiment by looking in silicon and assuming quarks would act as fractionally charged impurities. In that case, conventional charge carriers within the semiconducting silicon would then bind, though weakly, with any free quarks to give entities that could be revealed by exciting the material with light. Another bulk matter option is seawater, though not everybody has thought to use oysters as quark concentrators, as did Franken. A UK collaboration of physicists from the Rutherford Appleton Laboratory (RAL) near Oxford, and chemists from the University of Southampton and Queen Mary and Westfield College, London, reported on a seawater search in 1992. They used evaporation to coat tiny niobium balls with seawater residue, mainly salt, the idea being that free quarks might accumulate in seawater as a result of cosmic rays, and that evaporation would enrich the levels of any fractionally charged components. The coated balls were then tested for fractional charges using magnetic levitation between electric plates. The experimenters found no free quarks, and set an upper bound of fewer than 0.001 quarks per gram of seawater.

Some people have tested moon-rock, but to no avail. And three of the RAL seawater team, Peter Smith, Graham Homer, and Wally Walford, were part of an earlier collaboration that looked for free quarks in meteorites (pristine, undamaged material from the genesis of our solar system) but they found nothing. In 2002 Perl and his team reported on a variant of their oil drop work in which they introduced bits of five billion year old meteorite into their oil drops. Nothing.

A popular place to hunt free quarks is in cosmic rays, some of which have energy vastly higher than any particles from a man-made accelerator, so offer the possibility of seeing something that those machines cannot. Over fifty cosmic ray free quark searches have been published. These include the famous free quark "candidate" found in cloud chamber photographs by Brian McCusker of the University of Sydney way back in 1969, a one-off that failed to withstand the test of time. Twenty years on, a US–Japanese team searched for fractional charged particles in three years of data gathered at the bottom of the Kamioka mine, 300 km west of Tokyo. They used the Kamiokande II detector, normally devoted to watching and waiting for protons to decay. The detector comprises 2400 tonnes of water viewed by nearly a thousand photomultiplier tubes, standard detection devices that turn tiny flashes of light into electrical pulses. The detector, like several other proton decay detectors, lives at the bottom of a mine since the rock around it provides good shielding against excessive cosmic radiation. That way, when the photomultipliers do see something, it's more likely to be a proton decay event from protons within the bulk of the water. Or it might be a free quark passing through. But though the experimenters totted up 70 million events, mostly very fast muons, they saw no free quarks.

Of course, the big accelerators have been put to work hunting for free quarks too. Looking at electron–positron collisions at LEP, the ALEPH detector collaboration analyzed 180 000 decays of Z particles into hadrons or leptons, looking for signs of unexpected masses or charges. Two years later, in 1995, the OPAL team published the results of another search, this time of more than a million decays of Z particles into strongly interacting particles. Neither group found free quarks, although both pushed back the limits for the production of free quarks. Other experimenters have used the Tevatron to look for free quarks in proton–antiproton collisions right up to an energy of 1.8 TeV, but have found nothing.

In truth, any kind of fundamental entity having a fractional charge would be news. There might, after all, exist particles lacking color, so they are not quarks, but which have a charge that's less than the unit of charge on an electron. So free quark searches are perhaps better termed "fractional charge particle" searches. Which ever it might be, nobody has yet found anything that stands up to a second look. But experimenters must, and do, keep looking.

MORE QUARKS

Scientists don't need to go far to find u and d quarks – they are made of them. Material objects of this world are made of atoms whose nuclei are made of protons and neutrons, and these, in turn, are made of u and d quarks. It's just that they are hard to see.

These, the first two quarks, were therefore sitting on and in the laboratory bench waiting to be found. The third quark, the s quark, was discovered through a combination of divine intervention and romantic science: cosmic rays and mountain top experiments. Strange particles, and implicitly the strange quark, surprised everyone. By the time the fourth quark was discovered, however, a few particle physicists at least were ready and waiting.

Bjorken and Glashow, at the time working together in Copenhagen, introduced a new quantum number, which they called charm, in a 1964 paper aimed at extending the Eightfold Way by adding an additional basic building block. Six years later, Glashow, along with John Iliopoulos and Luciano Maiani at Harvard University, pointed out that a fourth quark, the charmed quark, provided a neat solution to an irritating problem in weak interactions, in which decays such as $K^0 \rightarrow \mu^+ + \mu^-$ and $K^+ \rightarrow \pi^+ + e^+ + e^-$ appeared to be allowed, but experimenters were sure they didn't commonly occur. Not only that, but a fourth quark meant that quarks and leptons possess what Glashow and his collaborators termed a "remarkable symmetry": the electron with the electron neutrino can be neatly bundled together with the up and down quarks, and the muon with its neutrino tied to the strange and charmed quarks. It turns out that without this bundling together of particles into families, including quarks with color, the Standard Model will not work – exactly the right particle mix is needed as a guarantee of renormalization.

The charmed quark was discovered four years later, in 1974, and announced simultaneously by two groups. Using the new SPEAR ring at SLAC, Burton Richter and his colleagues witnessed a sharp peak in the cross-section for electron–positron collisions producing hadrons, Figure 4.59. The peak was so sharp it could easily have been missed, lost between different collision energy settings of the accelerator as its power was stepped up. This spike they ascribed to a resonance, which they christened ψ. Across the other side of the United States, Sam Ting and his group, working at the Brookhaven National Laboratory's AGS, had also found a resonance. Their experiment looked at the combined

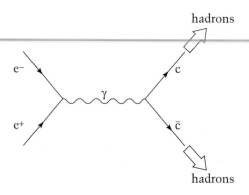

FIGURE 4.59 Charm discovery: a resonance in e^+e^- collisions at an energy of 3.1 GeV shows the creation of the charmed meson ψ, a $c\bar{c}$ pair.

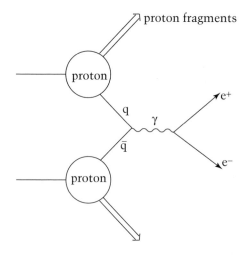

FIGURE 4.60 Charm discovery: a quark from one proton annihilates with an antiquark from the other. The result was termed the J resonance.

mass of electron–positron pairs emanating from proton–proton collisions, achieved by directing a beam of protons onto a beryllium target. A quark from one proton annihilates with an antiquark from the other proton to ultimately yield, via a photon, an electron–positron pair, or some other lepton pair. It's a mechanism known in the trade as the Drell–Yan process, Figure 4.60, after SLAC physicists Sydney Drell and Tung-Mow Yan. Ting's group found the same resonance as Richter's, and at the same energy, only they called it the J resonance. The two teams could never reach a compromise over the name, and particle physics has been lumbered with the appellation J/ψ ever since. The J/ψ is the simplest meson comprising one charmed quark and one charmed antiquark.

Ting and Richter received the 1976 Nobel prize in recognition of their discovery, a discovery that caught the particle physics community by surprise. That November in 1974 is sometimes known as the "November revolution," in view of the way particle physics was changed forever. Ultimately the incredibly narrow width of their new resonance could only be explained by the existence of a new, fourth quark, and the J/ψ's reluctance to decay quickly into a collection of pions found a natural explanation in terms of gluons and QCD. Indeed this was a piece of evidence that helped QCD gain acceptance in its early days. Experimentalists had a whole new vein to tap, including particles combining the new charmed quark with previously known quarks, and other states based on the simple charmed quark–antiquark pair, which helped to bolster belief in the still-young theory of QCD. The first of these charmed quark–antiquark possibilities was found at SLAC just ten days after the J/ψ was announced, an "excited state" comprising a charmed quark and antiquark now known as the $\psi(2S)$, with a mass of 3686.0 MeV. The first hint of a baryon containing a charmed quark came in neutrino–proton scattering at Brookhaven in 1975. And the SELEX experiment at Fermilab announced candidates for baryons containing *two* charmed quarks – important in the four-flavor quark model, Figures 4.61 and 4.62 – as recently as 2002. But the earliest unequivocal evidence of "naked charm," a charmed quark lacking a charmed antiquark companion, came with the 1976 discovery at SLAC of the first of the D mesons. With the sighting of the D^0, a meson made from a charmed quark and an up antiquark, or $c\bar{u}$, then the D^+ and D^- the same year, the reality of the charmed quark was beyond doubt.

As if to make absolutely sure that the 1970s go down in history as a golden era in particle physics, 1976 also saw the publication of another dramatic discovery. Analyzing around 35 000 SPEAR electron–positron annihilation events, Perl and his colleagues found a mere 24 that defied understanding in terms of the then-known particles. These few events were of the form $e^+ + e^- \rightarrow \mu^- + e^+ +$ missing energy, or $e^+ + e^- \rightarrow \mu^+ + e^- +$ missing energy – just what would be expected if the electron-positron annihilation were producing a pair of heavy leptons. What Perl and company had seen was an entirely new lepton, the τ or tau lepton. The process they had uncovered was $e^+ + e^- \rightarrow \tau^+ + \tau^-$, followed by taus decaying either to electrons or muons in roughly equal numbers, $\tau^- \rightarrow e^- + \bar{\nu}_e + \nu_\tau$ or $\tau^- \rightarrow \mu^- + \bar{\nu}_\mu + \nu_\tau$, with the corresponding decays for the positive taus.

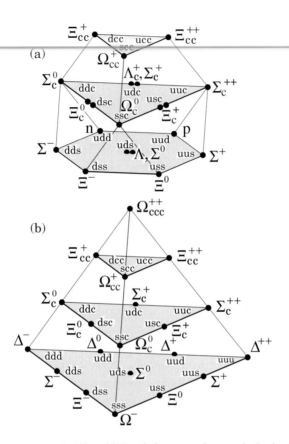

FIGURE 4.61 The additional charm quark extends the family of baryons: the "ground floors" of these larger multiplets are the more familiar particles built from just u, d, and s quarks. Credit: Particle Data Group.

The tau is a heavy relative of the electron and muon. Its mass is 1777.1 MeV, a figure that is about seventeen times the mass of the muon, 105.7 MeV, and vast compared to the featherweight electron, at 0.511 MeV. Like the electron and muon, it is a "true" fundamental particle, and is accompanied by its own neutrino, the tau neutrino, or ν_τ. Symmetry between quarks and leptons also implies that, just as the electron is part of a "family" embracing the up and down quarks, and the muon a family containing the strange and newly discovered charmed quark, the tau should be part of a family with two quarks. In other words, the tau's existence implies *another* quark–lepton family with *another* pair of quarks.

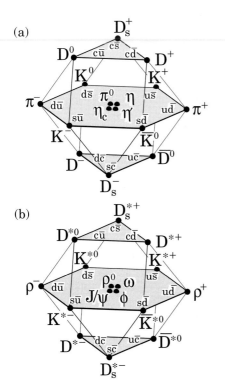

(a)

(b)

FIGURE 4.62 The charm quark extends the meson families. The middle levels are the more familiar meson families. Credit: Particle Data Group.

The world didn't have to wait long for the next quark, one of this new pair. Nobody had much idea what the mass of this new quark might be. Ting's technique for spotting charm seemed a good way to start the search, and Leon Lederman and his collaborators used Fermilab's 400 GeV proton synchrotron in what was in essence a re-run of Ting's experiment but at a higher energy. In 1977 they found a resonance at around 9.5 GeV, which they dubbed the upsilon, or Υ. This was beyond the energy range of SLAC's SPEAR, which was to forego the honor of finding the Υ in electron–positron collisions. On 30 June 1977, the day that the Υ discovery was announced at a Fermilab seminar, the PLUTO collaboration, whose detector at that time monitored collisions in the DORIS storage ring at DESY, submitted a proposal to upgrade the energy of the DORIS machine. Lucky them – as news of the Υ discovery spread, they hastily upped their target energy to embrace the Υ: DESY experimenters saw it the following year.

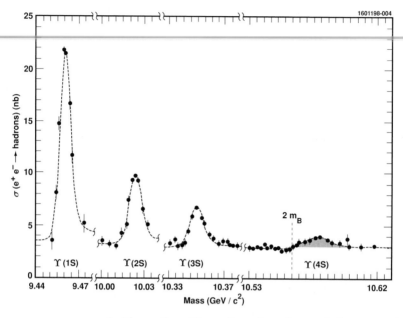

FIGURE 4.63 The upsilons. The upsilon, a bound state of a bottom quark and antiquark, has several excited states that can be accessed by tuning the accelerator energy and which are revealed as bumps in the cross-section plot. Credit: Courtesy Cornell University & Columbia University – Stony Brook (CUSB) Collaboration.

Beautiful bottom

The mass of the upsilon is 9460.4 MeV, about three times that of the J/ψ. And like the J/ψ, it too is a quark antiquark pair, only this time the quarks in question are the b or bottom quark, sometimes called the beauty quark. The first of several higher states based on the same pair were seen soon after the ϒ discovery, starting with an excited b$\bar{\text{b}}$ combination denoted ϒ(2S), in a simple parallel with the heavier cousins to the J/ψ, Figure 4.63. The first evidence for the delightfully termed "naked bottom" came in 1981, from the CERN ISR, the Intersecting Storage Ring proton–proton collider, and from the CLEO collaboration using CESR, the Cornell electron–positron collider. The CERN group saw signs of a "bottom baryon," the Λ_b, a combination of up, down, and bottom quarks, which at 5624 MeV remains the heaviest confirmed baryon, and the only confirmed bottom quark baryon. Its surprisingly short lifetime was a puzzle. For their part, virtually simultaneously, the CLEO collaboration spotted the first ever hint of B mesons, in which just one of the quarks is a b quark. The front line B mesons are the B^+, B^-, and B^0, with quark assignments of u$\bar{\text{b}}$, ūb, and d$\bar{\text{b}}$, respectively. All three have a mass of 5279 MeV, more than five times the proton mass, making the bottom mesons weighty animals indeed.

Experimenters have since unearthed a whole zoo of "heavy quark" mesons, in other words those containing c or b quarks, in addition to the basic D and B mesons, J/ψ and ϒ. There are mesons combining bottom and strange quarks, such as the B^0_s (with s$\bar{\text{b}}$), mesons combining charmed and strange quarks, such as the D^+_s and D^-_s, c$\bar{\text{s}}$ and c̄s respectively, and a whole array of excited states of the various basic quark combinations. In addition, experimenters have found a handful of charmed and bottom baryons.

At a seminar in Fermilab on 5 March 1998, CDF (Collider Detector Facility at the Teatron) scientists announced that they had found the B_c meson, a c$\bar{\text{b}}$ combination and the first particle combining charmed and bottom quarks to be discovered. It has a mass of about 6400 MeV, and a lifetime of around 0.46 picoseconds, but its main claim to fame is that it's the last of the "ordinary" mesons, for reasons that will become clear in due course, Figure 4.64. That's ordinary in the sense of mesons as simple quark–antiquark combinations: other mesons are based on these combinations. It has taken experimenters fifty years to find all the ordinary mesons.

With particles containing charmed and bottom quarks being so heavy, their decay possibilities into lighter particles are

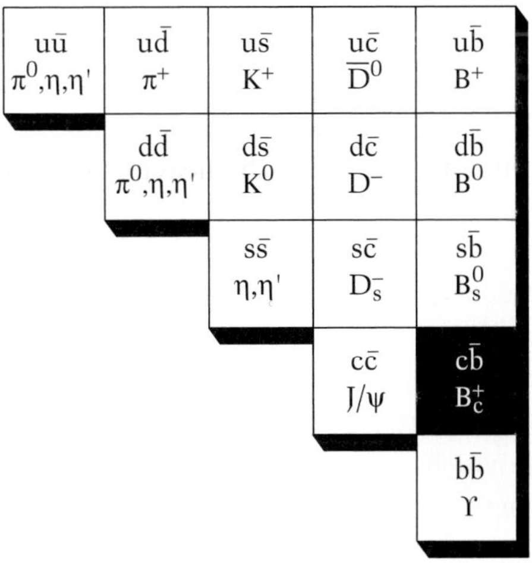

u$\bar{\text{u}}$ π^0,η,η'	u$\bar{\text{d}}$ π^+	u$\bar{\text{s}}$ K^+	u$\bar{\text{c}}$ $\overline{D}^0$	u$\bar{\text{b}}$ B^+
	d$\bar{\text{d}}$ π^0,η,η'	d$\bar{\text{s}}$ K^0	d$\bar{\text{c}}$ D^-	d$\bar{\text{b}}$ B^0
		s$\bar{\text{s}}$ η,η'	s$\bar{\text{c}}$ D^-_s	s$\bar{\text{b}}$ B^0_s
			c$\bar{\text{c}}$ J/ψ	c$\bar{\text{b}}$ B^+_c
				b$\bar{\text{b}}$ ϒ

FIGURE 4.64 The ordinary mesons, highlighting the last to be discovered, the B_c. Credit: Fermilab

mind-numbingly numerous and complex. For example, each of the "basic" charmed mesons, D^0, D^+, and D^-, and the "basic" bottom mesons, B^0, B^+, and B^- offer several dozen different routes through which they can decay, typically into an assortment of kaons (as K mesons are often called) and pions. In all, there are around a thousand decay modes for the B meson family alone. Much of this information is expressed in terms of decay rates, or decay probabilities, the decaying particles' version of cross-section. The decay rate, usually called the width and written Γ, is the inverse of particle lifetime. Decays into specific products are listed relative to the total decay probability, that is, the total width. So, for instance, the B meson decay $B \rightarrow \bar{D}^0 + l^+ + \nu_l$, where l^+ means any positive lepton and ν_l its corresponding neutrino, is listed as having a width of 1.6% relative to the total width for B^+ decay. One way of understanding this particular decay channel is as a measure of the likelihood of the weak conversion of a b antiquark into a c antiquark, plus a positive lepton and neutrino, in the environment existing inside the B^+ meson. This decay is just one of more than a hundred B^+ decay channels which have either been seen and measured, or which have been looked for and not (yet) seen.

Much of the huge task of accumulating this wealth of data has fallen to a clutch of experimental collaborations at Fermilab, and teams at the SPEAR, DORIS II, and CESR electron–positron colliders. They have made it their business to map out in detail the properties of these new particles and their inter-relationships. Fresh input has come from proton–antiproton collisions at the Tevatron, which can readily create B mesons and measure their lifetimes and masses, and from the study of Z decays at LEP and the SLC: for Z bosons decaying via $b\bar{b}$, the weakly decaying b quark-containing products comprise about 38% each of B^+ and B^0, 11% of B_s and 13% of Λ_b. The first signs of more bottom quark baryons, the Σ_b and Ξ_b, were announced by LEP teams at a conference in Manchester, UK, in 1995.

The b quark sector in particular is so rich that, despite the ability of machines such as CESR to produce B mesons at the rate of millions a year, along with comparable numbers of charmed mesons and tau lepton pairs, particle physicists want more. Enter the B-factories.

Factory machines are something of a parallel path in accelerator design. On one path is what might be called "discovery" machines, those such as the Tevatron, where the aim is to reach higher energy than ever before, and thus enter uncharted waters where new physics

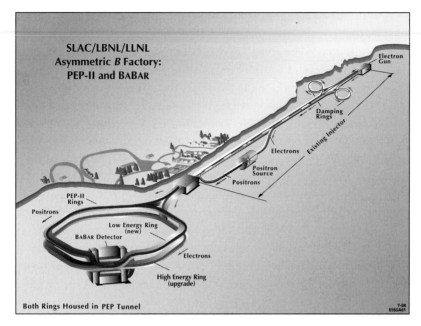

FIGURE 4.65 The Asymmetric B-Factory at SLAC. The injector provides both electrons and positrons. Credit: SLAC.

awaits discovery. In contrast, factory machines work at more modest energies, tuned to mass-produce certain particles. What they lack in energy they make up for in beam brightness and statistical precision. For experimenters at factories, new physics comes from exploiting the available precision to spot features such as unusual decay routes, or decay probabilities that disagree with theory. These are the kinds of things only revealed through the accumulation of millions of events, instead of mere handfuls.

The PEP upgrade at SLAC, the Asymmetric B-Factory, Figure 4.65, belongs to this new generation of factory machines. The asymmetry derives from the fact that the machine comprises two independent storage rings of *different* energies. The existing PEP collider was rebuilt to store a 9 GeV electron beam, and a whole new ring added above to store 3 GeV positrons, the SLAC linear accelerator plus damping rings providing the particle beams. The big idea of this unbalanced collision arrangement is that the more energetic electron beam imparts an overall flow to the collision fragments, making it easier to disentangle what happens.

Building an accelerator to do this is not so easy, so there must be a good reason why the machine designers went to all the trouble of working out how it could be done. There is. Mass-producing particles implies tuning the accelerator so that the collision energy exactly matches the energy required to produce the particles in question: this is resonance, where the cross-section is very large. The particles thus produced move slowly, since there's no leftover collision energy to turn into the collision products' motion energy. To supply that motion energy, and make them move faster, would mean winding up the accelerator energy a little, but then the cross-section would be reduced because now the accelerator is off-resonance. Why is it desirable to have the collision products moving quickly? Because, seen by the stationary detector, they decay more slowly than they would if they were moving slowly or not at all, the time dilation effect described by special relativity at work. The slower decay means that B mesons, which when stationary decay in just a trillionth of a second, live longer, so the detector has a better chance of seeing what's going on. The way to get the best of both worlds, the high production rate of collisions at the resonance energy, plus the dilating effects of high speed for the particles produced, is to use a lopsided accelerator.

Watching those drifting collisions at the SLAC Asymmetric B-Factory is a brand new detector, called BaBar. That's a play on $b\bar{b}$ that comes out sounding like the name of the elephant hero of the Babar children's stories: Babar the elephant has been adopted as the collaboration's logo. BaBar the detector sees hundreds of millions of b quark and c quark-containing mesons every year.

Sharing the SLAC facility's 1999 start date was Japan's own factory machine, KEKB or the KEK B-factory. This has collider rings of 3.5 GeV positrons and 8 GeV electrons, the lopsided energy likewise imparting a boost to the collision fragments, which are monitored by the new BELLE detector, Figure 4.66. Figure 4.67 shows a B meson event recorded at BELLE. B mesons are also a focus of an upgrade to the CLEO detector at Cornell, CLEO III, and the HERA-B experiment at DESY.

All of which points to a great deal of interest in the apparently humble B meson. One of the reasons for this intense interest is the fact that it should shed light on one of the big questions of cosmology: why does matter dominate antimatter in the Universe? The answer may lie in the recalcitrant B mesons' refusal to abide by a symmetry combination called CP symmetry.

FIGURE 4.66 The Great Gates of BELLE. The BELLE detector at KEKB, ready for "roll-in." The accelerator beam pipe passes through the central hole. The "gates" are the end caps of the muon plus kaon detector that envelops most of the detector when it's in position. Credit: Courtesy KEK–BELLE collaboration.

The C symmetry relates particles to their antiparticles. The C operation, called charge conjugation, means replacing every particle with its antiparticle, and vice versa. If C is a good symmetry, then everything still works as before: strong and electromagnetic interactions are respectful of this symmetry. They are also respectful of the P (for parity) symmetry, also called mirror reflection symmetry. The P operation amounts to reflecting everything in a mirror, for example interchanging left-handed and right-handed spins, and if P is a good symmetry then, again, everything works as before. For weak interactions, however, C and P are not good symmetries, as the neutrinos make plain: left-handed neutrinos and right-handed antineutrinos predominate. So C symmetry is not respected, since it would mean swapping a left-handed neutrino with a left-handed antineutrino. Nor is P respected, since under the P operation, a left-handed neutrino would become a right-handed neutrino. But put the two together, and the combination C and P relates the left-hand neutrino to the right-hand antineutrino, and all is well.

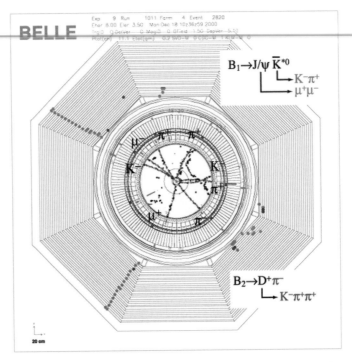

FIGURE 4.67 Two decaying B mesons, each weighing about as much as an entire helium atom, are captured in the BELLE detector. The insets show their respective decay paths. Credit: Courtesy KEK–BELLE collaboration.

Almost. Nature has thrown particle physicists another teasing twist. In an effect called mixing, a B^0 meson and its antiparticle, the $\bar{B}^0$, can convert into one another. In other words, starting out with a pure beam of one, say B^0 mesons, then downstream the beam will comprise a mixture of the both B^0 and $\bar{B}^0$, even though nothing has been done to the beam. This B meson mixing was first seen in 1987 by the ARGUS team (*A* *R*ussian–*G*erman–*US*–*S*wedish collaboration), using the DORIS II electron–positron collider. It's a hint of CP symmetry violation, or simply CP violation, though not proof of it. Similar mixing occurs in the K^0 meson system, which has in addition provided more direct evidence for CP violation. This took the form of a decay process that would otherwise be forbidden, seen in 1964 by James Cronin and Val Fitch working at Brookhaven National Laboratory, a result that earned them the 1980 Nobel prize. Experimentalists have now caught sight of unequivocal CP violation in B mesons, announced by both

BaBar and BELLE collaborations in 2001. Though CP violation is known to be real, however, it is a complicated business that remains poorly understood: B mesons provide the best chance of getting to the bottom of this exotic effect.

Some topics in physics have an intrinsic appeal, and CP violation is one of them. The reason is its deep-rooted philosophical implications. First off, CP violation defines a unique and absolute way of describing left and right. Imagine a distant alien nation, and the issue is to explain to them our convention of right and left. Earthling experimenters could, for example, tell them to look at a neutrino, and then explain that the neutrino is, according to Earthlings, a left-handed affair. Ah, comes the alien reply, how do we tell which is the neutrino and which the antineutrino? The answer to that must be phrased in terms of some CP violating process, for example the comparison of K meson decays into either $\pi^- + e^+ + \nu_e$, or the less common $\pi^+ + e^- + \bar{\nu}_e$.

Through CP violation, K and B decays differentiate between matter and antimatter, and it's this that excites cosmologists. So far as they can tell, matter rather than antimatter dominates the Universe. Explaining this asymmetry is seen as one of the fundamental challenges facing cosmologists in their quest for an understanding of our Universe, and how it was created: there is nothing in the Big Bang scenario for the creation of the Universe that offers any clues. Inserting some imbalance by hand is hardly an appealing solution. However, it's possible that some CP violating effects in the early Universe provided just enough of a shove towards matter to give the huge imbalance apparent today.

The seemingly arcane business of CP violation offers yet more. It turns out that a four-quark Standard Model would not permit CP violation – there wouldn't be enough quarks to provide sufficient flexibility to accommodate it. As early as 1973, a year before even the fourth, charmed quark was seen, Makoto Kobayashi and Toshihide Maskawa, both at Kyoto University in Japan, realized that adding more quarks to a four-quark picture was the only way of permitting that flexibility, and accommodating CP violation within what is now termed the Standard Model. Following Kobayashi and Maskawa, and assuming quarks come in pairs, particle physicists expect there are six quarks in all. The B-factories will be so sensitive that they might be able to detect modifications to the Kobayashi and Maskawa scheme due to the existence of still more quarks, should there be any. So B-factories are one relatively low energy way of tracking down circumstantial, indirect evidence for more quarks. There is one other quark, however, for which experimenters now have much more direct evidence: the top quark.

The top quark

The front page of the *New York Times* 26 April 1994 gushed:

The quest begun by philosophers in ancient Greece to understand the nature of matter may have ended in Batavia, Illinois, with the discovery of evidence for the top quark, the last of the twelve atomic building blocks now believed to constitute all the material world.

Grandiose stuff. At issue was the CDF collaboration's announcement of what appeared to be a dozen top quark events in proton–antiproton collisions at Fermilab's Tevatron. The following day, *The Times* of London, in a story relegated to the depths of the paper under the heading "overseas news," wrote:

If confirmed, it will complete the proof of the Standard Model, the physicists' present view of how the atom is constructed.

The CDF team submitted a 152-page preprint announcing their claim on 22 April. At a press conference soon after, they described their results to the world's media, pointing out that they had not seen enough top quark candidates to rule out the possibility of some kind of statistical blip. At that time, there was a 1 in 400 chance that they were seeing merely background events mimicking the top signature. In addition, at the same press gathering, the rival D0 collaboration, whose detector sits a third of the way round the Tevatron ring from CDF, reported that, as yet, they could muster no compelling evidence for the top quark. Time passed, data analysis continued, and statistics accumulated. The following year, in 1995, both groups were able to offer clear and unambiguous evidence of the top quark. The top quark, the sixth and possibly the last quark, had been found.

The mass of the top quark is an incredible 174.3 ± 5.1 GeV, making it nearly as heavy as a complete gold atom. That huge figure is worth a moment's reflection: an elementary particle as heavy as an entire big atom! It's this gigantic mass that explains why only Fermilab gets to see the top quark directly – only Fermilab has enough energy.

According to QCD, at Fermilab around 90% of the top quark and antiquark pairs are created via the annihilation of a quark with an antiquark from the original proton and antiproton, Figure 4.68(a).

(a)

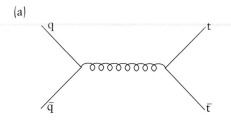

(b)

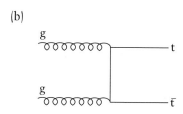

(c)

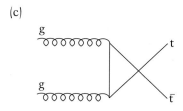

(d)

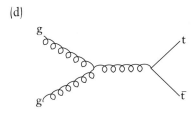

FIGURE 4.68 Simplest Feynman diagrams for the production of top–antitop pairs from proton–antiproton collisions: (a) quark–antiquark annihilation and (b)–(d) gluon–gluon fusion.

The remaining 10% come from gluon–gluon fusion, Figure 4.68(b)–(d). The top quark then prefers to decay weakly into a b quark and a W boson, according to $t \rightarrow W^+ + b$, the antiquark via $\bar{t} \rightarrow W^- + \bar{b}$. In turn, the W might spawn a lepton (an electron, muon, or tau) plus a matching neutrino, $W^+ \rightarrow l^+ + \nu_l$ and $W \rightarrow l^- + \bar{\nu}_l$, with l representing the lepton. Alternatively, the W might decay into a quark–antiquark pair, $W \rightarrow q + \bar{q}$. Top quarks can be created singly, not only in pairs, for example via $q + b \rightarrow q + t$, but in practice this route is far less common.

Whatever the mechanism of their birth, the probability of creating top quarks is very small. In data collecting spanning eight years,

CDF and D0 between them accumulated a modest couple of hundred or so top pair events. The CDF collaboration puts the cross-section for t̄t production at 7.5 ± 1.9 picobarns, D0 offer 5.2 ± 1.8 picobarns at fractionally lower collision energy. Put another way, proton–antiproton collisions produce around three million b̄b pairs for every t̄t pair. Or again, about one in every ten billion proton–antiproton collisions at Tevatron energies gives rise to a top quark–antiquark pair.

As experimenters accumulate more data and this cross-section measurement is refined, it offers a good test of QCD. Currently available at a level one step beyond the simplest, the perturbative QCD calculated cross-section for t̄t production at the Tevatron is 4.8 picobarns, within range of the measured value. Theorists trust perturbation theory for heavy quark calculations. This is because a heavy quark, and there are none heavier than the top, places the energy scale of the calculation well inside territory where the strong coupling is small, and perturbation theory meaningful, a point to be revisited in Chapter 5. Another factor making this particular calculation more reliable than some is that most of the t̄t pairs are created by quarks and antiquarks from the colliding proton and antiproton. Again, theorists are pretty comfortable with their description of protons and antiprotons as sources of freely interacting quarks and antiquarks, more comfortable than they are with their corresponding description for gluons. Some factors are less certain, however; for example, the exact value of the strong coupling parameter to be used. Then there is the thorny issue of correlating groups of tracks in the detectors with the creation of tops. There may also be some unknown decay options, new to science, which theorists will not have included in their estimations. These, if they exist, may be revealed when experimenters are able to measure the various top quark decay channels in the future.

Hunters of the top quark have an additional agenda. The top mass, combined with the W boson mass, imposes limits on what is arguably the most sought-after particle in the history of high-energy physics: the Higgs particle. This is because the W spends part of its life as combinations of Higgs particles and top quarks, a consequence of its quantum nature, and as a result the three mass values are interlinked. The Higgs, which is like no particle seen so far by physicists, is involved with the breaking of symmetries between particles, forming part of the machinery by which particles acquire their masses. Symmetry breaking and the Higgs particle were central to the correct prediction of the W and

Z masses in the electroweak theory of weak interactions. Clearly the Higgs particle is no minor player.

Yet the Higgs particle and the symmetry-breaking scheme of which it is part beg a number of questions. Actually finding the Higgs would help cast in concrete a facet of the Standard Model that remains slightly mysterious. The top quark has a special role here since, because the quark–Higgs interaction strength is proportional to quark mass, the top quark's large mass means it interacts more eagerly with the Higgs particle than does any other quark. This makes the top quark more significant than any other quark flavor in both the creation and decay of Higgs particles. In addition, accepted symmetry-breaking wisdom says that decays of top quarks into W bosons should show a particular angular spread, since top quarks are expected to produce a specific mix of polarized W bosons. Future data will allow experimenters to test for such an effect.

So far, experimenters have not found the Higgs particle. Electron–positron experiments at LEP show that a Higgs mass must exceed 65 GeV. Theorists expect its mass to lie somewhere between 100 and 1000 GeV, the range permitted by the current uncertainties in both the W and top masses, coupled with the weak dependence of the Higgs mass on the other two. Researchers are praying that the Higgs will reveal itself at CERN's LHC, if not sooner. Higgs aside, the LHC will produce many more top quarks than the Tevatron, something like seven $\bar{t}t$ pairs per second.

In many ways, finding the top quark was not a surprise. Ever since the discovery of the bottom quark back in 1977, particle physicists believed that its partner was out there somewhere. Experimentalists have actively hunted for the top, progressively pushing the mass limit higher and higher as they failed to see it at energies more modest than the Tevatron's. Even the mass of the top, when it was finally measured, wasn't a surprise. In 1995, a LEP working group, exploiting the link between the Higgs, W, and top masses in reverse, together with a sensible guess for the Higgs mass, calculated a value for the top quark mass. They predicted a top mass between 150 and 203 GeV, with a favored value of 178 GeV, impressively close to the measured value.

Rooting out the top quark at Fermilab marked the end of a seventeen-year search, an ending that is in itself a whole new beginning. Finding the top quark without doubt rates as an exciting discovery. Not finding it would have rated as an exciting disaster.

FIGURE 4.69 Aerial view of the Fermilab accelerators. The ring in the front is the Main Injector. The ring towards the back is the Tevatron itself. The rest of the Fermilab facility lies to the left of the neck where the two rings meet. Credit: Fermilab.

The atom smasher under the prairie

The machine that found the top quark, the Tevatron, is the world's most powerful particle accelerator. Its home is Fermilab, in Batavia, Illinois, 60 km west of Chicago. However, the Tevatron cannot create top quarks alone. It relies on less powerful machines to feed it. In fact, the Tevatron is the last in a string of five accelerators acting in concert, Figures 4.69 to 4.71.

The sequence starts inside a square metal box – the ion source – where electrons are added to hydrogen atoms, creating negative ions comprising a proton and *two* electrons. These are then accelerated through 750 000 V, in a version of the electrical accelerator that Cockcroft and Walton used to split the atom back in 1932. A 150 m long linac, the second acceleration stage, ups the ions' energy to 400 MeV. They then pass through a thin carbon film that strips off the electrons, leaving just protons to enter the third acceleration step. This is a

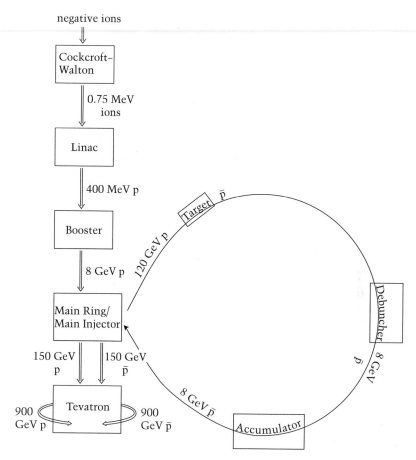

FIGURE 4.70 Flow diagram showing the main features of the Tevatron collider. The Main Injector replaces the old Main Ring.

"small," 500 m circumference proton synchrotron, called the Booster, which winds up the protons' energy to 8 GeV.

All of which is merely a prelude to the big machines. From the Booster, protons travel to the Main Ring, the old 6.28 km circumference proton synchrotron that cranks them up to an energy of 150 GeV. Finally, the protons are ready to meet the mighty Tevatron itself. The Tevatron – so-called because it can accelerate protons up to a peak of one TeV (tera electronvolt) or 1000 GeV – entered service in 1984 as the world's first superconducting synchrotron. It shares the Main Ring's tunnel, and is mounted directly beneath it. Its one thousand

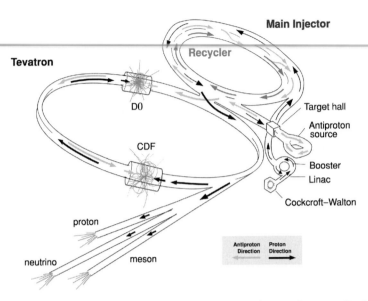

FIGURE 4.71 Sketch of the Tevatron accelerator chain. Credit: Fermilab.

superconducting magnets, about equal in number to the Main Ring's conventional magnets, provide stronger fields for less power consumption. It collides protons and antiprotons at two spots, watched by the two detectors CDF and D0, at an energy of 1.8 TeV, or 900 GeV per beam.

There's something missing though – the antiprotons. Creating protons is easy – just peel away the electron wrapping from hydrogen atoms. Creating antiprotons is a bit more challenging. Experimenters can make antiprotons by ploughing a proton beam into a metal target, and spawning antiprotons via nuclear interactions within the metal nuclei. The problem is that those antiprotons emerge in a disorderly stream, and not in some well-ordered beam suitable for injection into a synchrotron.

To tame them, engineers employ a cunning piece of beam manipulation termed stochastic cooling, invented by Simon van der Meer in 1972. In essence, the idea is to set the antiprotons circulating in a storage ring and, at some point round the ring, test the beam using electrical sensors, to see how spread out it is relative to the central circulation path. This information is then hastily relayed to an electrical "kicker" further round the ring ahead of the antiprotons. The kicker deflects those same sampled particles towards a perfect orbit. When this is done cycle after cycle, the antiprotons rapidly converge towards a well-behaved beam that is then stored. More and more antiprotons are added, until enough have been gathered to pass on to the next stage of the acceleration sequence. The "stochastic" in the name relates to the randomness of the sampling of the beam particles. The "cooling" refers to the parallel between the reduced spread in antiproton speeds, and the reduced speeds of molecules of a gas whose temperature is lowered. Stochastic cooling is one of those great enabling ideas that allowed experimental particle physics to bound forward. First trialed at the CERN's ISR, it was put to work in the Spp̄S, where it enabled Rubbia and his team to find the W and Z bosons. In fact, van der Meer shared the 1984 Nobel prize for physics with Carlo Rubbia for his stochastic cooling work.

Stochastic cooling features at Fermilab. Antiprotons are first manufactured by directing a 120 GeV proton beam from the Main Ring onto a copper target, from where they pass to the 520 m circumference Debuncher ring for cooling. The cooled antiprotons are then warehoused in the Accumulator ring until a useful number have been gathered. By the time an assortment of cooling strategies, including stochastic cooling, have been inflicted on the jostling antiprotons, their packing in the beam has increased a million times over. Finally, the cooled antiprotons are injected into the Main Ring, where their energy is upped to 150 GeV before they pass to the Tevatron itself, forming a counter-rotating beam to the waiting protons.

At least, that was how it was all done when CDF and D0 found the top quark. Since then, the Tevatron has had an upgrade. The centerpiece of the upgrade program is a whole new accelerator, a proton synchrotron called the Main Injector. Nestling up close to the Tevatron, the Main Injector is 3.3 km in circumference, about half that of the Tevatron, and its job is to replace the Main Ring. In fact, some parts of the Main Ring have been utilized in the new machine. The reason for building the Main Injector, which cost around 260 million US dollars and came on line in 1999, is largely to eliminate the bottleneck caused by the Main Ring, which simply couldn't handle enough particles. Relying on much more tightly focused beams, the 150 GeV Main Injector provides more intense beams to the Tevatron, and increases its brightness fivefold. Other enhancements include a further ring, the 8 GeV Recycler ring, which shares the Main Injector's brand new tunnel, and whose function is to store antiprotons left over at the end of the Tevatron's collision cycles. Overall, the combined improvements push the Tevatron's brightness up roughly tenfold.

The upgrade program leaves Fermilab better able to engage in a whole range of improved experiments, from searching for "neutrino oscillations," in which one type of neutrino transforms into another type, to exploring properties of B mesons, or to looking more closely for deviations from the Standard Model. One program certain to benefit is the study of top quarks. With its brighter beams, and slightly increased total energy of 2 TeV, the new Tevatron can produce top pairs at twenty times the rate of the old. Where the old Tevatron produced by the hundred tops, in "Run II," started in 2001, the revamped Tevatron will create them by the thousand.

To spot the top

The Tevatron is not alone in enjoying a makeover. The CDF detector, and across the ring the D0 detector, have also been upgraded. This had to happen: the brighter beams of the revamped Tevatron produce so many collisions they would inundate the older versions of the detectors.

The CDF detector, the older of the pair, is reasonably typical of a modern particle physics detector, and exemplifies their staggering statistics. The CDF collaboration is roughly 600 strong, with participants from about three dozen institutions around the US and across the world, led by two democratically elected spokespersons, together with other officers and an executive board. Then there's the sheer size of the machines: CDF weighs in at about 5000 tonnes, is 17 m long, and 12 m both wide and high – big enough to dwarf the average family home.

CDF is a general-purpose detector comprising several distinct component parts. Since it watches head-on smashes between equals, the collision debris can be spread all over. So those different detector components take the form of cylindrical layers enveloping as much of the collision point as possible, given there is a beam pipe in the way. This is in contrast to detectors designed for fixed-target experiments, whose mission is to look for debris projected into a cone around the direction of the incoming beam. Detectors like CDF are a kind of set of Russian dolls, with one type of detection element performing a specific task, surrounded by another element doing another job, in turn surrounded by another, and so on. In terms of function, however, there is a clear division into two camps. If CDF were an apple, its core is its tracking system. This core comprises a cylindrical magnet, the beam along

FIGURE 4.72 The CDF at Fermilab. The central tracking chamber is being installed. Credit: Fermilab.

its axis, encircling detectors that accurately measure the positions of emerging charged particles. Then the fruit surrounding that core is the particle energy-measuring "calorimeter" system, Figure 4.72.

The CDF detector watching for collisions at the revamped Tevatron is an upgraded version, CDF II, of the detector that was first to see the top quark. Its innermost layer, closest to the interaction point, comprises a wholly new version of the CDF's silicon vertex detector. It's a 1 m long cylinder comprising five concentric wrappings of silicon wafers, the five layers ranging in diameter from 2.4 to 10.6 cm. From the outside, it's the size of a waist-high length of drainpipe. The basic detector element, known as a silicon microstrip detector, exploits the physical properties of semiconductors – the stuff of transistors – to pin down a passing charged particle's position to an accuracy of about

10 microns. One of the silicon vertex detector's principal tasks is to locate the forks, or vertices, in particle tracks resulting from particle creations and decays close to the interaction point.

Finding the top quark would have been much harder without silicon microstrip detectors, which are ten times as accurate as the next best electronic detection methods. It would also have been harder had not theorists given top-hunting experimentalists some idea of what to expect of their quarry. As the top quark mass is greater than about 85 GeV – experimenters hadn't demonstrated this until the early 1990s – then, since the W mass is about 80 GeV and the b quark mass is about 4 GeV, the top quark is able to decay like $t \rightarrow W + b$. This decay route, closed to the other quark flavors since they are much lighter than the W, is very fast. That is because the top can spawn real, physical W bosons, whereas decays of other "heavy" quarks such as b and c involve only virtual W bosons, Figure 4.73. So, the other quarks have to wait

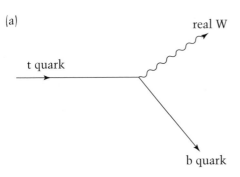

(a)

real W

t quark

b quark

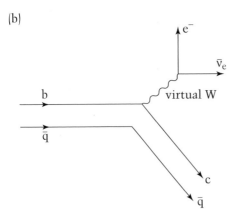

(b)

e^-

$\bar{\nu}_e$

b

virtual W

$\bar{q}$

c

$\bar{q}$

FIGURE 4.73 (a) Top quarks can decay to give a real W boson and a b quark. (b) A b quark, shown here as part of a B meson, decays via a virtual W.

for a suitable quantum energy fluctuation before they can decay via a virtual W, whereas the top can go right ahead and decay without delay. This it does in about 10^{-24} s. In contrast, the b and c quarks live around a thousand billion times longer, with lifetimes of around 10^{-12} s.

Pity the top quark! It decays so quickly that a freshly baked top hasn't the time to turn into a meson in the way that a new b quark can become a B meson. Top quarks are so unstable they will never form versions of the usual mesons and baryons, which is the reason why sighting the B_c in 1998 is billed as the discovery of the last regular meson, despite the existence of the top. The top is unique. All other matter particles are either stable, long-lived, or become part of composite particles. The top, by contrast, is doomed to a fleeting, lonely existence.

Sentimentalism aside, the unique properties of the top have implications for how experimentalists seek them out. With such a short life, there's no chance of distinguishing between the point where the top quarks decay and the original interaction point where the proton and antiproton collided. The W itself has a lifetime comparable to that of the top, so W bosons resulting from top decays will also appear to decay at the original interaction point. However, the bottom quark is relatively slow to decay, and will travel about half a millimeter before giving up and converting into a lighter quark. This means that in a jet of particles associated with a bottom quark or antiquark there will be a vertex marking its decay. So particle jets showing a vertex close to the original collision point are likely to be b quark jets. In fact, experimentalists actively look for this property, an example of what they call tagging – hunting for a particular process via the particles or properties of particles that are created. It's important to be able to identify b and $\bar{b}$ quark jets: most of the top quark events seen at CDF comprised two jets arising from the decay of a single W boson into a quark and antiquark pair, two more jets, one each for the b and $\bar{b}$ quarks arising from t and $\bar{t}$ decay, and a positron (or some other lepton) coming from the other W decay, its partner neutrino being invisible.

The first part of the detector to witness this starburst of particles, and help to resolve it into particular jets hinting at top quarks, is the silicon vertex detector. In principle at least its elemental unit, the silicon microstrip detector, is a simple device. A wafer of silicon, doped with carefully controlled levels of an additive, for example phosphorus, has fine parallel lines of a second material, for example boron,

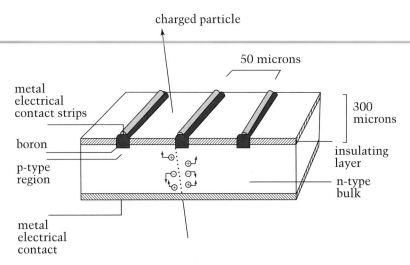

FIGURE 4.74 Silicon microstrip detector. Charges created in the semiconductor material by the passage of a charged particle separate, giving pulses detected via electrical contact strips.

implanted onto it. These lines, the microstrips, are a few microns wide and typically 50 microns from the center of one strip to the center of its neighbor. Metal electrodes are attached to the base of the silicon, and to the fine lines, Figure 4.74.

Silicon atoms bind to four of their crystal neighbors. Phosphorus has five electrons available for bonding, so when it finds itself trying to fit in to a silicon crystal, one of those electrons is surplus to requirement and easily freed. This freed electron can carry charge around the material, and because it's negative the doping is described as n-type. Boron is the reverse of phosphorus. It has just three electrons available to form bonds, so those boron atoms that leak into the silicon have a gap where otherwise an electron should be. This gap, a missing negative electron, behaves like a mobile *positive* charge, called a hole, so this is p-type doping. In the junction region, the two types of doping between them give rise to an electric field.

When a charged particle, say a charged meson, passes through this junction region, it leaves in its wake a trail of fresh electrons and holes electrically dislodged from the crystal. These separate under the influence of the field established by the impurity atoms, the holes going towards the p-type strips, the electrons away from the p-type strips. The result is an electrical pulse, generated in around twenty billionths of a second. A single passing charged particle causes current pulses in several strips, allowing its position and trajectory to be calculated. The high resolution of silicon detectors allows a primary, or creation, vertex to be distinguished from a secondary, or decay, vertex for particles such as charmed and bottom mesons. The spacing between vertices translates into particle lifetimes.

Silicon microstrip detectors, developed through the 1980s as fast, accurate trackers for exploring the then-new charmed particles, entered service that same decade in fixed-target experiments. The first colliding beam experiment to make use of them was the MARK II detector at SLAC in 1990. The first hadron collider experiment to use them was none other than CDF, in 1992. The upgraded CDF has a second-generation silicon vertex detector, designed to cope with the higher proton–antiproton collision rates of the upgraded Tevatron. To measure an event, it blasts out nearly ten times as many electronic signals as its predecessor, over 400 000 in total. This bolt of data has to be swallowed by the waiting electronics in just 10 μs.

Two more silicon layers, between 20 and 30 cm from the beam pipe and protruding beyond the ends of the vertex detector, provide additional cover for particles emerging at shallow angles relative to the beam. Next comes a particle-tracking layer called a drift chamber, in the form of a thick cylindrical jacket embracing all the layers of silicon, and extending out to a 1.3 m radius. Drift chambers rely on detecting charged particles by their ionization of the gas within the chamber. The ionization products drift, under the influence of an electric field supplied by charged wires in the chamber, to sensing wires that pass electrical pulses to electronic counters. The timing of a pulse, relative to a reference time, which for CDF is supplied by the accelerator clock, gives a time interval that translates into a distance moved and hence a position. Drift chambers are a key element of particle detectors: Georges Charpak, who invented the forerunner of the modern drift chamber back in 1968, received the 1992 Nobel prize for his work. In CDF II's drift chamber, there are more than 30 000 sensing wires, five times as many as the old detector. This larger number reduces the distance ionization charges must drift before encountering a sensing wire, making the new chamber several times faster. Both old and new versions of the CDF drift chamber record the position of a particle to an accuracy of around two tenths of a millimeter.

Surrounding the silicon layers and the drift chamber is a powerful superconducting solenoid magnet, 3 m in diameter and 5 m long. It's so large that the family car could sit inside with room to spare. The magnet is there to supply a magnetic field, along the direction of the beam, to bend the trajectories of the charged particles on their journey away from the collision point. The curvature of a particle's track then yields its momentum, the sense of curvature the sign of its electric charge.

The solenoid, drift chamber, and silicon detectors together comprise the tracking system. Outside the tracking system, the rest of the apple beyond the core, lies an energy-measuring jacket nearly 2 m thick. This comprises the calorimeters, whose job is to measure the energy of emerging electrons, photons, and jets of particles. They are destructive: whereas particles transit the tracking system essentially unhindered, most come to a halt in the calorimeter layer, where their energy is converted, via cascades of particles spawned in the material of the calorimeter, into flashes of light.

Closest to the collision action are electromagnetic calorimeters, alternating thin layers of lead and scintillator material that converts some of the energy of transiting particles into detectable light. Electrons and photons undergo bremsstrahlung and pair production in the lead, giving showers of photons, electrons, and positrons. These showers produce light flashes in the scintillator that are piped to photomultiplier tubes. The hadron calorimeters, which rely on a similar principle, are built of thicker layers of steel and scintillator material. Strongly interacting particles, which need not be charged, strike the steel nuclei and initiate a cascade of hadrons. At least some charged particles visible to the scintillator result, the inherent uncertainties of the cascade process for hadrons reducing the accuracy of these units compared to their electromagnetic counterparts.

Not much gets through the calorimeters, which of necessity destroy the very particles they set out to measure. One notable exception is muons. Their greater mass suppresses their urge to lose energy via bremsstrahlung, so they do not dump their energy via particle showers the way that electrons do. Instead, they pass relatively unscathed to an outer layer of muon-detecting drift chambers.

Picking out muons is useful because, for example, they can signal W or Z bosons, and the weak decay of heavy quarks. Muon counters also help avoid the problem of swamping the detector with data. The upgraded CDF will be exposed to several million particle collisions every second. The trouble is that the measurements have to be written onto computer tape, where they are stored for later analysis. The tape can only swallow around 50 collisions-worth of data per second. So the CDF II detector, in common with other particle detectors, employs a "trigger," some decision-making system for separating the experimental wheat from the chaff. The trigger is responsible for deciding that an event looks interesting, and firing up the entire detector to measure and then record it. Other collision events – the vast bulk – are simply ignored.

It's always a ticklish business, setting up an experiment to ignore something like 99.999% of the data to which it's exposed. The challenge facing CDF experimenters is to both avoid missing interesting events and avoid triggering on boring ones. At the simplest level, if the detector sees two 1.5 GeV muons it will trigger, for example. A 12 GeV electron will also do the job. The calorimeters will demand action when they see large energy flows away from the beam direction, which might be energetic electrons or jets of particles. High-momentum tracks emerging at large angles, detected via the tracking system, will also pass muster. Unique to CDF II is its ability to "trigger on b quarks," through the silicon vertex detector's ability to spot the telltale signature of a b quark decay. The trigger system has a total of about 25 µs to decide whether to ignore a particular collision or direct the detector's full attention to it.

Even after ignoring almost everything, CDF II can record about 300 million events a year, yielding in total about 300 terabytes of data over a two-year run. That's enough to fill tens of thousands of regular computer hard disks. The data are processed at the rate of six million events per week by "farms" of computers, mostly programmed using C++, a computer language now widely used in particle physics.

So that's what it takes to spot the top, including the one in Figure 4.75: an army of experimental physicists, supported by legions of technicians and engineers, a detector bigger than a house weighing thousands of tonnes yet built to micron precision, lots of fast and intricate electronics, and enough computational power to make the average computer jockey weep. And millions of dollars. And a powerful particle accelerator. That's all.

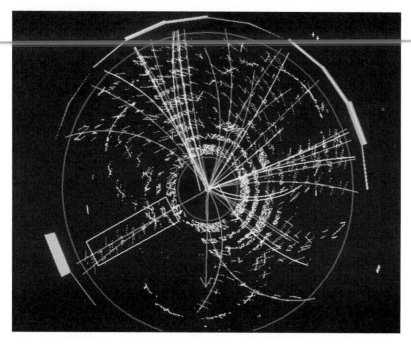

FIGURE 4.75 A CDF top event captured on a computer screen. Credit: Fermilab.

The six-quark universe

So there are six quarks: up, down, strange, charmed, bottom, and top. It's tidy to bundle them all together into a table, Table 4.6, and instructive to compare their masses. To do this is to be more than a little cavalier with the concept of quark mass, which comes with caveats attached, and which will be discussed further in Chapter 5. A fun way to illustrate the masses is to draw balls whose volumes are in the same ratio as the various quark masses, Figure 4.76. This is not supposed to indicate that the quarks differ in physical size in this simplistic way, but to illustrate the huge range in quark masses. How can these mass values, and the enormous range of masses, be explained? Good question.

Another good question to ask is are there any more quarks to be found? Their names await them, b′ and t′, charges $-1/3$ and $+2/3$, respectively. With these quarks would come matching leptons, making a fourth family. For these days, postulating a new quark, or new lepton along the lines of the tau, or a new neutrino means postulating an entire family, or generation as they are sometimes called. But there is no kind

Table 4.6 *Properties of the six quarks. By convention, flavor numbers (charm, etc.) have the same sign as electrical charge. The mass figures are approximate.*

Quark attribute	Quark flavor					
	u	d	s	c	b	t
Charge	+2/3	−1/3	−1/3	+2/3	−1/3	+2/3
Mass (GeV)	0.004	0.007	0.135	1.3	4.2	174
Isospin	+1/2	−1/2	–	–	–	–
Strangeness	–	–	−1	–	–	–
Charm	–	–	–	+1	–	–
Bottomness	–	–	–	–	−1	–
Topness	–	–	–	–	–	+1
Baryon number	1/3	1/3	1/3	1/3	1/3	1/3

of consensus that there really should be more quarks. In times past, introducing the charmed quark solved a known problem, the lack of a type of weak decay, and inventing the idea of six quarks made some sense of CP violation. Martin Perl stumbled on the tau lepton, so in due course the rest of its family would have been deemed necessary. But there is no widely acknowledged riddle that is resolved by adding still more quarks, or more heavy leptons, or additional neutrinos. So, for the time being at least, we have a six-quark, three-family universe, Figure 4.77.

None of which dissuades theorists from postulating more quarks, of course. And experimenters, in turn, continue to look for them. It might seem improbable that a further quark would reveal itself now, given that the top has only just been found. But some theorists have suggested that a b′ quark might have a smaller mass than the top, so would be accessible to current accelerators. Certainly experimenters have looked using the Tevatron and LEP, but seen nothing. They will look again with the revamped Tevatron, and with the LHC when it comes on stream. And who knows, the B-factories, with their high precision, might yield signs of another quark, if such a thing exists.

Is it possible to reverse the argument, and look to experiment for a limit to the number of families? The answer is yes, and it's been done, in particular by studying Z production in electron–positron collisions at SLAC, using the linear collider, and at LEP. The Z boson can

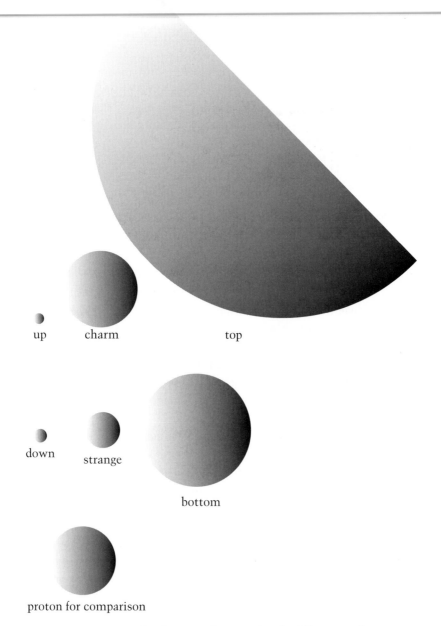

up   charm   top

down   strange

bottom

proton for comparison

FIGURE 4.76 Quark masses. Representing the different quarks as
spheres having volumes in the same ratio as the quark masses
emphasizes the huge mass of the three heavier quarks. This diagram
does *not* suggest that real quarks actually differ in size in this way.

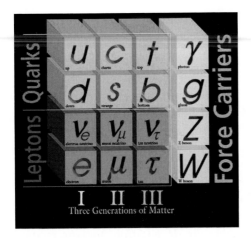

FIGURE 4.77 The three families, or generations, of leptons and quarks, plus the force carriers of the Standard Model. Credit: Fermilab.

decay in many ways, for example into leptons such as electrons and positrons, or muons and antimuons, but it prefers to decay into quarks, and thence to hadrons. Adding up the various measured decay options, then comparing them to the total decay probability, reveals a significant shortfall. This shortfall matches the Standard Model prediction for Z bosons decaying into exactly three neutrino species, decays which rate as invisible due to the problems of seeing neutrinos. So the millions of Z boson decays picked out at the SLC and LEP point to three neutrino species. Or rather, the results point to three *light* neutrino species, allowing for the prospect of a fourth neutrino that has a mass in excess of half the Z mass, in which case it would not feature in Z decays. So, *if* there are no such heavy neutrinos – and, in the main, the working assumption remains that all neutrinos have tiny mass – and *if* the number of neutrinos matches the number of quark pairs via the family patterns, then it's a pretty safe bet there are only six quarks and three families.

Naively, the number of different quarks doesn't matter to QCD, which simply explains the workings of the strong force binding them together, irrespective of their flavor. Family matters, and numbers of quark flavors, are more issues for the weak interaction department of the Standard Model, and for what lies above and beyond the Standard Model.

Quantum chromodynamics may be flavor blind, but it is not entirely flavor ignorant. Though in a simple sense a single quark flavor would satisfy QCD, it would lead to a very odd strongly interacting world. For instance, most of the mesons would have no place in a single-flavor universe, likewise the baryons, both groups reduced to excited states based on a single quark type. Much of the industry that seeks to explain meson and baryon structure ultimately in terms of QCD, with different quarks having different masses, would be redundant. The proton as it is understood today would no longer be, replaced by some baryon comprising a single quark flavor, a bit like the Δ^-, Δ^{++}, or Ω^- for example. There would be no neutron, and presumably no atomic nuclei to explain via the strong force.

When, for example, theorists interpret a proton colliding with an antiproton at the Tevatron, their description of each as a bag of essentially freely interacting quarks depends to some extent on how many different quark flavors there are. In addition, making contact between the experimental results, and theoretical calculation, depends on the number of quark flavors, since the reaction could proceed via different quarks and antiquarks from the colliding species. The same is true of virtually every experiment that is to be compared to some QCD-based prediction.

But there are even more fundamental links between QCD and the number of quark flavors. A gluon can flex its quantum muscles and, for a fleeting moment, turn itself into a quark–antiquark pair. More quark flavors offer the gluon more choices, and in this way the number of quark flavors is interwoven with the fabric of QCD in a way that cannot be ignored. According to QCD, the strength of the strong interaction actually depends on the number of quark flavors. Even the basic distance scale determining QCD's sphere of influence is linked to the number of flavors: quantum chromodynamics might be flavor blind, but flavor features in the one number of QCD.

5   The one number of QCD

Traditional Tibetan tea is an acquired taste. It's made very strong, brewed for a long time, and then diluted down. After being mixed with salted butter in a cylindrical, brassbound wooden churn, it's then reheated, and finally drunk.

The butter is made from yak milk, which contributes somewhat to the surprise. But an important factor in the taste is that the water, though boiling, is not hot enough to make tea of the kind more familiar to the English gentleman in his rose-covered cottage. The point is that Tibet is on a high plateau, high enough that the local atmospheric pressure is much reduced relative to that at sea level. As a result, water boils at a lower temperature, low enough even that bugs are safe and tourists with weak stomachs are not.

Though the boiling point of water drops with decreasing pressure, it rises with increased pressure. In other words, the temperature at which the water phase transforms into the steam phase, called a phase transition, rises with pressure. However, this simple relationship does not go on indefinitely. At a pressure of 218 times atmospheric pressure, water boils at a temperature of $374\,°C$. At higher pressures, there is no boiling point: the gas and liquid phases become indistinguishable. The temperature and pressure where the distinction between liquid and gas disappears is termed the critical point of water.

Another example of a critical point, and a favorite of condensed matter physicists, is magnetization in iron. At high temperatures, iron has no innate propensity to become magnetized. However, if iron is cooled, when it reaches a temperature of $770\,°C$, it becomes magnetized. At this temperature, individual atomic magnetic elements within distinct regions, called domains, spontaneously line up, their mutual aligning forces finally overpowering thermal disruptions. Different domains within the sample will, in general, have different magnetic orientations, but subject bulk iron at room temperature to a magnetic field and at least some of these domain magnetizations respond by reorienting, or by swelling at the expense of others, and the iron becomes strongly magnetized. At the critical temperature, $770\,°C$, the ordered "ferromagnetic" phase, familiar in room-temperature iron, merges into a disordered "paramagnetic" phase, common in metals such as aluminum. There are many other instances of such so-called critical phenomena, effects in which the distinction between two phases of matter vanishes.

As Kenneth Wilson (Figure 5.1) himself once pointed out, two neighboring water molecules don't know if they are in a teapot or in the middle of the Pacific Ocean. What goes on at the length scales where the behavior of individual atoms is noticeable is of little consequence at the length scales of a teapot or ocean, which are billions of times larger. Yet there is one instance where the way a molecule interacts with its neighbor really does impact on the entire contents of the teapot, where the smallest length scales right through to the largest all have a role to play, and that is at the critical point.

For a container of water held at a pressure of 218 atmospheres, as the temperature rises and closes in on 374 °C, the "water" becomes a mix of liquid drops and gas bubbles of all possible sizes. At the critical point, these bubbles and drops, nested inside one another, span the complete range of length scales, from the size of molecules to the size of the container. They form a complete spectrum of density fluctuations throughout the container. Every molecule is felt by the whole.

Accounting for the contributions from such a totality of length scales is a challenge. Usually, tiny fluctuations average out in some way, and physicists have long known how to deal with such averaging problems. But for those phenomena in which all length scales contribute, where averaging gives the wrong answers, they need something new. That something is the renormalization group.

Applied to critical phenomena, the renormalization group links different "magnifications" of the system, as though a powerful camera were panning back to include larger and larger features, as finer details are lost from view. The renormalization group maps the evolution of fine-grained images into successively coarser grained ones – it relates physics at different scales. At the critical point features such as density fluctuations or regions of magnetic alignment appear on all length scales; panning in and out with the camera does not alter the overall look of the scene. In other words, from the distribution of liquid drops and gas bubbles in a fluid at the critical point, or the clusters of similarly magnetically aligned regions in a lump of iron at the critical temperature, it's not possible to figure out if the magnification is set on

FIGURE 5.1 Kenneth Wilson in his office in Ohio State University, 1998. Wilson's renormalization-group-related insights have had far-reaching consequences, both in quantum chromodynamics and elsewhere. He was awarded the Nobel prize in 1982 "for his theory for critical phenomena in connection with phase transitions." Credit: Courtesy Kenneth Wilson.

high or low: both appear the same, statistically at least; just by looking down the eyepiece you couldn't tell what the microscope knob was set to. This is an invariance, since nothing changes if the distance ruler is changed. This scale invariance is a symmetry, as anticipated by the use of the word "group" in renormalization group.

The renormalization group applied to critical phenomena allows theorists to predict measurable properties that rely on all length scales of the system. For water, for example, the change in volume in response to a change in an applied external pressure, the compressibility, becomes infinite at the critical point, and close to the critical point varies according to a particular mathematical relation. For a simple iron-type magnet, the change in magnetization of the material in response to an applied magnetic field follows the *identical* mathematical relation, and likewise becomes infinite at the critical point. These two examples illustrate the two dominant characteristics of critical phenomena. Firstly, close to the critical point, they show characteristic dependences on the relevant quantities describing the system, with infinite spikes in physical parameters actually at the critical point. Secondly, they exhibit a critical point universality, with apparently very different physical systems, in this case the liquid–gas transition and magnetization in iron, behaving in identical ways.

These insights into critical phenomena via the renormalization group, embracing the critical point behavior of fluids and magnetic systems, of transitions into the superconducting state, of the total mixing of oil and water at elevated temperatures, and many more, form one of the great triumphs of modern theoretical physics. On the face of it, however, none of this seems to have much to do with the strong force. But there *is* a link. The link is the renormalization group, and invariance under changing length scales.

ELECTROMAGNETISM GROWS STRONGER . . .
The family encyclopedia gives a figure of something like $1.602\ 177 \times 10^{-19}$ C for the charge of the electron. Though this is perfectly correct it is, in a sense, an oversimplification. Charges and masses of interacting particles don't have rock-solid values that can be engraved in stone. The values of these quantities turn out to depend on who's looking, or more exactly on how hard they look. For electrons and QED, this is little more than a curiosity. For QCD, it's a cornerstone.

David Koltick and his colleagues in the TOPAZ collaboration used the now-defunct TRISTAN electron–positron collider at KEK, near Tokyo, to check the QED curiosity angle by looking at the effective charge on an electron probed at high energy. This they did this by comparing the two processes, $e^+ + e^- \rightarrow \mu^+ + \mu^-$ and $e^+ + e^- \rightarrow e^+ + e^- + \mu^+ + \mu^-$, Figure 5.2. The first involves just a single photon, and

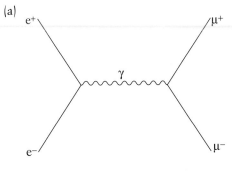

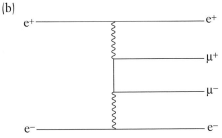

FIGURE 5.2 The dominant Feynman diagrams for (a) $e^+ + e^- \rightarrow \mu^+ + \mu^-$ and (b) $e^+ + e^- \rightarrow e^+ + e^- + \mu^+ + \mu^-$.

the momentum transfer via this photon is high since the full, combined momenta of the electron and positron flow through it. So in this case the interaction between the electron and positron is short range. The second process involves two photons, and by contrast the momentum transfer via either of these photons is small; for a loose analogy, think of two rivers that combine to form a single mighty river, and compare the current there to that flowing along a little creek linking two rivers flowing side by side. Compared to the first process, the electron–positron annihilation, this second interaction is relatively long range.

By comparing the two scattering rates for the two processes, the TOPAZ team extracted a value for the electromagnetic coupling strength α at high energy. Condensed matter physicists, on the other hand, have some very precise techniques for determining α at low energy: the current accepted value, ignoring the ubiquitous errors, is about 1/137.036 – the value of α is quoted in this way, as 1/137, for largely historical reasons. In their 1997 paper, the TOPAZ group reported a value of 1/128.5 for α at a momentum transfer of 57.77 GeV^2, which agrees, to within experimental error, with the value of 1/129.6 from theory.

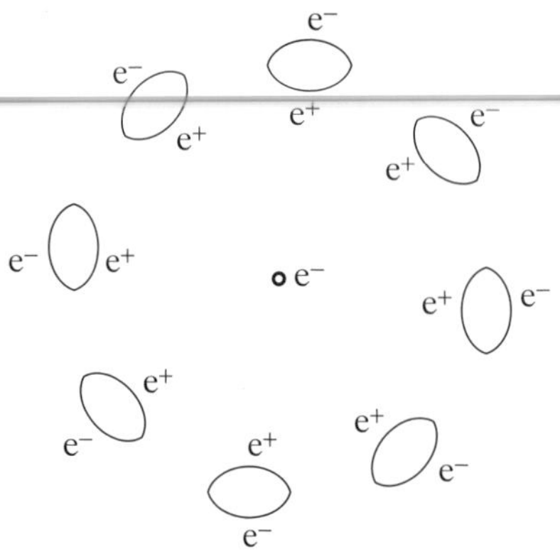

FIGURE 5.3 Schematic representation of how a bare charge, the electron at the center, is screened by virtual electron–positron pairs from the vacuum, shown as loops in the diagram.

What the TOPAZ team saw was the strengthening of the electromagnetic coupling between two charged particles as those particles approach one another more closely. Theirs was by no means the first experiment to witness this effect, but it's a clean and unambiguous test. The physical picture envisages a bare charge that surrounds itself with a cloud of virtual photons – the same physics as modifies Coulomb's law, manifest as the Lamb shift in hydrogen spectra. At any one moment, some of those virtual photons will turn into virtual particle–antiparticle pairs, for example electrons and positrons, with the oppositely charged virtual particle closer to the bare particle, on average, Figure 5.3. The net effect is therefore to screen the bare charge with a cloud of virtual particles, and it's the screened electric charge that's at play at low energies. But, attacking an electron with a high-energy probe means penetrating deeper and deeper into the screening layer, revealing more of the bare charge, so the apparent strength of the electromagnetic force increases.

One might imagine a despotic dictator so extreme in his (or her) views that his circle of councilors, somewhat repelled by their leader's outrageous philosophy, does his bidding, though at the same time tempering his schemes. Further levels of moderation enter with the layers of civil servants who implement the dictator's wishes, so that what the public sees is a diluted image of the dictator. Just as the dictator's advisors and civil servants screen the "bare" horror of the man himself, so that the public sees a weaker version, virtual particle pairs screen a bare charge, so that from the outside the electron charge looks relatively weak. Penetrate the screening of either, and the strength of the bare core lies exposed.

The screening featuring in the TOPAZ experiment is the same screening mechanism that in the renormalization program links a bare unmeasurable particle to one having a finite measurable charge and mass. Crucially, in making this link the renormalization program introduces a renormalization scale, marking the rather arbitrary boundary between "short" and "long" distances. However, the physics cannot care about the exact position of this boundary, where the short distance cut-off lies, so theoretical predictions must remain unchanged when the scale is altered – a length scale invariance. So where has this renormalization scale gone? It's absorbed by key quantities in the theory such as particle charge and mass. These quantities must therefore change as the boundary between short and long distances – the renormalization scale – varies. By exploiting the invariance of the physics with respect to this changing length scale, the renormalization group shows explicitly *how* these key quantities change, establishing a precise relationship between charges and masses at different renormalization scales, and therefore at different energies. Here, what the renormalization group shows is just how much the charge on an electron, say, as seen by another incoming charged particle, depends on how intimately the two interact. This evolution of the coupling with energy also reveals that the expression "coupling constant" is a misnomer, since it is not constant at all: physicists have coined the phrases "effective coupling constant" or, better, "running coupling constant," to flag this energy-dependence.

Though the renormalization group allows theorists to calculate the coupling at different energies, there's something missing; experiment has to provide the first value of electric charge, presumably measured at low energy. It's a little like knowing how often people are leaving a crowded room, and wanting to know how many remain, but not knowing how many people were in the room to start with: an extra piece of information is needed to give an absolute value for an answer, and set the overall scale. Thus, for people leaving the room, the moment the original number is made known, the absolute number remaining is easy to find. Likewise for the electron charge, once the scale of the charge values has been set by a low-energy measurement, for example the electric charge determined using techniques from condensed matter physics, the renormalization group provides a way of calculating the running coupling constant at higher energies. This expression is what the TOPAZ group used to calculate their theoretical value of the coupling to compare with their experimental value.

It is also clear that there is some unfinished business in QED. The coupling just grows and grows for ever-larger values of momentum transfer, which means that there comes a point where it is no longer small enough for perturbation theory to make much sense, since perturbation theory is supposed to give an answer as a power series of ever-decreasing terms. In other words, as mentioned in passing in Chapter 3, perturbation theory eventually fails even for QED. But this happens at an energy so high – around 10^{26} GeV – that it's totally off the dial of any particle accelerator. In fact, at this energy the conventional notion of space-time itself no longer makes sense, so as a failing on the part of QED this really is pretty academic.

. . . WHILE THE STRONG FORCE GROWS WEAKER

No conference on QCD is complete without an account of the latest measurements of the strong coupling parameter, α_s.

Of the many particle reactions physicists have explored in their hunt for a reliable and accurate value of α_s, one of their favorites remains the decay of tau leptons. These can decay into a collection of neutrinos, antineutrinos, muons, and electrons, the tau leptons themselves having first been produced by electron–positron annihilation, Figure 5.4. But tau leptons are also heavy enough to decay, via W bosons, into neutrinos and strongly interacting particles, Figure 5.4(c). This is perhaps a little surprising at first, but the fact is that quarks, the only particle in nature sensitive to all of nature's forces, are perfectly able to interact weakly with W bosons. Theorists like tau lepton decay as a place to measure the strong coupling parameter because they can calculate it up to Feynman diagrams containing three loops. This about equals the greatest level of refinement available in QCD calculations. Another reason theorists like it is that the bits they are not sure how to calculate, lumped together in the too-hard basket and labeled "non-perturbative effects," are reckoned to be too small to upset the basic conclusion of the experiment.

Several experimental groups, for instance the CLEO collaboration at the CESR electron–positron collider, have published measurements

(a)

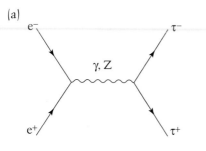

(b)

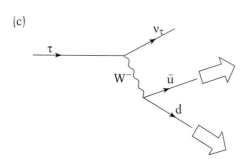

(c)

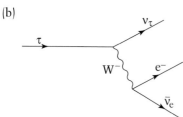

FIGURE 5.4 (a) Production of tau leptons in electron–positron collisions. The tau leptons can decay via other leptons (b), or into quark–antiquark pairs and thence hadrons (c). By including higher order diagrams based on (c), comparison of the lepton versus quark decay routes gives a value for the strong coupling.

of the strong coupling parameter based on tau decays. The result comes in at $\alpha_s (m_\tau) = 0.35 \pm 0.03$ – no units, because α_s is a simple number. The subscript s stands for "strong" to distinguish α_s from the α of QED. And compared to the α of QED that's a huge error on α_s. The other noticeable feature of the α_s result is the (m_τ) specification. This is because the strong coupling parameter depends dramatically on energy. So it's necessary to say what the energy scale is, and in this instance the appropriate figure is the tau lepton mass, 1.777 GeV.

The need for the (m_τ) label becomes abundantly clear when the α_s result from tau decay is compared to the results from other experiments. These are too numerous to list in full, but another well-traveled road is the study of collisions giving rise to distinct showers of particles called jets. For instance, the ZEUS and H1 collaborations at HERA have

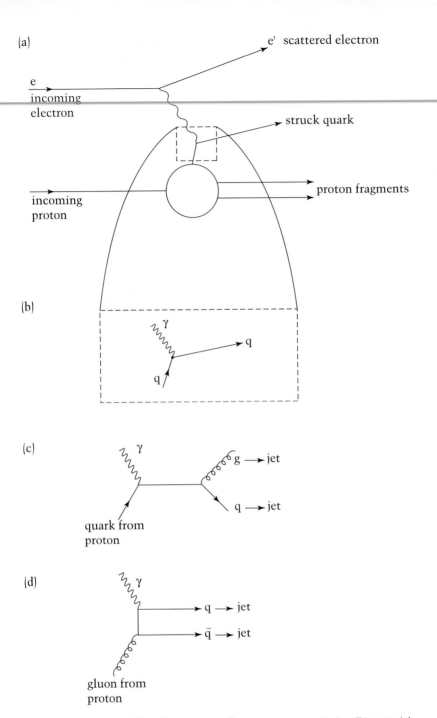

FIGURE 5.5 Three jet events in electron–proton scattering. Diagram (a) shows conventional deep inelastic scattering. The subprocess $\gamma q \to q$ is indicated by the dashed box, and reproduced in (b). For three-jet events, this subprocess is replaced by those shown in (c) or (d), which both create two jets in addition to a proton fragment jet.

looked at jets emanating from electron–proton collisions as a way to extract a result for the strong coupling, Figure 5.5. One shower of particles is the leftover smashed proton, the two others corresponding to either the quark and gluon of Figure 5.5(c) or the quark and antiquark of Figure 5.5(d). Their combined result is $\alpha_s \, (m_z) = 0.1179 \pm 0.006$. That number looks very different indeed from 0.35, and reflects the dependence of the strong coupling on energy. So it's important to include that energy information when quoting an α_s value.

Experimentalists have accumulated a plethora of α_s measurements. When those measurements are collated, several features emerge. One is that a wide variety of approaches yield consistent answers – it would be a problem for QCD if they didn't, to be sure. Another is that the strength of the interaction does not depend on quark type, to be expected since the strong force is about color, not flavor. Finally, as energy increases, the strength of the strong coupling really does diminish in the way predicted by asymptotic freedom.

Turning this established link between energy and coupling strength on its head, measurements of the strong coupling are usually quoted in terms of the corresponding value at a certain energy scale, chosen by convention to be the Z boson mass, m_Z. This allows experimenters to compare measurements of the strong coupling taken from very different experiments. For example, the result from tau lepton decays, $\alpha_s \, (m_\tau) = 0.35 \pm 0.03$, evolves into $\alpha_s \, (m_Z) = 0.121 \pm 0.003$ at the higher energy, consistent with the HERA results above. Averaging over many experiments, a good choice for the value of the strong interaction coupling is $\alpha_s \, (m_Z) = 0.117 \pm 0.002$.

Experimenters have shown that for both QED and QCD the strength of the interaction depends on energy. However, though both share a similar mechanism for this energy-dependence, the two respond in opposite ways to changes in energy. In QED, screening provides a simple physical picture of how the coupling *increases* as the particle under test is probed more closely. In contrast, for QCD the coupling *decreases*. This is explained in terms of an antiscreening effect. The virtual particle pairs surrounding a bare color charge actually make the color charge on a quark appear stronger than it is, and when a probe quark penetrates through the layers of antiscreening, the core is revealed to be weaker, not stronger. An electron induces an opposite charge around itself: by contrast, a quark appears to induce around it a color charge of the same type. The root of the difference lies in the fact that the force carriers of the strong force, the gluons, themselves carry color. So the gluons that appear in the cloud of virtual particles around a bare quark contribute directly, whereas the photons in the QED case do not. It's these gluon contributions, larger and acting in opposition to the contributions from quark–antiquark pairs, which carry the day. It turns out that if nature had supplied several more varieties of light quarks, then the quark's contribution would win out, and the Universe would be a very different place.

There is no simple physical analog of this antiscreening effect, but the story of the dictator may help. Imagine that a benign dictator is surrounded by councilors, as before, and they in turn by civil servants. This time, however, all communications between parties is via devilish messengers – the gluons. With their own agenda, clandestine associations and secret handshakes, they impart their own influence, dominating that of councilors and civil servants alike to the extent that, from the outside, the dictator appears an utter despot. Close up, he's a relative pussy cat.

In QCD, the link between the strong coupling parameter, α_s, and energy is established using the renormalization group, aided by perturbation theory. The resulting expression for α_s is one of the key results in the whole of QCD. At first glance, this result looks similar to the corresponding equation for QED. But then the eye picks out a strange looking portion in the denominator. It is a $33 - 2n_f$ piece, where the n_f means number of quark flavors. Remarkably, perhaps, the number of types of quarks actually makes an appearance in this basic result. This factor controls the sign of the strong coupling, as calculated to the simplest level, since if the number of quark flavors is 16 or less, then $33 - 2n_f$ comes out positive. But if the number of flavors is 17 or more, then this factor is negative, in which case the theory would no longer possess the key property of asymptotic freedom. This is why the Universe would be very different if nature had laid on more light quark species. The 33 comes from the gluons, and the fact that the color group is $SU(3)$: a different symmetry group would give a different number. The $2n_f$ comes from the quarks, and reflects the number of different virtual quark–antiquark pair possibilities into which gluons may transform during the course of an interaction. Thus QCD with asymptotic freedom demands 16 or fewer quark types, a limit of sorts, though still a long way above the limits imposed on the number of quark types from other sources.

There's another subtle point that emerges in the relationship between strong coupling and energy. Theory shows how the coupling strength changes with energy scale, but it does not set the zero of the scale. This issue was present for QED, and there its resolution was simple – measure the coupling at "zero energy," and then relate all others to that value. In other words, use the readily available and very precise techniques for determining the charge of an isolated electron.

But in QCD, there's a problem. There's no way of measuring the strength of the color charge on an isolated quark in the way that experimenters can measure the electric charge on an isolated electron. Indeed given that the coupling becomes very large at low energy, there isn't even a way that a low-energy measurement can help calibrate a result derived using perturbation techniques valid only at high energy. All that physicists can do is to measure α_s at some reasonable energy to fix the overall scale, just as knowing the charge on an electron fixes the sliding scale of the electromagnetic coupling in QED.

Actually, it's more convenient in QCD to relate the strong coupling and energy in terms of some parameter, whose value is known just as soon as experimenters have made a measurement of α_s. This parameter, or equivalently the strong coupling α_s measured at some specific energy, is *the* one number of QCD, the single fundamental constant in the theory. It must be determined from experiment – everything else that really matters comes from the theory.

By convention, this unknown parameter is called Λ, the Greek letter lambda, and it has dimensions of mass. One might have thought that the "fundamental constant" of QCD would have a suitably illustrious name all of its own, but it seems destined to remain innominate: particle physicists simply refer to it as Λ, or the "Λ parameter," or "QCD scale parameter."

In principle, determining this fundamental constant Λ is easy. For instance, in the H1 and ZEUS measurements of the strong coupling in electron–proton jets mentioned above, both groups simply compared their scattering data to the QCD prediction of the same process, and allowed the value of Λ to drift until it gave the best agreement between theory and data. This in turn fixed the strong coupling parameter, α_s.

In practice, however, extracting a rock-solid value for Λ is hard. The main reason is that at higher energies it takes more and more energy to effect a reduction in the coupling. In other words, it becomes harder and harder to penetrate the antiscreening layer surrounding the color source. This means that a huge range of energies is required to extract an accurate value of Λ. At low energies, this telescoping effect is less of an issue, but then the experiment is entering an energy range where perturbative QCD is less reliable. The accepted value of Λ is about 200 MeV.

For something that started life as an anchor-point constant, the Λ parameter has hidden depths. The master equation of QCD knows nothing about the Λ parameter. And ignoring the quark masses – a perfectly acceptable and common high-energy approximation – there is nothing in the master equation of QCD that makes any mention of a mass, or a length or intrinsic size. The (massless) master equation of QCD knows nothing of the size of the world to which it is supposed to apply. In this naive way the theory is scale invariant: it could be about particles the size of planets for all the difference it makes.

When the master equation of QCD runs the gauntlet of quantization and renormalization, however, something remarkable happens. Renormalization introduces a length scale – a momentum cut-off or equivalently a renormalization scale. The fact that the physical content of the theory is invariant with respect to this scale factor yields the *change* of the strong coupling with energy. But to get an *actual* value for the strong coupling means including the renormalization scale in the guise of the lambda parameter. The quantum theory of QCD thus acquires a scale-dependence in the form of the lambda parameter: the scale invariance of the master equation of QCD is *sacrificed* in the transition to a full quantum theory. The value of 200 MeV for lambda corresponds to a distance scale of around one fermi, a bit less than the wavelength associated with a pion and roughly the size of a proton. In other words, the value of lambda puts the theory firmly into the subnuclear ballpark. The quantum theory of QCD is not a theory of planets or billiard balls.

If the energy scale gets close to the value of lambda, then the strong coupling parameter is large, and perturbation theory no longer holds good. In other words, lambda dictates where useful perturbation theory ends. This is the low-energy end of the graph of coupling versus energy, Figure 5.6. At high energy, the coupling is small and perturbation theory viable, so all kinds of useful perturbation theory calculations – for example, those of deep inelastic scattering – are on offer. This is the realm of perturbative QCD. Most QCD is perturbative

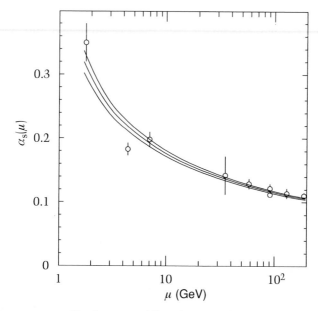

FIGURE 5.6 The distinctive fall in the strength of the strong coupling parameter α_s as energy increases. The upper and lower curves indicate a statistical spread about the central line. Credit: Particle Data Group.

QCD. The low-energy world, below about 1 GeV, is the hidden valley of QCD, less accessible and certainly boasting rougher terrain. However, one group in particular has managed to cut some kind of a path into this jungle. They are the practitioners of so-called lattice QCD. Their goal is to tackle this "non-perturbative" region of QCD head-on by using vast computers to solve the equations of QCD with simulation methods. Lattice QCD is the subject of Chapter 10.

Somewhere in this darkness of non-perturbative QCD lies confinement, the antithesis of high-energy's asymptotic freedom. The energy is low, so the coupling is large and this explains why nobody sees free quarks and gluons. They are too tightly bound, confined into place.

A TREACHEROUS NUMBER

As fundamental parameters go, lambda is pretty ghastly. One might have hoped for a simple, well-defined and easily measurable number, but no such luck. Extracting lambda from experiment could hardly be more difficult.

To begin with, lambda suffers from the energy telescoping effect mentioned above, but this is only the beginning of a long and sorry tale. Commonly used calculations make use of, at most, Feynman diagrams having just one or two closed loops. But at this level of complexity, the relation linking the strong coupling and lambda is not unique. Experimentalists attempting to deduce a value for lambda must, therefore, explain carefully how their strong coupling and lambda values are connected.

A related twist has its roots in the fact that there's no way of including all possible Feynman diagrams of every conceivable complexity in a real calculation. The poor theorist has to stop somewhere, usually at diagrams containing two loops or even, occasionally, one tier beyond to diagrams with three loops. The issue is that, although a physically measurable quantity, such as a cross-section, cannot depend on the theorists' choice of renormalization scheme, an actual calculation can. So although a never-ending string of terms calculated in two different ways might ultimately give the same answer, stopping short and comparing the resultant two expressions reveals differences that arise from the choice of renormalization scheme, since this choice impacts on the details of the calculation. To illustrate the idea, the number one may be broken down in two ways, as $1 = \frac{1}{2} + \frac{1}{4} + \frac{1}{8} + \frac{1}{16} + \cdots$ or as $1 = \frac{1}{2} + \frac{1}{6} + \frac{1}{12} + \frac{1}{20} + \cdots$, using the two mathematical series $\frac{1}{2^n}$ and $\frac{1}{n(n+1)}$, respectively, with n a whole number greater than zero. The right-hand sides of both these strings of fractions add up to one if *all* the terms are included, and the first term of each string is the same. But if each string were to be cut off, say, at the fourth or perhaps the fifth term, or some other number of terms, and then summed up, the two strings would give two different answers. For QCD, what this truncating means is that for all but the simplest calculations, the equations that two theorists write down as their predictions for some specific particle process can differ, depending on which renormalization scheme each used. This is one reason why theorists have agreed a convention to try and reduce the complications of comparing calculations both with one another and with experiment.

Then there's the problem of the number of quark flavors. The energy of the interaction dictates the ease with which different quark flavors can feature as fleeting, virtual particles created then destroyed. Up, down, and strange quarks are so light they are readily produced. Charmed quarks are easily created during some higher energy interactions, and at high enough energies the bottom and even top quarks also play a role. The only quarks making a significant contribution are those whose mass is less than the energy scale of the process in question. This means the value of lambda depends on how many quark flavors contribute, revealed by the $33 - 2n_f$ factor in the relation linking the strong coupling to lambda. Usually the figure for n_f is four or five: the corresponding values of lambda differ significantly.

Quoted values of lambda therefore carry a label saying how many quark flavors are involved, four or five, and strictly speaking a label identifying the renormalization convention used. And lambda's value is known only imprecisely, at around 200 MeV. For five quarks and using theorists' accepted convention, $\Lambda \approx 200$ MeV, where "$\approx$" stands for "roughly equals," is the way the "one number" of QCD is written down. It's not very pretty.

QUARKS GET LIGHTER

How heavy is a quark? Though this might seem a simple enough question, in fact the answer turns out to be anything but simple. Quark masses are just as slippery as the coupling "constant" of QCD: the mass of a quark depends on how you look at it, and is a ticklish issue because of confinement.

With quarks confined to live inside protons, pions, and other hadrons, experimenters are denied the luxury of *free* quarks whose mass they can measure in the way they are able to measure that of an electron, for example. Because quarks are never free, quark masses are revealed only through their effect on some interaction. So quark masses are not unique, but depend on how they are defined. Any quoted value of a quark mass should really come with a footnote, setting out the regime used to extract the mass figure.

For example, one way of estimating quark masses is to use a simple quark model of hadrons that can reproduce the pattern of hadron masses, and extract values of quark masses by fitting the model to the measured hadron masses. Because each quark lives inside a hadron "bag" filled with interacting quarks and gluons, a kind of strongly interacting, sticky molasses, they behave as though they have an effective mass that need not be identical to their "true" mass. This effective mass is called the constituent mass. Roughly speaking, the constituent masses of the up and down quarks are about a third of the mass of the proton, so come out at around 300 MeV each, the strange quark rather more, with a constituent value of about 550 MeV.

How to reconcile these quark model masses with the other set of oft-quoted quark mass values? In Chapter 4, the up and down quarks were ascribed masses of 4 MeV and 7 MeV, respectively, very different from the 300 MeV or so for the constituent quark masses. These smaller values, extracted from pion-decay data, are termed current quark masses. They arise from the way in which small but finite quark masses spoil the additional symmetry that a massless version of QCD would posses, a point to be discussed further in Chapter 9. A constituent quark is a current quark surrounded by a gluon cloud, and it's this cloud that carries most of the mass and which gives rise to the difference between the two sets of mass values.

And there is more. The quark mass also features as a parameter in the master equation of QCD, but is not predicted by the theory. This mass parameter is not the physical mass, however. The physical mass is the mass that would feature in the full-blown expression giving the probability for a quark to move unmolested from place to place, the two-point Green's function. This version of the mass also has a name: it's called the pole mass: it can be measured for electrons, for instance, and is the value give in data tables for the value of the electron mass, but is the very thing that cannot be measured for quarks. The best particle physicists can do is to include quark mass in the master equation and, working in the high-energy regime where perturbation theory is valid, subject it to the renormalization program. The quark mass then depends on energy just as the coupling parameter depends on energy. In this way, quark masses are not fixed, but are "running" masses that decrease as the energy is increased. This running of quark masses is as much a part of QCD as the running of the coupling parameter.

Particle physicists have taken the running quark mass for granted for many years. However, the first experimental evidence for the running quark mass was announced at a QCD conference in Montpellier, France, in 1997, by Liverpool University's Salvador Martí i García and his colleagues. They used data gathered over two years by the DELPHI experiment at LEP for the decay of Z bosons into showers of hadrons. Jets from b quarks contain some particles originating from outside the original electron–positron interaction point, a consequence of the relatively slow decay of B mesons, Figure 5.7. By contrast, particles in jets from light quarks can more often be traced back to the original

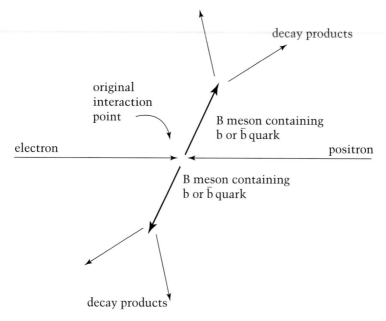

FIGURE 5.7 The relatively slow B meson decay gives particle tracks whose origin lies away from the original interaction point.

interaction point. So significant numbers of particles originating from outside the original collision point provide a signature of b quarks. This allows the two types of jet, b quark and light quark, to be counted separately. In addition, b quarks give rise to fewer gluons, a consequence of their greater mass. The difference between the probability of producing b quark jets, and that for producing light quark jets, reflects this, and is therefore a measure of the b quark mass. By comparing jet probabilities, and using refined QCD calculations, published that year, for three-jet production in electron–positron collisions, Martí i García and his colleagues deduced a value of 2.67 ± 0.50 GeV for the b quark mass. Their figure, extracted at an energy of 91 GeV, the mass of the Z boson, is just 60% of the quoted value of the b quark mass at 1/20 of the energy, based on studies of the upsilon meson. This fall in b quark mass with increasing energy is in line with the running mass prediction of QCD.

6   The gregarious gluon

Physics can be a dirty business. On the one hand is the elegance of mathematical theory, the experimenters' shiny vacuum chambers, flashing electronics, and precision computing. On the other hand is the real, mysterious, and untidy world. The list of actual physical systems that continue to defy complete explanation is almost endless, from the intricate workings of stars to the flowing of sand from a bucket. Sometimes, however, experimenters uncover graphic evidence of a deep simplicity in nature. In particle physics, one of the best places to witness this simplicity is in the phenomenon of jets. Figure 6.1 shows an example of a particle jet.

JETS AND SWISS WATCHES
A proton hurtling around the Tevatron ring encounters an antiproton in a head-on smash. Particles thus created could, in principle, fly off randomly in all directions, but they don't. Instead, they emerge as two, three, or even more distinct cone-shaped sprays of particles, in addition to the remains of the proton and antiprotons themselves. A similar thing can happen for colliding electrons and positrons at LEP and elsewhere. This is more remarkable still since, in contrast to the composite proton and antiproton, here the protagonists are elementary. These emerging particle sprays are the "jets" already encountered in Chapters 4 and 5.

   The prototype jet event is electron–positron annihilation into two jets of strongly interacting particles. This proceeds via the creation of a quark–antiquark pair, with each of the pair subsequently responsible for a hadron jet, Figure 6.2. And this is one of the attractive features of jets – a simple diagram to explain their formation is instantly recognizable as the core of the relevant Feynman diagram. Not only that, the jets themselves are the nearest experimenters are likely to get to seeing quark and gluon tracks in their detectors, the jets themselves tallying with the underlying quark and gluon processes. Multi-jet events in electron–positron collisions are today's quintessential particle physics images.

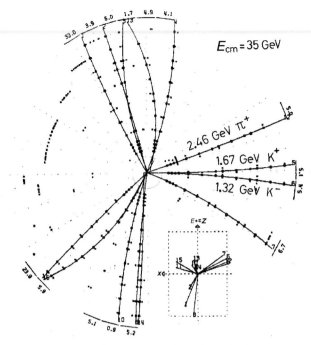

$E_{cm} = 35$ GeV

2.46 GeV π^+

1.67 GeV K^+

1.32 GeV K^-

$E \div Z$

$X \triangleleft$

FIGURE 6.1 A 1979 three-jet event from the PETRA storage ring, evidence for the existence of gluons. Credit: DESY, Hamburg, Germany.

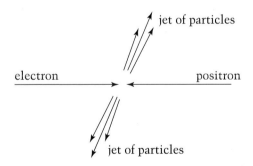

jet of particles

electron   positron

jet of particles

FIGURE 6.2 The prototype jet process: an electron and positron annihilate into two jets of particles.

There is, of course, a twist. How do the bare quarks and gluons from the depths of the initial reaction dress themselves, becoming the particles seen by experimenters? Having at least a reliable model of how this happens is essential, since otherwise there's no way to compare the

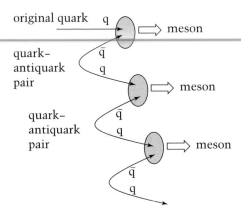

original quark q

quark–
antiquark
pair

quark–
antiquark
pair

FIGURE 6.3 Quark fragmentation. An original quark sparks a
quark–antiquark pair from the vacuum, combining with the antiquark
to form an observable meson. The new quark continues the process.

outcome of calculations with the particles that experimenters actually
measure.

The basic idea is that, as the initial quark and antiquark, say, try
to separate, they are unable to do so in a straightforward way. This
is because free quarks and antiquarks are not allowed. Instead, the
mechanism of confinement says that quarks and antiquarks should
stay bound together in composite particles. So as the quark and antiquark from a collision try to fly apart, confinement asserts itself. As
the pair try to move apart, it's energetically favorable for each to conjure a fresh quark–antiquark pair from the vacuum. So as the quark
leaves the collision scene, it combines with the antiquark of a fresh
quark–antiquark pair from the surrounding empty space, the vacuum,
creating a meson, Figure 6.3. This then leaves a new quark, which
in turn spawns another quark–antiquark pair before combining with
the antiquark to give another meson. The result is a chain of meson
creations, each with less energy than the previous one due to the
energy absorbed in creating the quark–antiquark pairs. The chain terminates when the final quark hasn't enough energy to pop another
pair from the vacuum, at which point it combines with a similarly
feeble antiquark left over from the antiquark meson-creation chain.
This balances the color books so that not one of the finished products
is left with a net color charge. The process by which quarks and gluons turn into hadrons is described as hadronization, or fragmentation.

It's the last, and thus far missing, ingredient of a real cross-section calculation.

During fragmentation a quark, say, shares its momentum between the new meson and the daughter quark. This is expressed as a probability distribution, called a fragmentation function, given in terms of the fraction of the parent quark's momentum passed on to the meson. These fragmentation functions are a kind of structure function in reverse: whereas structure functions describe the probability of finding quarks and gluons of some momentum within a nucleon, fragmentation functions describe the probability of creating a meson having a particular momentum given a parent quark or gluon. Like structure functions, to calculate fragmentation functions from first principles would mean knowing more about how to deal with the low-energy, non-perturbative QCD world than theorists do at present.

Likewise, fragmentation has attracted its own community of scholars dedicated to its elucidation. They were quick to extend the simple, but inherently flawed, "independent jet model," outlined above, adding both baryon creation and splitting of gluons into quark–antiquark pairs. The enhanced model is used to understand proton–antiproton collision data. Another widely used model, the cluster fragmentation model, envisages radiated gluons in turn giving quark–antiquark pairs that combine with others, not necessarily from the same parent gluon, to form clusters lacking any net color. These color singlet clusters of quarks, antiquarks, and gluons then transmute into observable jet particles, mainly pions with surprisingly few strange mesons. The third popular model visualizes the color interaction forming a narrow tube, or "string," between a quark and an antiquark as they fly apart. In this "string model," the string snaps into pieces that become particles, a little like a bar magnet with a north and south pole which, when snapped, gives two bar magnets each with a north and a south pole. Gluons introduce kinks in the string that are manifested in the angular distribution of the final detected particles. This approach gives a very satisfactory account of the angular spread of particles in three-jet events in electron–positron collisions. However, the fact remains that hadronization is a log jam – while particle physicists are forced to rely on hadronization models, there will always be limits to what they can understand about the physics of particle jets.

To paraphrase MIT's Robert Jaffe, hadronization means that colliding together a proton and an antiproton, then studying the debris, is like smashing together two Swiss watches and finding only more Swiss watches, with no sign of cogs, springs, or other fragments. Likewise particle physicists don't see quarks and gluons, only fully formed composite particles – more Swiss watches.

THE WORLD OF JETS

"When you're a jet, you're a jet all the way" so they sing in *West Side Story*. The trouble is, it's not always so easy to see that jets really are jets, let alone to actually count them. For example, it's common enough for some collision to produce a distribution of particles that might be two fat jets, or one fat and two thin jets. The distinction is important, since the number of jets produced tracks directly back to the core process of the interaction.

The result is a whole collection of exotic-sounding quantities that researchers have invented over the years to attempt to quantify "jetiness," and to disentangle one jet from another. Sphericity, for example, combines the momenta of each and every observed particle, giving a single number that is zero if all particles can be aligned along an axis, in other words a fully two-jet-like event, or one if the particles emerge evenly in all directions. Sphericity, originally suggested by Brodsky and Bjorken, was how MARK I experimenters at SLAC's SPEAR electron–positron collider pinned down the first quark jets back in 1975. The opposite to sphericity is "thrust" which, instead of looking at particle momenta perpendicular to a selected axis, focuses on momenta parallel to an axis: a two-jet event with back-to-back jets has a thrust of one, and a spherical event a thrust of zero. Where sphericity and thrust are ways of identifying whether or not an event as a whole is jet-like, researchers have also evolved other strategies to distinguish, for example, a fat jet from two thin jets. One way of identifying a jet is to see if more than a certain amount of energy is deposited into some cone emanating from the collision point, this energy being measurable in a detector's calorimeters. Another way is to devise some energy and flight direction-related number that can be calculated for every pair of particles. The pair giving the smallest value is taken as a cluster, and this process is then repeated treating this cluster as a single particle. Eventually, after enough repeat calculations, all particles are assigned to clusters, with those clusters remaining at the end being the jets.

A great place to see jets is in an electron–positron collider. An electron and positron annihilate to give a virtual photon (or a Z at sufficiently high energies) that can decay into a quark–antiquark pair. These should then materialize into two back-to-back jets of hadrons, their direction mirroring that of the original quark and antiquark, and this is exactly what experimenters see. They also see that, as they pump up the collision energy, the jets become increasingly jet-like, in other words the relative spread of each shower about the overall shower direction diminishes, making each jet more pronounced, in line with expectation.

Tightly collimated jets allow experimenters to measure the angle that each jet makes relative to the colliding beam. If jets are from quarks and antiquarks that are spin 1/2 entities, like muons and antimuons, this angle should follow the same spin-dependent pattern as that of muon–antimuon production via electron–positron annihilation. SLAC results in 1975 confirmed that this is indeed the case, providing powerful evidence in support of spin 1/2 quarks.

With enough energy, the next layer of complexity is revealed, the radiation of a gluon by one of the emerging quarks or antiquarks. The result is a three-jet event, as CERN theorists John Ellis, Mary Gaillard, and Graham Ross realized in 1976. The tell-tale three-pronged event was first seen at the PETRA storage ring at DESY in 1979, Figure 6.1. This was the first visual evidence for gluons and marks the experimental discovery of the gluon, though there followed an unseemly squabble over which of the four PETRA detector groups got there first.

Compared to two jets, calculating the probability of a three-jet event involves an additional appearance of the strong coupling parameter, in line with the new quark–gluon–quark junction, Figure 6.4. Complications aside, the ratio of the two- to three-jet cross-sections therefore gives a value of the strong coupling parameter relatively free of theoretical uncertainties, a route that both SLC and LEP experimenters in particular have taken to measure the strength of the strong force. Three-jet data also show that the gluon has a spin of one, the spin 1 gluon predictions giving convincing agreement with experiment, whereas a spin 0 gluon version is clearly ruled out of court.

And with three or even more jets, it becomes possible to look for differences between jets arising from quarks or antiquarks, and those originating as gluons. Again in line with expectation, experimenters find that gluon jets contain more particles and are fatter than quark jets, two consequences of gluons' greater effective color charge causing more gluon radiation. Gluon jets are also found to have a total electric

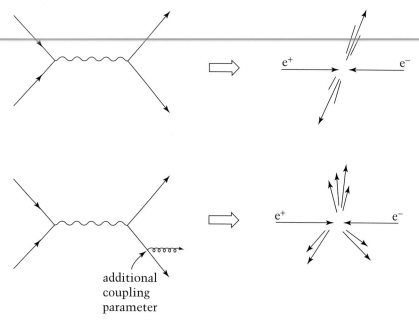

FIGURE 6.4 Two- and three-jet events.

charge consistent with zero, whereas jets from quarks have a charge different from zero.

Another clue to the difference between quark and gluon jets is that, for a three-pronged event, the gap between the quark and antiquark jet directions is less populated with particles than the gap between either of these jets, and the gluon jet, Figure 6.5(a). This effect can be traced back to a destructive interference effect, called color coherence, involving "soft," that is low-energy, gluons emitted from a quark or antiquark. A consequence of color coherence is that successive gluons are emitted from a quark at ever-decreasing angles, Figure 6.5(b), giving an angular ordering that has a counterpart in QED for the emission of photons.

Higher energy colliders, first TRISTAN at KEK and then in particular CERN's LEP and SLAC's SLC, make for one of the best experimental tests of QCD. Crucial to the whole structure of the theory is its non-Abelian nature, which translates into the prediction that gluons interact with other gluons, as distinct from Abelian QED in which photons do not interact with other photons. In other words, QCD predicts the three-gluon vertex. To spot this means measuring four-jet events,

(a)

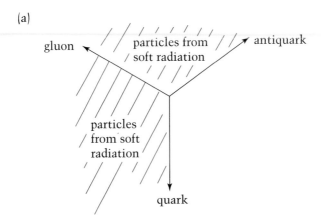

(b)   emitted gluons

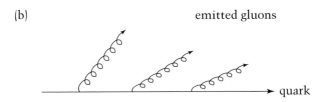

FIGURE 6.5 Color coherence: viewed along the beam axis, the electron–positron three-jet event has a pattern of soft radiation particles dominated by color coherence (a) – an interference effect that gives rise to angular ordering (b).

and then being able to count events involving the three-gluon vertex against a backdrop of those of "more mundane" origin, Figure 6.6. The first hint of a four-jet event involving gluon self-coupling came in 1989, from the AMY collaboration at KEK. By the early 1990s, LEP was mass-producing three-gluon vertex events, enough to allow a detailed study, together with searches for direct evidence of the "color group," already discussed in Chapter 4. The secret to spotting the three-gluon vertex is to explore carefully the angles between the jets, since a gluon spawning two further gluons has a distinctive spin composition, revealed in the angular spread of the resultant jets.

Electron–positron colliders have not annexed the jet market entirely. A proton colliding with an antiproton can, according to QCD, proceed via a quark or antiquark from one interacting with a quark or antiquark with the other through the exchange of a gluon, resulting

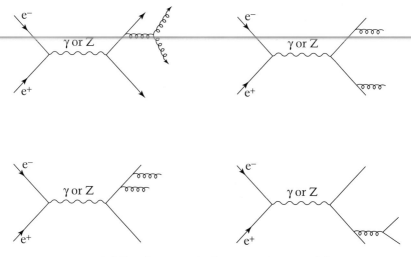

FIGURE 6.6 Four-jet events in electron–positron annihilation.

once again in the creation of jets, Figure 6.7(a). Another possibility is for a quark from one to scatter off a gluon from the other, Figure 6.7(b); there are many possible scenarios, some involving three- and four-gluon vertices. Seeing these jet events clearly, however, is really only possible when the jets are energetic and produced at large angles relative to the incoming beams. It wasn't until 1982 that unambiguous evidence for such jets was found, by the UA2 collaboration at the then-new $Sp\bar{p}S$ at CERN. At still higher Tevatron energies, jets are even more sharply delineated, with plenty of events having two, three, four, or even more jets. The number of two-jet events seen at the Tevatron, their energy and angular distribution, and a host of other jet cross-sections all support QCD predictions.

Hadron–hadron jets do bring with them their own special challenges, however, since the two colliding species are effectively bags of quarks, antiquarks, and gluons rather than simple, elementary particles. The total collision probability is dominated by relatively feeble collisions; only the most violent collisions give what practitioners term the "hard scattering" that firstly gives rise to jets of particles at large angles relative to the beam and secondly relates in some simple way to elementary quark and gluon scattering processes. Complications then include the fragments of the smashed proton and antiproton, the beam jets, which can confuse the disentangling of the collision debris into

(a)

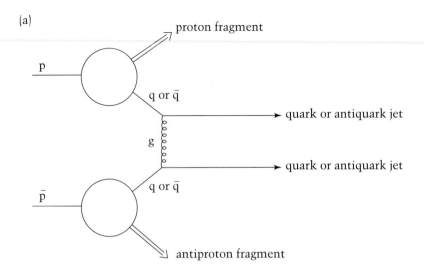

(b)

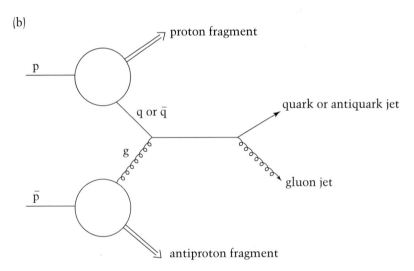

FIGURE 6.7 Two-jet events from proton–antiproton collisions, showing (a) scattering via a gluon or (b) scattering of a quark or antiquark with a gluon.

quark and gluon jets. Another issue is that calculating collision probabilities means blending in probability distributions of quarks, antiquarks, and gluons within both the proton and antiproton, in other words making use of the structure functions already introduced in

Chapter 4. This is a whole new Pandora's box, to be opened up in Chapter 8.

GLUONIC TEASES

Any jet involves the radiation of gluons. Incorporating all of these radiated gluons exposes some of the subtleties and pitfalls of working with QCD. One such sting in the tail ensues when a gluon is radiated from a quark in exactly the same direction as the quark, in other words collinear with it. Another, related, case is when the radiated gluon is so feeble, so soft, that its energy is effectively zero. The consequence of either of these apparently innocuous scenarios is the *bête noire* of perturbative field theory, an infinity. Because both sources of infinity are to do with low energies, these are examples of what theorists' term infrared divergences. Both collinear gluons and soft gluons threaten to reduce calculations to a rubble of infinities. And in both cases the root of the problem is that gluons have no mass.

Incredibly, these infinities cancel out with yet another infinity lurking just along the calculational road. The simplest Feynman diagram for electron–positron scattering into quarks and antiquarks, $e^+ + e^- \rightarrow q + \bar{q}$, is shown in Figure 6.8(a). The next most complicated diagrams, to be incorporated in a more refined calculation, include some with closed loops involving gluons, Figure 6.8(b). These closed loops have already been a source of angst in the form of ultraviolet divergences, subjugated via the renormalization program. But that leaves unresolved trouble at the other end of the energy spectrum: those same loops give infinities as the momentum round the loop drops to zero, a problem that is again traceable to gluons having no mass. The fact that these infinities, and those linked to soft and collinear gluons, share a common cause paves the way for the mutual annihilation of two disasters: by adding together contributions from gluon loop diagrams, Figure 6.8(b), and real gluon radiation diagrams, Figure 6.8(c), the infrared infinities cancel.

This, surely, has to be a fudge? The loop diagrams apply to the physical process $e^+ + e^- \rightarrow q + \bar{q}$, the gluon emission diagrams to an apparently *different* process $e^+ + e^- \rightarrow q + \bar{q} + g$. The resolution of what appears to be a paradox is that, for soft and collinear gluons, in a real experiment with a limited ability to resolve particles, the process $e^+ + e^- \rightarrow q + \bar{q} + g$ is completely indistinguishable from $e^+ + e^- \rightarrow q + \bar{q}$. The key to making the infrared divergences vanish is to

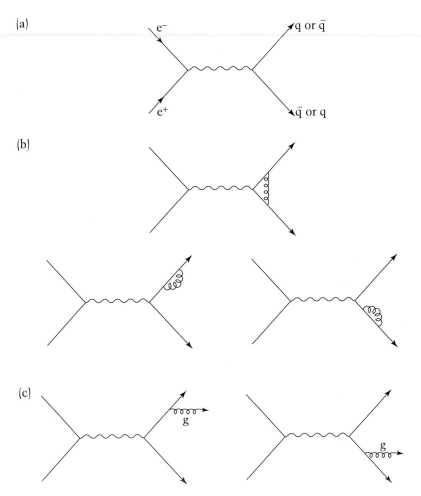

FIGURE 6.8 Simple gluon emission in electron–positron annihilation.

sum over indistinguishable final states, a recipe that also works for combinations of more complex Feynman diagrams. It's a strategy that corresponds to removing what is really an artifact introduced by using perturbation theory. It's just as well this problem, which has a progenitor in QED, does have a resolution: virtually all calculations in QCD are potentially threatened by infrared divergences. So if these cancellations didn't happen, theorists' calculations just wouldn't make much sense.

A further subtlety, likewise evident in calculations of $e^+ + e^- \rightarrow q + \bar{q}$, is also a rare instance of the free lunch in QCD. To calculate the cross-section for this process in full would mean calculating an infinite number of Feynman diagrams, starting with those in Figure 6.8 and embracing diagrams of ever-increasing complexity. But the simplest loop diagrams have a familiar look to them: corresponding diagrams appeared in the discussion of renormalization, in the context of QED. In fact, the QCD loop diagrams are part of the set of diagrams that contribute to the key relation between rising energy and the falling QCD coupling strength. So, if in a simple cross-section calculation theorists use that running coupling parameter in place of a fixed coupling constant, they are straight away incorporating a whole string of related diagrams. In fact, they are including a sum over a subset of all possible diagrams, and that subset is itself infinite. This is the renormalization group at work: the free diagram summation is just enough to remove any dependence of the cross-section calculation on the renormalization scale.

However, the cross-section, or whatever other process is under study, has not had all of its Feynman diagrams mysteriously evaluated and added together. To go further, somebody will have had to explicitly work through diagrams having more loops, giving a more refined expression for the running coupling parameter. At the next level of complexity in a cross-section calculation, this improved expression sums a further infinite swathe of terms. Still more loops yield more refinement in the coupling parameter expression, and corresponding further summation of terms. So, as free lunches go it isn't much. It's more an efficient use of that which is already known.

GLUEBALLS AND OTHER EXOTICA

Gluons interact with one another. So what's to stop them sticking together into a blob containing nothing but gluons? The answer, according to theorists, is nothing: gluons should bundle together to form a new kind of neutral meson called a glueball. The glueball has probably been "discovered" more times than any other particle in high-energy physics.

A glueball not in New York

At the CESR electron–positron collider in Ithaca, New York, experimenters have investigated the properties of a heavy meson with the ugly name $f_J(2200)$. It's a meson with problems. First reported as a relatively long-lived state in J/ψ decay back in 1986, from the very outset its reality has been questioned, and it is yet to graduate to a "confirmed without doubt" status. Its spin is probably 2, but could be 4, nobody is certain, and several of its purported decay routes are listed merely as seen, rather than measured. But it's the relationship between the ways this putative particle is born and then dies that physicists find particularly intriguing.

At CESR, CLEO experimenters have picked over the collision debris looking for pairs of pions, where each pair contains a positive and a negative pion. The $f_J(2200)$ meson can decay into just such a pion pair, and a spike in the number of pairs having a total energy matching the mass of the $f_J(2200)$ would herald the $f_J(2200)$'s creation in electron–positron annihilation. But even though the experimenters saw pions, they found no sign of the $f_J(2200)$.

Which, it turns out, is good news for glueball hunters. The reason is that a regular meson, made of electrically charged quarks and antiquarks, can experience both electromagnetic interactions, via photons coupling with its charged constituents, and strong interactions, via gluons coupling with the color charges on the quark and antiquark. In contrast, a glueball, as a bundle of gluons, contains no electrically charged constituents so is insensitive to photons, though it feels the full brunt of the strong force. This imbalance between strong and electromagnetic forces is measured in terms of "stickiness," the ratio of gluon to photon interaction strengths: a large stickiness value implies a relatively feeble electromagnetic response, and suggests the presence of a glueball.

The CLEO results provide a limit on the electromagnetic interaction of the $f_J(2200)$. To reach a stickiness figure, the CLEO experimenters combined their results with those from the 4.4 GeV Beijing Electron–Positron Collider (BEPC) at BHEP, the Institute of High Energy Physics, Beijing. Here, the BEijing Spectrometer (BES) team *has* seen the $f_J(2200)$, but via an initial J/ψ that then decays to give a photon and a $f_J(2200)$, which in turn expires to yield a pion pair. This final link in the decay chain provides a measure of the strength of the interaction of the $f_J(2200)$ with gluons. Fed into the stickiness calculation, the combined results give the kind of high-stickiness value to be expected from a glueball. These combined results also show how the $f_J(2200)$ appears in electron–positron collisions via a J/ψ, though not directly, hinting that it is a different type of object from an ordinary quark–antiquark pair. Other properties also suggest it is something special: it has never been seen in proton–antiproton collisions, yet the BES collaboration has seen the $f_J(2200)$ decay into a proton–antiproton pair. And at LEP, three-jet events, which rely on a gluon jet, show evidence for an object that could be the $f_J(2200)$ that is absent from two-jet events.

Supporters of QCD-by-computer are quick to point out that their simulations of spin 2 glueballs predict an object having a mass very close to that of the $f_J(2200)$. Couple this with the strands of experimental evidence, and it looks increasingly likely that the $f_J(2200)$ is a glueball. Yet so far there is still uncertainty, and the $f_J(2200)$ is classed merely as a glueball candidate. Whether or not it really is a glueball remains a matter of debate and opinion; an example of how, in science, things are not always so black and white as they might sometimes appear. Despite years of searching, particle physicists are not certain that any of the plethora of glueball candidates are the real thing, even though the glueball's discovery has been announced several times. For example, the spin zero $f_0(1500)$ meson, whose existence is on rather firmer footings than the $f_J(2200)$, was promoted to a leading glueball candidate in 1996. That was when physicists realized old data revealed the $f_0(1500)$ as likewise a product of J/ψ decays, and showed good glueball stickiness: "Forgotten particle fits glueball profile," ran one headline.

And there's the rub – what exactly is the "glueball profile"? In other words, what *should* experimenters expect to see in their detectors when a glueball passes by? It will have no electrical charge, which straight away makes it harder to see. It lacks many of the conserved properties that serve as a guide in many particle interactions, and estimations of its mass rely on computer simulations. And if that wasn't bad enough, untangling and identifying different meson states is incredibly difficult, especially when, as is the case with glueballs, they can "mix" with others having the same quantum numbers. At best, glueball hunters can expect to see a combined signal from a glueball and at least one other particle species.

An exotic find

Across New York State, on Long Island, experimenters at the Brookhaven National Laboratory's AGS proton accelerator have explored the twilight zone between glueballs and "regular" mesons. And they have found more than just a "candidate."

At experiment number E852, circulating protons struck a target placed in their path, reacting via the strong interaction to create pions.

These were drawn off, via a system of magnets, and directed onto a liquid hydrogen target. A pion can collide with a target proton to give a number of reactions: of special interest at E852 was the reaction π^- + p $\rightarrow$ η + π^- + p, which the team measured by identifying the decay of the eta meson into a pair of photons. Combining the masses and energies of the eta and the final pion revealed a peak in the number of observed events. This resonance peak flagged the creation of an intermediate particle of mass around 1300 GeV, which subsequently decayed to give an eta and a pion. It's a meson called the $a_2(1320)$, a relative of the pion, which contributes most of the spike in the particle count rate.

But this was not the end of the story, since the angular distribution of the eta meson was uneven, and best explained if a second particle, having different spin and angular momentum properties, was created alongside the $a_2(1320)$. What makes this new entity, the $\pi_1(1400)$, so exciting is that it makes no sense in terms of the tried and trusted quark model. It is something more, something called an exotic. "We thus conclude that there is credible evidence for the production of a . . . exotic meson," wrote the E852 group in their 1997 paper. Within a year, CERN's Crystal Barrel Collaboration, whose detector is a barrel-shaped array of 1380 light-sensitive cesium iodide crystals, had turned "credible" into "direct." They used CERN's LEAR antiproton accelerator to collide antiprotons onto a deuterium target, focusing their attention on the reaction $\bar{p}$ + n $\rightarrow$ η + π^- + π^0, and finding clear evidence for the $\pi_1(1400)$ via an eta–pion resonance. The first exotic meson was confirmed.

The quark model is quite clear about which quark–antiquark mesons are allowed, and which are not. Including quark spin, and relative motion between the quark and antiquark, the model predicts a spectrum of mesons having particular allowed angular momentum and symmetry properties. However, introducing gluons admits the existence of "hybrid" mesons. These states, for example $q\bar{q}g$ – a new kind of quark–antiquark–gluon state – modify the rules and can possess "exotic" properties, forbidden by the regular quark model but an aid to experimenters seeking to track them down. So-called bag models, to be discussed in Chapter 7, predict a mass of about 1400 MeV for the $q\bar{q}g$ hybrid, very close to the mass of the $\pi_1(1400)$: computer simulations predict higher masses. There are other exotic possibilities, for example four-quark states such as $qq\bar{q}\bar{q}$, or a four-quark meson–meson combination, qq̄–qq̄. The $\pi_1(1400)$, originally named the ρ̂(1405), is presumed to be a qq̄g hybrid, the simplest case having the right properties. A second related meson, the $\pi_1(1600)$, emerged soon after.

Even more recently, 2003 saw the announcement of an exotic particle built of five quarks. This object, christened the Θ^+, has a mass of 1540 MeV and is made of two up quarks, two down quarks, and one strange antiquark, uudds̄. Russian theorist Dmitri Diakonov and his collaborators predicted its existence back in 1997, and evidence for its existence turned out to have been lurking in the available data of several experimental groups. Had somebody told them where to look sooner, the 35-plus year search for a "pentaquark" might have been over sooner.

The discovery of exotic mesons and multi-quark states, and what many researchers take to be overwhelming evidence for the existence of glueballs, reveal new forms of matter that present theorists with a fresh challenge, since these creatures are all beyond the pale of the "normal" quark model. Yet understanding even the simplest particles, let alone the exotics, within the framework of QCD turns out to be a tough assignment.

7 Quarks and hadrons

An atom is, crudely speaking, a core of positive electric charge surrounded by a bunch of negatively charged electrons. The charge on the nucleus is due to protons, which are equal in number to the orbiting electrons, giving rise to the obvious but important fact that, overall, an atom is electrically neutral. Strip off an electron, and what's left is an ion with an overall positive charge. Stripping an electron away from an atom is easy, using heat, light, or electrical discharges, for example; ions are commonplace.

Why, then, are things so much more difficult when we descend deeper into the structure of matter and, instead of thinking about whole atoms, focus on the structure of the protons and neutrons that comprise an atom's core? The idea is that protons and neutrons are made of quarks, combined in such a way that the total *color* charge of a proton or neutron is zero, in analogy with the electrical neutrality of atoms. But whereas it's a simple matter to chip one or more electrons from an atom to create an electrically charged ion, the corresponding task of prizing a quark from a nucleon to leave behind a "color ion" is not just difficult, it's impossible, or so it seems.

The reason, of course, is confinement. The *postulate* of confinement says that only color-neutral, or "white," particles are to be observed in nature. So a proton with a quark simply ripped away is not allowed, nor is a quark thus removed allowed to stay naked and free. Yet this is only a postulate, designed to account for the total non-observation of free quarks and gluons. It has never been comprehensively proved. Given the enormity of the claim, that the central players in the theory of the strong force are forever hidden from view, the fact that it remains unproven is pretty unsatisfactory. Confinement is like a committee that always meets behind closed doors and whose members have never been seen or heard: the issue of confinement rates as one of the most significant unresolved challenges in science.

There is an added dimension, over and above the need to prove a central postulate of the theory. Quantum chromodynamics seems to work nicely for special cases such as certain violent collisions between protons and the production of jets, but these are merely part of the package. The original problem, easily forgotten, is to understand matter at the most fundamental level. Quantum chromodynamics is a theory of the strong interaction that explains the force between elementary, indivisible entities – the quarks – in terms of gluons. Really, though, before theorists can consider their job of explaining the strong force to be complete, they need to be able to use QCD to explain how quarks and gluons combine into the protons and neutrons that make up the cores of the atoms from which our material world is built. They also need to be able to predict the properties of nucleons – and other observable particles – from the theory. Ultimately, the complete picture would explain how protons and neutrons, in turn, stick together to give the nuclei that are the cores of atoms.

Now this is a problem, since quarks inside observable particles, be they nucleons or mesons or any other hadron, are in a relatively low-energy environment. Here, the coupling between them is strong, rendering QCD perturbation theory inapplicable. So both explaining real particles as interacting collections of quarks, and proving quark confinement, are non-perturbative issues, where "non-perturbative" translates into "low-energy, relatively long distance – about one fermi." At the moment, tackling the strong force at these kinds of distances and energies means using computer simulations (a subject called lattice QCD, postponed until Chapter 10), devising clever models, and forging sneaky schemes for separating out the merely hard from the impossible.

CONFINEMENT

One simple approach that nevertheless yields some understanding of how particles work, how they are related to one another, and perhaps some pointers to confinement, is to breathe a little QCD into the quark model, and use the particle zoo as a guide.

The simple confinement comic strip, evocative of the quark fragmentation picture, runs something like this. Imagine that the quark and antiquark of a pion, say, are joined by a string of some kind. The quark and antiquark rotate one about the other at colossal speeds, held a fixed distance apart by the string, their centrifugal motion keeping the string taught. Not only will it have to be a pretty strong type of string but, because confinement is taken to mean that the force between a quark and antiquark remains constant as they are pulled apart, as the

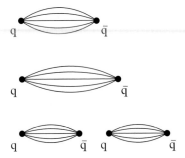

FIGURE 7.1 Cartoon of string confinement. As the quark and antiquark are pulled apart, eventually it becomes energetically favorable to spawn a further quark–antiquark pair from the vacuum.

string is stretched it does not weaken. Instead, the energy added to the system to try and pull it apart is converted into more string, and the string simply becomes longer.

The string does not elongate indefinitely, however. There comes a point when it is energetically favorable to snap, a quark popping into existence at one of the newly exposed string ends, and an antiquark at the other – an idea already encountered in our quark fragmentation discussion of Chapter 6. So now there are two pions, each comprising a quark and antiquark connected by string. For baryons, a slightly more involved version of this string cartoon links the three quarks, but the idea is just the same, Figure 7.1. To make contact with QCD, the "string" is to be pictured as woven from "fibers" that are lines of color force.

What of the gluons? Very close together, the quark and antiquark behave as though they are relatively free; models based on a heavy quark and its antiquark partner supporting a single gluon exchange and a Coulomb's-law-type of approach. As more energy is pumped into separating the quark–antiquark pair, however, the energy goes into popping more and more particles, including gluons, from the vacuum. The swarming gluons cause an attractive force between the gluons mediating the quark–antiquark attraction, in effect sucking the gluon color force lines together. This is a direct consequence of the fact that gluons themselves carry color, and so interact with one another. The more the quark and antiquark are pulled apart, the more gluons enter the fray, and the more the lines of force linking the pair are drawn together. The interaction is no longer feeble, and is directed along a tube of force, yielding the "string" linking the quark and antiquark. The concentrating of lines of color force into a tube is something new, over and above the anti-screening effect, discussed in Chapter 5, that increases the apparent color charge on both the quark and antiquark. Antiscreening, more significant when the quark and antiquark are closer, gives a strengthening of the interaction with increasing separation. The channeling of color force into a tube linking the two, which comes into play as they are pulled still further apart, gives a force that is roughly constant with increasing separation, and which is ultimately held responsible for the pair's confinement.

For some, that is enough of a picture. For others, there is a more fundamental version of the confining string story, one that many believe provides an even closer insight to what really happens in confinement. The model is borrowed from condensed matter physics and the phenomenon of superconductivity, the total loss of electrical resistance of many materials when cooled below a certain material-specific temperature. The new ingredient in a superconductor is a "fluid" of weakly paired electrons which, like dew forming on grass on a summer's evening, condenses out from the electrons of the electron "gas" pervading a metal at higher temperatures. The electron pairs in this fluid, or condensate, as a higher density state created by reducing the energy of some gas is usually termed, are called Cooper pairs after Leon Cooper, one of the trio of American scientists who won the 1972 Nobel prize for their description of superconductivity. The basic idea of a Cooper pair is that one electron deforms the lattice in which it finds itself, and a second electron sees this deformation and adjusts accordingly so as to lower its energy. So a pair of electrons interact indirectly with one another via the lattice, giving a spin 0 entity, a boson, with a charge that's double that of the electron.

Cooper pairs are pretty strange animals. The two electrons of a pair might be hundreds of atom spacings apart, so huge numbers of pairs overlap. This, and the fact that, as bosons, they are freed of the shackles of the Pauli exclusion principle applicable to electrons, means that Cooper pairs all behave in unison – they undergo Bose–Einstein condensation, just like the ultra-cold rubidium atoms of Chapter 3. It's this collective behavior that's responsible for the vanishing electrical resistance, and stands in dramatic contrast to the conventional picture of electrical conduction. In conventional conduction, individual electrons of the electron gas in metals battle their way through the material, bouncing off charged ions at every step, like a commuter rushing through a crowded street trying to reach the metro station. In contrast, conduction in a superconductor bears more resemblance to the synchronized movement of parading troops.

Superconductors have all kinds of amazing properties. Relevant to the confinement story is an effect discovered by Walther Meissner in 1933. Meissner found that a superconductor inside a magnetic field excludes the field from its interior. This so-called Meissner effect is due to electric currents circulating in the surface of the superconductor, which generate magnetic fields that are exactly equal and opposite to the applied field.

Superconductors fall into two categories. A metal such as mercury is termed a type I superconductor because it displays the classic Meissner effect, excluding the magnetic field throughout its bulk. However, there are some materials, such as niobium, that are classed as type II superconductors. These allow filamentous magnetic, non-superconducting regions to thread the bulk material. A vortex of circulating electrical current around each such line creates a quantized magnetic field within. These materials therefore exhibit only a partial Meissner effect, the magnetic field excluded from most of the material, but localized in narrow tubes within, where it is held in place courtesy of the surrounding superconducting material. It's the idea of filaments of field, the field squeezed into tubes by its surroundings, that excites QCD theorists' interest.

Developing the parallels between type II superconductors and confinement in QCD means adding a further ingredient. For a meson, a string linking the quark and the antiquark starts at one partner of the pair, and ends at the other. If tubes of trapped magnetic field are to somehow model this meson string, then there must be a source at one end of the magnetic tube and a sink at the other end. In other words, there should be a lone magnetic pole, say a magnetic north pole, at one end of the tube, and the opposite kind of pole, say a magnetic south pole, at the other. Now this is a problem, since nature seems to have passed up on the chance of manufacturing lone magnetic poles, even though single poles, or monopoles, would have made Maxwell's equations even prettier. Real magnets, so nature decrees, be they bar magnets or solenoids, shall always have two poles, a north and a south. In contrast with these magnetic two-poles, or dipoles, the source of an electric field is a single pole. That is, a positive charge, the electric field source, can exist in isolation, as can a negative charge, the sink for an electric field.

No matter. It's easy enough to invent magnetic monopoles, for the purpose of building a model, and imagine that they live inside a superconductor. Then the magnetic monopole can be the source of a tube of magnetic field, and its opposite number, the antimonopole, can be the sink where the magnetic tube ends. The energy of such a magnetic tube increases as its length increases, and the magnetic field is both quantized and trapped in a tube configuration, all of which are desirable features in a confinement model.

Despite its encouraging features, the model is still not quite right as a model of confinement. The reason is that, in the real world, electric charges come first, so to speak, and magnetic effects second. That is, the force between an electron and a positron, say, depends in the first place on their electric charges. Magnetic forces are a consequence of moving electrical charges, and it's in this sense that magnetic effects are secondary. Crossing to QCD, and the counterpart of electrical charge is color charge, through which quarks and gluons feel one another's presence.

To develop the superconductor model, the way forward is to interchange electric and magnetic fields. This interchanging is an example of what mathematicians call a duality transformation and the result is called the *dual* superconductor: a single pole is then the source of *electric* field and a single antipole is the sink, with the tube of *electric* field maintained by a surrounding sea of *magnetic* monopoles. As a model of QCD confinement, this then becomes a quark and an antiquark, the source and sink of chromo-electric field, linked by a tube of chromo-electric field, the field squeezed into a tube by a surrounding sea of chromo-magnetic monopoles. This, then, is the dual-superconductor model of quark confinement.

The real content is that the confining mechanism preventing quarks and antiquarks from ever becoming free is due to an environment of chromo-magnetic monopoles. The "empty space" surrounding the meson or baryon is superconducting, in the sense of being a condensate of chromo-magnetic monopoles, and via the Meissner effect the chromo-electric field is expelled from the surrounding monopole sea and squeezed into a narrow tube linking the quark–antiquark pair.

Another child of the 1970s, the dual-superconductor model owes its genesis to the efforts of many people, especially Holger Nielsen and Poul Olesen from Copenhagen, Bruno Zumino at CERN, Stanley Mandelstam in California, Alexander Polyakov in Moscow, Yoichiro

Nambu in Chicago, and Gerard 't Hooft in Utrecht. It remains popular amongst theorists, who know that QCD is comfortable with the existence of chromo-magnetic monopoles, in theory at least, even though none have ever been seen by experimenters. But is confinement really due to a Universe filled with monopoles? Nobody yet knows.

MODELS OF HADRONS

The simplest and best-studied atom of them all, hydrogen, has a single-proton nucleus and lone orbiting electron. That electron can inhabit a number of different energy levels in the atom, and quantum mechanics does a comprehensive job of describing those levels. Excite the atom in some way and the electron climbs the rungs of the energy level ladder. As the electron drops back down after a time, the atom emits light or some other radiation, the discrete frequencies mapping to the energy levels of the atom. Studying the emission (or absorption) of light and other radiation is called spectroscopy, and is a powerful way of identifying atomic species in some unknown sample, the spectrum offering a kind of "light fingerprint."

It's possible to imagine an "atom" in which the proton is replaced with a positively charged entity of the same mass as the electron, in other words a positron. Unlikely though it may seem, given that, as particle–antiparticle opposites, electrons and positrons are in the habit of annihilating, such an "atom" is feasible. It's called positronium.

Positronium comes in two breeds. In one, the longer lived variety called orthopositronium, the spins of the electron and positron point in the same direction. In the other, parapositronium, the spins are opposed. Even though positronium lives such a short life (around 1.25×10^{-10} s for parapositronium, a thousand times longer for its spin-aligned variant) it's possible to accurately measure its spectrum of energy levels. The energy levels of positronium mimic those of the hydrogen atom, though with the spacing between the principal levels halved, a consequence of replacing the heavy proton with a lightweight positron.

Viewed at a finer resolution, the main energy levels of hydrogen are in fact composed of several lines of slightly different energy. Several lines where one might do is a consequence of electron spin. Positronium has similar additional "fine structure," due to the electron "magnet" finding itself in a magnetic field arising from the relative motion between it and the positron. There are still more spectral lines to consider. This "hyperfine structure," again present in atomic spectra, is due to the fact that, because *both* the electron and positron have spin and so behave like tiny magnets, they interact with one another magnetically in a way that depends on their relative alignment. Parapositronium, with spins opposed, turns out to have the lower energy of the two, with orthopositronium a mere one thousandth of an electronvolt above it. The measured spectrum of positronium agrees very well with the calculated version: positronium is something physicists understand really well.

What makes positronium relevant is its potential as a model for mesons. One option for understanding the meson spectrum would be a full-blown QCD calculation, but nobody knows how to do that. A simplistic, yet fruitful, alternative is to use positronium as a template: replace positronium's electron with a quark, its positron with an antiquark, and photon-based interactions with those involving gluons, and see if the result resembles a meson. The questions then are, do meson masses show a positronium-like pattern of levels, and how much of the simple quantum mechanics applicable to positronium can particle physicists import into the physics of mesons?

There will be some important differences. Firstly, because in a meson the force binding the quark and antiquark together is so large compared to the electromagnetic forces in positronium, the jumps between energy levels in mesons should be correspondingly huge. The electrostatic force between the electrical charges on the electron and positron in positronium has a strong interaction analog, the "chromo-electric" force between two static color charges. So the Coulomb law electrostatic description of positronium should be replaced by a Coulomb's-law-like chromo-electric force between the color charges on the quark and antiquark. In addition, quarks have spin, so like electrons will behave as tiny magnets able to interact electromagnetically, though it's not this small contribution that's important. What is important is that spinning color charge gives rise to the "chromo-magnetic" force, a color analog of the usual magnetism ascribable to spinning electric charge. This chromo-magnetic force acting between quarks and antiquarks, which will depend on their relative spin alignment, should feature in a family of meson states for which the energy level ladder of positronium is an analog.

A good place to see the chromo-magnetic force at work is the mass patterns of the baryons. The simple quark model leaves some questions unanswered. For example, the classic mass relations of the early quark model fail to account for the difference in mass between a member of the baryon octet and a member of the decuplet having the exact same quark composition. A nice example is the proton, mass 938.3 MeV and quark configuration uud, which is roughly 76% of the mass of the Δ^+, 1231.6 MeV, which nevertheless contains the same quark flavors. The difference can be tracked to the fact that quarks are chromo-magnets. The Δ^+, a member of the baryon decuplet, has all three quark spins aligned. The proton, on the other hand, has two of the three quark spins opposed. This results in different energies due to chromo-magnetic interactions in the two cases, and neatly explains the proton–delta mass difference. The other classic example is the π meson, mass 140 MeV and built from u and d quarks and antiquarks, which is vastly lighter than the ρ meson, mass 770 MeV. Again, both share the same flavor make up. The mass difference comes about through the chromo-magnetic interaction, and the fact that the quark and antiquark spins are antiparallel for the π and parallel for the ρ.

Cramming quarks into a space as small as a pion, which is much smaller than positronium, leads to an additional complication, however. Via the uncertainty principle, the uncertainty in quark momentum is consequently large which, for lightweight quarks, implies near-relativistic quark speeds. The fact that pions – and other hadrons composed of light quarks – are relativistic systems invalidates the strategy of parceling up their internal interaction into distinct spin parts, relative motion-dependent parts, and an electrostatic part. The positronium approach, which amounts to virtually ignoring special relativity, would therefore appear doomed. But the November revolution of 1974 changed all that. Nature has taken pity on particle physicists and provided them with the J/ψ meson, a bound state of a heavy charmed quark and charmed antiquark. Because the charmed quark and antiquark are so massive, they sit close together, ensuring that the binding between them is relatively modest. They are also heavy enough, compared to the energy holding them together, that they are quite slow moving, so to a fair approximation the effects of relativity can legitimately be ignored. In addition, with heavy quark systems, it's more reasonable to assume all the gluons, and the quarks and antiquarks comprising the mesons' "sea," go about their business very rapidly compared to the lumbering, heavy (valence) quarks, so the sea's net impact can be smeared out to give a kind of average interaction holding the quark and antiquark together. Things are even rosier for the upsilon meson Υ, which is made of still-heavier bottom quarks that are even less relativistic.

The J/ψ is one of a whole brood of mesons which, aping positronium, are dubbed charmonium states, or simply charmonium. All are built of the same charmed quark and charmed antiquark pair, but have different masses and different quantum labels. The least massive is the η_c, at 2979.8 MeV, compared with 3096.9 MeV for the J/ψ. The η_c has its quark and antiquark spins in opposition, whereas those of the J/ψ are aligned parallel, so the relationship between the η_c and J/ψ mimics that between parapositronium and orthopositronium. And because its components' spins are opposed, the η_c has the wrong quantum numbers to be made directly in electron–positron collisions. It emerges, for example, in the electromagnetic decay process $\psi(3686) \rightarrow \eta_c(2980) + \gamma$. Below a mass of 3730 MeV, there's a string of relatively long-lived $c\bar{c}$ states, including the χ_{c0}, χ_{c1}, and χ_{c2} mesons, and the confirmed $\eta_c(3594)$. The figure of 3730 MeV is significant because charmonium states above this value have so much mass that they can decay into charm-carrying D mesons, the lightest of which is the D^0 with a mass of 1864.6 MeV. Those states too light to decay rapidly into D mesons can be measured very accurately, and the spectrum of their masses tabulated and set alongside that of positronium. The same is true for the lower mass states of the Υ meson system which, based on the bottom quark, is called bottomonium. Mesons based on top–antitop pairs would be even better, if the top quark lived long enough to give them life: bottomonium is as good as it gets.

Positronium turns out to be a pretty good prototype for both the charmonium and bottomonium states below their respective thresholds for D and B meson production. The spectrum of meson masses, annotated with their various quantum labels, bears a similarity to the spectrum of positronium energy levels, though the actual energies involved are very different. The basic physics of the charmonium and bottomonium systems is the same, and results from charmonium studies have been used to predict some bottomonium states. In fact, the program also lends weight to a description of mesons in terms of a quark and antiquark, and an understanding of the relative meson masses in terms of a chromo-magnetic interaction between spinning, colored quarks and antiquarks.

For the lighter states at least, a Coulomb's-law-like interaction, commensurate with single gluon exchange, is sufficient, but the higher mass states demand something more. By working backwards, and shoe-horning the model to match observed particle masses, theorists have found a modified interaction energy that performs better for heavier particles. This additional portion of the interaction energy doubles as the quark–antiquark separation doubles. And this is a vital clue.

One way of making confinement happen is to ensure that the energy needed to separate a quark and an antiquark pair is infinite. And one way to ensure this is for the interaction energy between them to increase in step with their separation. Pulling them apart then involves inserting so much energy it becomes energetically favorable for the system to "rearrange" and, instead of splitting into a free quark and antiquark, initiate further quark–antiquark pairs from the vacuum, as discussed above. Charmonium and bottomonium suggest a single gluon exchange at short distances, rising to a confining type of interaction for larger quark–antiquark separations relevant to the more excited states.

An interaction energy rising in step, in other words linearly, with separation is the same thing as a constant force: it's as though there were a string joining the quark and antiquark, a string whose tension remains constant as the pair are pulled apart. The tension in that string, which can be deduced from matching the model to the charmonium and bottomonium states, comes in at around fourteen tonnes. That's the weight of a truck pulling on a string less than the size of a proton. The strong force really is strong.

Baryon bags

So much for mesons built from heavy quarks. What about ordinary baryons, and in particular protons and neutrons? These are a tougher assignment still, since there are three main (lightweight) players, not just two as in mesons, and tackling even a "simple" problem with three moving, interacting parts is much harder. A full-blown direct mathematical assault on baryons is inefficient.

The trick is to try and build QCD models of baryons. One of the earliest attempts at applying QCD to nucleons is the MIT bag model, proposed in 1974 by Kenneth Johnson and his colleagues at MIT. The idea is simple: take confinement at face value, and model the baryon as a bag, or bubble, from which quarks cannot escape. An analogy comes from a kettle in which the water is about to boil. Bubbles of water vapor form in the liquid, and inside each bubble the water molecules interact only feebly. In a similar way, the bag model envisages a baryon – or meson – as a bubble in space, the baryon bubble populated with freely moving quarks. But creating hadronic bubbles demands something a little more drastic than a kettle. A particle accelerator, for example, is one way of dumping enough energy into a sufficiently tight space to blow a bubble of quarks. The energy needed to create a bubble, called the bag constant, is about 55 MeV for every cubic fermi of bubble volume. Crucially, this inflation energy is proportional to the bubble volume, so that an infinite amount of energy would be needed to simply keep on inflating the bubble. What stops the bubble from collapsing and relinquishing its energy is the pressure within, due to the quarks racing around inside. Color confinement amounts to the constraint that color sources, in other words quarks and gluons, and lines of color force, never cross the bubble surface, but are always contained within. And, of course, the total contents of the bubble are color-neutral. The mathematics boils down to solving Dirac's equation for massless quarks incorporating the special conditions at the bubble surface.

"Baggers" have chalked up successes, explaining the relationship between hadron mass and total spin, and the basic chart of hadron masses. They can also offer a picturesque model of processes such as a meson fragmenting. Then again, some of their achievements are not unique – the string model, introduced above, in particular shares some of their turf. Strings and bags are a little crude and unphysical, and face a number of difficulties, but they have one great asset: they capture a key thread of the plot in terms of a problem that theorists can actually *solve*, and for a theorist this is good thing.

OF PENGUINS AND BROWN MUCK

Figuring out how a proton works in terms of its component parts is tough because those components are going about their business in what is, by QCD standards, a relatively big box. It's a comparatively genteel environment where quarks and gluons hug each other tight: pure perturbation theory will not do for QCD here.

The question is, are there special circumstances where at least part of the action is amenable to perturbative QCD? Is it possible to hive off the short-distance high-energy aspects, where theorists can justify using their powerful Feynman diagram perturbation theory, from the less tractable longer distance, lower energy parts? The answer is yes, and the piece of mathematical machinery that shows how it's done labors under the epithet "operator product expansion," or OPE for short.

The idea of the OPE is to split the physical phenomenon under scrutiny according to length scales. Features characterized by length scales exceeding some cut-off assume the status of "background" processes of some kind. Features possessing a length scale shorter than the cut-off contribute "likelihood measures" that determine the relative significance of the larger scale effects.

An example of an interesting process possessing two different length scales is deep inelastic scattering of electrons on protons. Here, the exchanged probe photon has a wavelength that's tiny compared to the overall size of the proton. So the two disparate length scales are the photon wavelength and the proton size.

In close-up, deep inelastic scattering is then recast in terms of scatterers that bind together in some mysterious way to constitute a proton, and likelihood factors that amount to distributions of scatterers within the proton. The likelihood factors, which depend on the probe photon's wavelength, become so-called structure functions describing the distribution of quarks inside the proton. The mystery surrounding the way the scattering centers constitute the proton is a reminder that to get the full story would mean having a complete low-energy theory of how quarks bind together into protons, something we do not have. Still, the division remains useful, and those structure functions, to be revisited at length in Chapter 8, reveal a great deal about the proton, even if they cannot be calculated from scratch.

Though it might appear at first sight that the OPE is a simple splitting ploy, it's much more than that. The clue to part of the extra content is that the division is based on length scales, where the choice of division into short and long is arbitrary, yet the physics is the same. This is renormalization group country. Indeed the likelihood measures are slave to the renormalization group. This means that once they are calculated at some vanishingly fine length scale using perturbation theory, the camera can "pan out" to give some insight into what happens when the length scale is inflated, perhaps to match that of an experiment.

War offers an analogy. Imagine that the armies of each side are structured in the conventional way, with an intermediate scale unit, the platoon comprising say 30 men, made up of perhaps three squads of 10 men each. In turn each platoon, led by its own lieutenant, is part of a larger company, which is ultimately part of a battalion. The lieutenant can then interpret his own little patch of war using his "platoon length scale" information to determine the relative importance of broader issues of a national length scale, for example the technological sophistication of the two warring sides, public sentiment at home, and the existence or otherwise of ally nations. As a specific, a lieutenant will see "on the ground" the effect of his own troops being equipped with M16s facing an adversary armed with ancient hunting rifles.

The platoon analyses the war at a small scale – it could be smaller, say squad or even individual soldier level – and the photon analyses the proton at a short scale. The backdrop to war – technology, public opinion, and allies for example – mimics the scattering backdrop provided by a complete proton. It's even possible to imagine a military version of the renormalization group, which shows how to scale up the lessons of the platoon to the level of a battalion, offering a perspective on the conflict of interest to generals and cable news channels.

The operator product expansion actually predates QCD. It was introduced by Kenneth Wilson in 1969, ironically as part of an attempt to invent an alternative to field theory. One of its first important outings was courtesy of Columbia University's Norman Christ and his colleagues in 1972, and by Gross and Wilczek two years later, who used the OPE and the renormalization group to discuss Bjorken scaling in deep inelastic scattering and – significantly – departures from perfect scaling, another topic to come under the microscope in Chapter 8.

The Russian revolution
A closer look at the way photons disturb the vacuum in which they find themselves shows how to work up the OPE into a powerful tool for estimating properties of strongly interacting particles. An electron might scatter off another electron, for instance, a simple enough electromagnetic process pictured in terms of a Feynman diagram showing a single exchanged virtual photon. Through quantum fluctuations, that virtual photon can spend part of its time as a quark–antiquark pair, however. This is vacuum polarization, discussed in Chapter 3: the classic vacuum polarization Feynman diagram of Figure 3.8(b) gives a correction to what is otherwise a straightforward QED calculation for the $e^- + e^- \rightarrow e^- + e^-$ scattering cross-section.

That's pretty much the story when the photon's wavelength is very short, and the quark–antiquark emergence from the vacuum and subsequent adsorption back into it occur over very short distances and times. But what happens if the spacing between the quantum birth and death is a little longer? This is the experience of the quark–antiquark pair associated with the virtual photon in the related process of electron–positron annihilation, when a positron and electron meet and destroy one another. With the increase in overall length scale, quark–gluon interactions, suppressed at short distances by asymptotic freedom, become increasingly important. Now the quark and antiquark can materialize into real particles, for example ρ or ω mesons, or even heavy quark J/ψ or Υ mesons, revealed as resonance peaks in the electron–positron annihilation cross-section.

The question facing theorists is how to extrapolate from the well-understood case, the straight quantum fluctuation and its quark-loop diagram, to larger distances and something more representative of actual strongly interacting particles. This was the question three Russian theorists, Mikhail Shifman and Valentin Zakharov at Moscow's Institute of Theoretical and Experimental Physics (ITEP), and Arkady Vainshtein at the Budker Institute of Nuclear Physics in Novosibirsk, asked themselves. The answer, they realized, lay in the OPE.

When the quark and antiquark have short wavelengths then perturbation theory is on firm ground. The trouble starts when their wavelengths are allowed to grow. Then the quark–antiquark pair become enmeshed in a molasses of gluons and further quarks and antiquarks all ultimately spawned from the vacuum. Here is the beginnings of a division according to length scale that is central to the OPE. Shifman and his collaborators argued that, at longer lengths, these quark and gluon effects can be bundled together as (approximately) constant background influences that cannot (yet) be calculated. The short distance quark–antiquark combo acts as a kind of sensor that responds to the relatively longer distance backdrop of vacuum quarks and gluons. At short lengths, regular perturbation theory describes the quark and antiquark, giving the likelihood factors in the OPE.

In essence, a short-distance snapshot of a hadron is recast as a series which, step-by-step, embraces more and more of the (nonperturbative) background. Actually it's not just a single hadron, but a whole slew of hadronic states, starting with a ground state such as the ρ or J/ψ, and including more excited states and other objects based on the same quark–antiquark combinations – after all, nobody told the emerging quark and antiquark which particle they were supposed to form. And the result amounts to a constraint on the aggregated

FIGURE 7.2 Mikhail Shifman (top left), Valentin Zakharov (top right), and Arkady Vainshtein (left), pictured here in the Soviet Union at around the time of their work on QCD sum rules. Credit: Courtesy Mikhail Shifman.

hadron parameters. It's a sum rule, in other words. This is the QCD sum rule approach, first developed by Shifman and his colleagues in 1978 (Figure 7.2).

So the starting point in the QCD sum rule method is a place where QCD is known to work because asymptotic freedom holds. The result shows that when QCD is pushed into places where it creaks under the strain, where quark–gluon interactions become stronger, vacuum-based non-perturbative contributions emerge in a systematic way.

A likely story – how can rewriting one intractable problem as another apparently intractable problem, relying on vacuum black magic to provide answers literally from thin air, move the plot forward? Actually it does. First off, as we shall discover in Chapter 9, there's much more to the vacuum that mere empty space. Secondly, though the averaged-out effects of quarks and gluons that live in that empty space cannot be calculated from cold, useful values come from experiment, for example pion decay, which measure processes linked in some way to vacuum properties. By way of example, the QCD sum rule approach yields a ρ meson mass of about 750 MeV, compared to a measured value of 770 MeV, and a decay constant value of 213 ± 20 MeV compared to an experimentally determined value (via the decay $\rho \rightarrow e^+ + e^-$) of 216 ± 5 MeV. In fact the QCD sum rule prescription has handed theorists the power to calculate properties such as quark masses, decay properties of mesons and baryons, magnetic moments, and much more besides. The original paper by Shifman and his colleagues, which consumed an entire volume in *Nuclear Physics B* in 1979, has now garnered more than two thousand citations in the scientific literature, making it one of the classic papers of QCD.

Shifman and his colleagues were having a pretty productive run in the 1970s, quite an achievement given the awkward circumstances they endured. Incoming journals from the West suffered long delays due to censorship, and getting permission to publish in a Western journal could take months. Even the conventional practice of submitting a pre-publication version of a result, a preprint, at the same time as the paper was submitted to the journal, was a challenge in its own right, according to Shifman. The rules dictated that a paper was not allowed to appear both as a preprint and in a journal, which rather flies in the face of the way the rest of world sees the process. So Russian authors tinkered with titles and the listings of authors names, amongst other things, to fool the censors into thinking that the preprint and the partnering paper were actually different. Occasionally papers were effectively smuggled out via rare Western contacts. The paper setting out QCD sum rules was an example of a paper whose corresponding preprint actually emerged as a series of preprints, following one after another, to beat the length impositions.

Added to all of that, they of course faced the usual problems confronting any author submitting papers for journal publication: the reviewing process. Though Shifman and his collaborators published their sum rules work in *Nuclear Physics B* in the end, they were tempted to run with an alternative, stung by the difficulties they encountered with an earlier paper that had languished at the hands of the journal and its reviewers for two years. Then again, that earlier paper featured penguins.

Weak windows on the strong world

All is not well in the strange world of strange K meson decays. In particular, the K^+ can decay into a pair of pions according to $K^+ \rightarrow \pi^+ + \pi^0$. There's not a lepton in sight, only strongly interacting particles in the finish, for which reason this is termed a non-leptonic decay. The neutral kaon is also seen to decay into two pions, $K_S \rightarrow \pi^+ + \pi^-$. A quirk of neutral K mesons is the source of the "S", which stands for "short-lived": neutral kaons have a short-lived form that survives about a thousandth of the time of the long-life or "L" form. The long-life version commonly decays into *three* pions.

Never mind three pions, however, it's the two-pion case that's more interesting. Experimenters know that the decay $K_S \rightarrow \pi^+ + \pi^-$ is about 450 times more likely than the decay $K^+ \rightarrow \pi^+ + \pi^0$. What makes this apparently arcane fact interesting is that the two cases, both involving weak decays and pretty much the same set of quarks, should be roughly the same. Indeed as long ago as 1964 Julian Schwinger estimated that the K_S option should be just 9/4 times as likely at the K^+. The discrepancy between these two numbers, 450 and 9/4, is so big and so well-known it has a name, though it's a name requiring a certain amount of preamble: in the $K^+ \rightarrow \pi^+ + \pi^0$ case, total isospin changes by 3/2 whereas in the $K_S \rightarrow \pi^+ + \pi^-$ case isospin changes by just 1/2. So the "change of isospin by 1/2 case," written $\Delta I = 1/2$ (where the ΔI part is simply shorthand for "change in isospin I") is far more likely than the 3/2 case. The $\Delta I = 1/2$ route, $K_S \rightarrow \pi^+ + \pi^-$, dominates. This is the "$\Delta I = 1/2$ rule." Understanding it is still something of an open problem.

The thing is, there's no cunning symmetry scheme to explain this rule. The trick is to instead acknowledge the role of the *strong* interaction. Strong interaction effects enter in two ways. Firstly, the decay depends on W bosons, and since these are so heavy they act over short distances where their exchange is subject to short-range gluon effects. Secondly, over longer distance scales the emerging quarks will in turn form up into the observed pions, and strong interaction effects enter here too.

Putting a positive spin on this, however, and looking at the issue from the other end shows that weak interations, for instance in kaon decay, are actually a way of accessing information about the *strong* forces at work within hadrons. Weak decays offer a window on the strong workings of the hadron.

Kaon decay is an especially dramatic and significant example. To try to understand the impact of short-distance strong-interaction effects in what appear at first glance to be purely weak processes, Mary Gaillard and Benjamin Lee at Fermilab, and Guido Altarelli and Luciano Maiani in Rome, turned to the OPE. Their pioneering efforts, published in 1974, launched an industry.

The hefty W marks out a short-distance scale by virtue of its mass. The idea is to bundle such short-distance detail, which might also include heavy quarks, into various "constants" of a revised theory involving just the three light quarks, photons, and gluons, yielding an effective field theory in the spirit of Chapter 3. In fact, what looks like a retreat to Fermi's old theory of weak interactions is actually a generalization. The magic new ingredient is the renormalization-group-enriched OPE. This shows how to build the effective theory properly, without losing the impact of those features that no longer appear explicitly, such as the W boson. QCD provides a useful description of the strong interaction contribution at these short distances, since here the strong coupling parameter is small by virtue of asymptotic freedom. The "constants," calculated using perturbative QCD, then evolve into those applicable at the inflated length scale of the final effective theory courtesy of the renormalization group.

Gaillard, Altarelli, and their respective co-workers did indeed find that strong interactions enhance the K_S decay relative to the K^+. That's good. Unfortunately, the enhancement they found was nowhere near enough. Over in Moscow and Novosibirsk, Shifman, Vainshtein, and Zakharov realized that something was missing, that theorists elsewhere had not fully appreciated the impact of the recently discovered charmed quark. Incorporating the charmed quark in turn led them to an entirely new species of diagram in their version of the effective theory for K meson decay. These were the penguin diagrams. Figure 7.3(a)

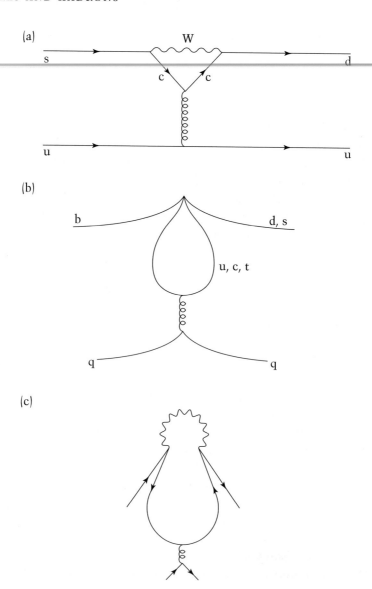

FIGURE 7.3 The three penguins: (a) shows a penguin diagram as used by Shifman and his collaborators in their work on kaon decay; (b) is a penguin diagram linked to B meson decay as envisaged by Ellis and his colleagues. The W boson line visible in (a) is closed up to yield a point vertex in this effective theory; (c) is a recasting of (a) that helps justify the name "penguin". Credit (c) only: Courtesy Johan Bijnens and World Scientific.

shows a penguin diagram along the lines of that presented by Shifman and his collaborators in 1975, with a quark loop linked to another quark via a gluon. They found that the new penguin diagrams contribute only to the $\Delta I = 1/2$ route, going one step further to explaining the enhanced K_S decay. Even this turned out to be insufficient, however, and the $\Delta I = 1/2$ rule remains an enigma. But the world now has penguins, whose reality was first established through B meson decay and whose significance is manifest in numerous weak decay processes: penguin diagrams are vital in understanding CP violation in B mesons, first seen at BaBar and BELLE in 2001.

But why penguins? In 1977, CERN theorist John Ellis was working with Mary Gaillard, Dimitri Nanopoulos, and Serge Rudaz on the properties of B mesons, the b-quark-based Υ having just been discovered at Fermilab. They invoked a similar mechanism to that of the Russian group, along with their own distinctive version of the corresponding diagrams, Figure 7.3(b). One evening Ellis, Rudaz, and CERN student Melissa Franklin went to a bar, where Ellis and Franklin played darts, betting that if he, Ellis, lost he would have to insert the word "penguin" into his next research paper. Franklin had to leave, and Serge Rudaz actually finished the game on her behalf, beating Ellis in the process. John Ellis felt obliged to honor the bet, but for a while couldn't see how. The answer came to him when one evening, leaving CERN, he dropped by to visit some friends where he smoked an illegal substance. Later that night, working again on his paper, in a moment of revelation he saw that the diagrams looked like penguins. The name went into the paper, and has stuck ever since. Clearly penguins are rare in Switzerland.

Mesons with heavy hearts

That Υ discovery handed theorists a truly heavy quark to play with. In a B meson, for example, the bottom quark (or antiquark) is several hundred times the mass of its accompanying light (anti)quark. In such cases, the hydrogen atom is a better model than positronium. This is because in hydrogen, the proton is a couple of thousand times heavier than the electron and, to a reasonable approximation, is just an immobile source of positive charge that hugs the electron to itself. In a similar way, the bottom quark in a B meson, and to a lesser extent the charmed quark in a D meson, is a static source of color charge that binds the accompanying, nimble light quark to itself. Theorists describe such mesons as hydrogen-like.

At the simplest level, therefore, the mass of the heavy quark should no longer feature. Sure enough, the properties of the D and B mesons are related, the fact that they have different flavors of heavy quark making little difference. This is analogous to atomic physics, where different isotopes have the same chemistry, even though their nuclear cores have different masses. The spin of the heavy quark also bows out, and again there's a lead from atomic physics. Flipping the spin of the hydrogen atom's proton corresponds to a well-known feature of hydrogen's radio-frequency spectrum, the so-called 21-centimeter line, and an energy jump just one millionth that of a typical electron transition. In much the same way, flipping the spin of the heavy quark in a hydrogen-like meson, which would be manifest through the chromo-magnetic force, has a negligible impact. The reason, in both cases, is that the (chromo) magnetic moment of the heavy partner is suppressed by its large mass, so flipping its spin orientation translates into only a tiny energy hop on the ladder.

In the late 1980s Mikhail Voloshin in Moscow, Shifman, and shortly after Nathan Isgur from the University of Toronto, and Caltech's Mark Wise, scented a new symmetry. They saw that allowing the heavy quark mass to run off to infinity gave a simplified QCD that is symmetric with respect to changes of both the flavor and the spin of the heavy quark. These are two new symmetries that full QCD does not possess. The program sparked by this insight, developed by many people including Edward Shuryak, Politzer, and others, crystallized in 1990 through the efforts of Howard Georgi into a potent calculational tool called "heavy quark effective theory" or HQET. Heavy quark effective theory is a new effective field theory describing the strong interaction of heavy quarks with lightweight partners, built around the new symmetry. Using this approach, theorists can work out exact results for processes such as decays for mesons containing an infinite-mass heavy quark. They can then let the actual mass of the heavy quark, still large so that the symmetry *almost* holds firm, creep back in as a symmetry-breaking correction to the infinite-mass version. Heavy quark effective theory is a powerful technique, in fact one of a family that exploits the properties of a heavy and light quark–antiquark duo, for dealing with what is truly a very tough problem. The reality is that the heavy quark is surrounded by an incalculable mess of highly relativistic quarks and gluons that Isgur christened "brown muck." The expression appears in the technical literature, a colorful antidote to the gory mathematics and an indicator of how bleak the true problem really is.

In the case of K meson decays, it's the W boson that defines a natural short distance scale. Here it's the heavy quark's turn. Predictably, underneath HQET lurks the operator product expansion once again. And once again HQET, as the name implies, is another example of an effective theory in the mold of Chapter 3.

Heavy quark effective theory deliverables include relationships between mesons having different heavy quarks, and in particular links between masses of members of the B and D meson families. Others are insights into the decays of heavy mesons, where for example a b quark converts to a c quark via a weak decay, giving overall processes such as $B \rightarrow D + l + \nu$, where l represents a lepton such as a muon or electron. This is a *weak* decay, to be sure, but the decaying quark is confined within a meson, meaning that strong interaction effects cannot be ignored when turning measurements of B decay into information about b to c quark transitions, for example. So particle physicists use HQET to disentangle the raw weak and strong interaction effects in heavy meson decays. There is a huge interest in applying techniques such as HQET to interpret the wealth of B meson data emerging from the new B-factory facilities.

8 Quarks under the microscope

Having a theory of the interaction between quarks is very nice. Having an explanation of some of the patterns between, and properties of, exotic particles such as upsilon mesons is nice, too. But what of the original problem, to understand the workings of the atomic nucleus, or of the "revised original problem," what makes the proton tick? Neither the quark model nor QCD tell us all we would like to know about the life of the main building block of matter, the nucleon.

One way to learn more is deep inelastic scattering, introduced in Chapter 4, in which experimenters use particle beam "microscopes" to see quarks and gluons inside protons and neutrons, a program that, in addition, offers important tests of QCD. Deep inelastic scattering has featured on the experimental agenda ever since the late 1960s, when SLAC experimenters first showed that the scattering centers within the proton are point-like entities. With continually improving data, experimenters can now disentangle the contributions made by different quark flavors, including charmed and strange quarks that, naively speaking, should not even be present in the nucleon. They can see antiquarks. Exercising skill and determination they have begun to reveal the gluon content. Even the secrets of the nucleon's internal spin structure are being made public.

SEA QUARKS AND VALENCE QUARKS

Shining light into protons and neutrons reveals their internal structure only if the probing light has a wavelength suitably shorter than the nucleon's size. Such experiments are described in terms of structure functions, which at one level are simply fudge factors encoding any structure that the target might have. With quarks added to the mix, structure functions are identified with quark momentum distributions in the nucleon.

The structure functions – equivalently, quark distribution functions – encapsulate much of what is known about the inner workings of the proton and neutron. They represent the division of a nucleon's momentum amongst its quarks in a way that parallels the

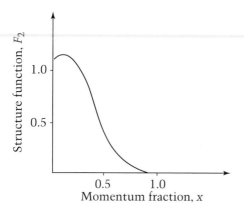

FIGURE 8.1 Schematic of the structure function F_2, a measure of the proton's quark and antiquark content.

division of a nation's wealth amongst its populace. Calculate each individual's wealth as a fraction of the nation's total, then plot their fractional wealth along a horizontal axis, with the number of individuals
worth a given amount showing vertically. Understanding the workings
of the nucleon from structure function data is tantamount to understanding a nation's economy from such a wealth plot.

Just as few people's wealth runs to a large fraction of the nation's
total, though plenty are closer to a tiny fraction of the nation's whole,
so few quarks have a large fraction of the proton's momentum, and
rather more have a smaller fraction. Figure 8.1 shows the approximate
form of the total quark (plus antiquark) content of the proton, which
is identified with the more important of the two structure functions
from deep inelastic scattering, F_2. At the very outset, this shows that
the proton is more than just a bag of three isolated quarks. For if it
were, then each quark would have a third of the proton's momentum,
rather than a whole spectrum of values, and the plot of Figure 8.1 would
reduce to some kind of spike centered on a momentum fraction of
one third, Figure 8.2. Clearly, the proton is more complicated than the
simple parton model suggests.

The three valence quarks that carry the proton's quantum numbers, thereby giving the proton its identity in terms of baryon number,
charge, isospin, and so forth correspond to the uud assignment of
the simple quark model. Since the valence quarks are held together
to make a proton, they must interact in some way, presumably via
gluons, and this permits momentum sharing amongst the trio. So the
momentum of the proton is not divided out equally, one third each to
the three quarks, but has some kind of smeared-out distribution, shown

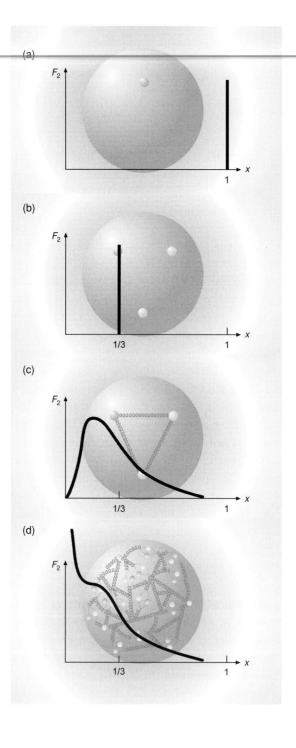

schematically in Figure 8.2. In turn, intermediate gluons are sources of quark–antiquark pairs. These particles will typically carry relatively small fractions of the proton momentum, and will appear predominantly at the low-momentum part of the quark content plot. This is because each such quark or antiquark has only a fraction of the gluon's momentum, which in turn has only a fraction of the original quark momentum, diluting down the original proton momentum. It is this picture that gives rise to the image of valence quarks immersed in a dynamic, low-energy "sea" of gluons, quarks, and antiquarks.

A helping handedness

If this picture makes any sense, then the proton should contain, via the sea, antiquarks, strange quarks, and charmed quarks, none of which would be present according to the simplistic parton model. Simply scattering electrons off protons is not enough to resolve these different possibilities, however, since alone it reveals only the total quark plus antiquark content. How can experimenters see exotic flavors and antiquarks within the proton?

Help is at hand, and from a perhaps unlikely source: the neutrino. One property that makes the neutrino rather special is its lack of electric charge. In a *neutrino* deep inelastic scattering experiment, a neutrino in becomes a negative electron or muon out, depending on the type of neutrino, Figure 8.3, and the electric charge is balanced via the intermediate W boson probing the target proton. Strange and charmed quarks aside, charge conservation means that only down quarks or up antiquarks can participate. Likewise for antineutrino scattering, only down antiquarks and up quarks feature.

FIGURE 8.2 Structure functions: if the proton was made of a single quark, then the structure function would have a single value at $x = 1$, x being the fraction of the proton's momentum carried by the quark (a). For a proton containing three equal, free quarks, the structure function would instead show a single value at $x = 1/3$ (b), since the momentum is shared equally between the three. If, however, quarks interact with one another via gluons in a simple way, the proton's quark distribution is smeared (c). However, those gluons are in turn a source of sea quarks and antiquarks, contributing a sharp rise in the structure function at small fractions of the proton momentum (d). Credit: DESY, Hamburg, Germany.

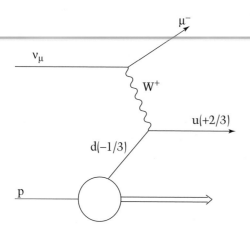

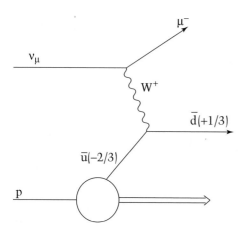

FIGURE 8.3 Muon–neutrino deep inelastic scattering from protons. Only d valence quarks, or d, s, ū, and c̄ from the sea can contribute. Quark charges are shown in parentheses.

A second crucial property that makes the neutrino special for studying nucleon structure is its handedness. Left-handed neutrinos predominate in nature, and a left-handed neutrino will interact only with left-handed quarks or right-handed antiquarks, a consequence of the fact that weak interactions violate mirror reflection symmetry (parity) as "badly" as possible. A third, parity-violating, structure function has to be enlisted to accommodate handedness.

Handedness, and the ability to interact via a charged boson probe, endows neutrino and antineutrino scattering with flavor and antiquark sensitivities denied to electron and muon scattering. These properties help experimenters disentangle contributions from quark versus anti-quark, and from one flavor versus another. In particular, the important third mirror-symmetry-breaking structure function itself can be measured via the difference in scattering probabilities between neutrino and antineutrino on a mixed proton–neutron target. In terms of the simple parton model, the result should equal the difference between quarks and antiquarks. Since quarks due to the sea should match the antiquarks in the sea, the sea contributions cancel, leaving a result that should simply equal the number of valence quarks. This is an example of a sum rule, a relationship based on momentum-summed quark distributions. This particular rule was first derived by David Gross at Princeton and Chris Llewellyn Smith at CERN way back in 1969 – an era when sum rules were popular and parton ideas still fermenting. The Gross–Llewellyn Smith sum rule is the most accurately measured of the bunch: for example, neutrino experiments by the CCFR group (a collaboration of scientists from Chicago University, Columbia University, Fermilab, and Rochester University) provide a test of the rule and give the number of valence quarks to within 10% of 3, once QCD corrections have been taken into account. So neutrino scattering confirms that the number of valence quarks matches that of the simple quark model.

Experimenters can also check out a further basic ingredient of the quark picture by comparing quark distributions from neutrino scattering with those from electron scattering. To within a simple numerical factor, the two are the same. This, firstly, provides a sanity check on the whole scheme, and hints at the idea that the distribution of the nucleon's constituents is independent of the type of particle microscope used to view them. Secondly, and more significantly for the simple quark model, the numerical factor relating the two equates to quarks having the expected fractional charges. Including the simple relationship between the first two structure functions, the Callan–Gross relation, deep inelastic scattering provides evidence for scattering centers that are point-like, spin 1/2 entities having electrical charges that are multiples of one third.

The first evidence of the quark sea, in the form of antiquarks, came in 1973 from the Gargamelle team's comparison of neutrino and antineutrino scattering from mixed proton and neutron targets. Neutrinos can provide further evidence of the sea by spotting "foreign"

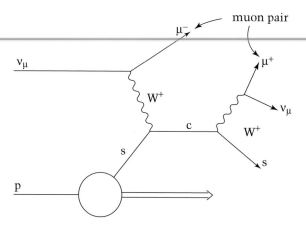

FIGURE 8.4 Neutrino deep inelastic scattering from a strange sea quark creates a pair of muons.

quark species, for example strange quarks. A neutrino scattering from a nucleon may hit a strange quark, Figure 8.4, and, assuming it's a muon neutrino, become a muon. Meanwhile the W boson interacts with the strange quark to create a charmed quark, which decays to give a *positive* muon (i.e., an antimuon), a neutrino, and a strange quark that ultimately spawns a spray of hadrons. The signature of this process is the simultaneous creation of two muons of opposite signs – experimenters including the CCFR collaboration at Fermilab have looked for this "dimuon" signature, and found it. Once the contribution from the d quark is cut away, a complication because the d quark can also interact in an analogous way to give a charmed quark, what remains is the signal due to strange quarks. The strange quarks are distributed within the nucleon in a way that parallels other, lighter, sea quarks, though there is fewer of them: strange quarks and strange antiquarks are assumed to have an equal presence.

It's even possible to probe for charmed quarks within the nucleon. A muon neutrino interacting with a charmed quark via a neutral Z boson remains a neutrino, Figure 8.5. The charmed quark, knocked from its roost, decays to give a strange quark, a positive muon, and a muon neutrino. The signature of this process is a "wrong sign" muon. A neutrino in is responsible for a positive muon, an antimuon, out: if the muon simply interacted by W exchange then lepton number conservation says that a muon should result, not an antimuon. To do

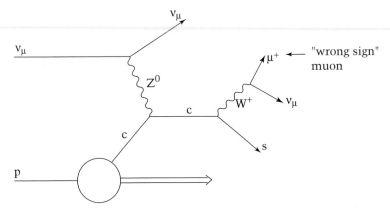

FIGURE 8.5 Neutrino scattering can spot charmed quarks in the proton via a "wrong sign" muon, a final μ^+ instead of a μ^-.

this experiment, however, means using a beam of either neutrinos or antineutrinos, rather than a mix of the two, since otherwise it's not clear if the exiting muon is a "wrong sign" one or not.

Traditionally, neutrino experiments have been a tough assignment. It's impossible to create and accelerate a neat well-controlled beam of neutrinos along the lines of an electron beam. And neutrinos are reluctant to interact, making it hard to see what happens to the beam anyway. Modern neutrino experiments use bright beams, created for example from the debris of a high-energy proton beam smashing into a fixed target, the high speed of the original protons translating into a relatively well-collimated neutrino beam. Another trick is to use a suitable massive target-cum-detector, to maximize the chance of seeing what the neutrinos do. These days, data samples of a million events are possible, enough for good statistical accuracy.

It's good statistics that puts the CCFR experiments at the top of the list. In its most recent incarnation, the CCFR collaboration smashed 800 GeV Tevatron protons into a beryllium target to create a secondary beam of kaons and pions. These decay in flight to give neutrinos. The resultant beam contained a mixture of neutrinos, mostly muon neutrinos but with a few muon antineutrinos, and a handful each of electron neutrinos and antineutrinos, with energies anything up to 600 GeV. The 36 m long detector comprised 690 tonnes of iron plates interspersed with scintillators and drift chambers, with the second half a muon spectrometer for detecting and tracking muons. A muon neutrino striking a nucleon in the iron becomes a muon that is detected in the muon detector. A quark knocked from the nucleon in the process turns to hadrons, detected by the active layers between the iron slabs. An upgraded version of the experiment, called NuTeV (for "Neutrinos at the Tevatron"), began running in May 1996 with an improved version of the same detector, and a beam that is *either* neutrino *or* antineutrino, achieved by selecting the appropriate secondary kaons and pions via their electric charge. NuTeV spotted unambiguous signs of charmed quarks within the nucleon, in addition to charting out the strange sea; the strange content of the nucleon is much greater than the charm content.

The sea: asymmetric and thick with glue

Because protons and neutrons are related to one another via an interchange of up and down quarks, in other words they are related through isospin symmetry, it's natural enough to assume that the up antiquark content of the proton is the same as the down antiquark content of the neutron, $\bar{u}_{proton} = \bar{d}_{neutron}$, and vice versa, $\bar{d}_{proton} = \bar{u}_{neutron}$. However, there is no symmetry principle demanding that a proton's up antiquark distribution should match that of its down antiquarks. That is, there's nothing that says $\bar{d}_{proton} = \bar{u}_{proton}$ should be true, although particle physicists have always assumed it is, for want of experimental evidence to the contrary.

Now they know it's not true. In 1998, Fermilab's E866 collaboration showed to their surprise that down antiquarks vastly outnumber up antiquarks. The E866 team fired 800 GeV protons into stationary liquid deuterium and liquid hydrogen targets, detecting collisions in which quarks from the incoming proton annihilated with antiquarks in the target nucleons to give muon pairs according to the Drell–Yan mechanism, Figure 8.6. The ratio of scattering cross-sections for the two different targets translates directly into a ratio of down antiquarks to up antiquarks in the nucleon which, when combined with existing results, yields the difference between the two. The results show that at large momentum fractions, the two make an equal but tiny contribution. But for quarks and antiquarks having less than a fifth of a proton's momentum, down antiquarks outnumber up antiquarks.

This is the first time flavor asymmetry had been witnessed across the antiquark distribution within the nucleon sea. It also represents a

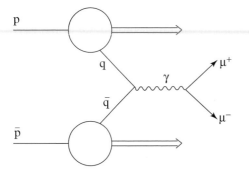

p

q

q̄

γ

μ⁺

μ⁻

p̄

FIGURE 8.6 The Drell–Yan mechanism in proton–antiproton collisions. Alternatively, the antiquark might arise from the proton sea.

violation of another sum rule, this one derived in 1967 by Cornell University's Kurt Gottfried, who at the time was looking for simple predictions that would test the elementary quark model. The Gottfried sum rule says that the difference in scattering probability between electrons on neutrons, and electrons on protons, should yield a simple fraction. The rule depends on the particular quark fractional charges, and the proton's uud make-up versus the neutron's ddu. It also assumes either no antiquarks, or at least that, within the nucleon, up antiquarks and down antiquarks have an equal presence. Violation of this simple rule, first glimpsed in 1991 by CERN's NMC collaboration in a more restricted regime, both adds weight to the evidence of a quark–antiquark sea, and throws up some surprising properties of that sea that theorists are yet to explain.

Tacit here is the assumption that the antiquark distributions of the E866 experiment, and those of NMC, obtained by scattering muons from hydrogen and from deuterium targets, are one and the same thing. The fact is that, generally, quark, antiquark, and gluon distributions from hadron–hadron collisions, such as those in the E866 experiment, are taken to be interchangeable with those measured by muons, electrons, or neutrinos striking nucleons. But this glosses over an important assumption. This is that the strong interaction of one hadron with another can be neatly partitioned into distinct portions describing, firstly, the parton distributions, then the interactions of free partons, and finally the fragmenting of the collision products into sprays of particles seen in real experiments. Only then does it make sense to use the quark and gluon description measured in one process as input into calculations of another. However, this assumption, called factorization, is *not* always justified. It could be, for example, that in a proton–proton collision, the bulk of the quarks within the beam proton distorts the distribution of quarks in the target proton. Or either may influence the dressing of struck quarks leaving the collision point. Theorists agree that there is a point where these types of effects become significant, and that then the factorization assumption breaks down.

But where the description of interactions such as proton with proton can be justified in terms of these distinct factors, the quark and gluon distributions are effectively universal. Accurately measured distributions from one experiment can be used to make predictions elsewhere. This is important. For instance, understanding top quark production at the Tevatron, and estimating how many top quarks will be created at the LHC, both need quark and gluon distributions as input.

As mentioned in Chapter 4, one of the reasons theorists are especially comfortable with their top quark predictions is that top quark production at Fermilab is dominated by annihilations between quarks and antiquarks. This means that the gluon content of the proton is relatively unimportant, at least for top quark production at Fermilab energies. And this is a good thing, because actually making accurate measurements of the gluon distribution in the nucleon is incredibly hard, bad news when it comes to predicting processes such as bottom quark production at current energies, or top quark production at the forthcoming LHC, both of which rely more heavily on the gluon component. The root of the problem is that gluons have no electrical charge and no flavor label, so the usual rules exploited for spotting quarks and antiquarks do not apply.

But revealing the gluon distribution *is* possible. HERA experimenters, for example, have measured the gluon sea in the proton via jets produced in electron–proton collisions. In these collisions, a photon from the incoming electron fuses with a gluon from the proton, Figure 8.7. The outgoing quarks each fragment into jets of pions and other strongly interacting particles. Measuring the probability of creating two such jets, in addition to the proton remnant, gives the proton's gluon content, once the contribution from jets arising from quarks in the proton has been accounted for. The HERA measurements, the first of which were published in 1995, were the earliest direct measurements of gluons that have small fractions of the parent proton's momentum, the region where gluons start to dominate. In fact, gluons having more than a tenth of the proton's momentum are scarce, but for lower momentum fractions the gluon population explodes. This

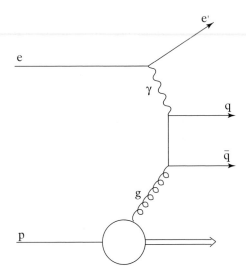

FIGURE 8.7 Boson–gluon fusion in electron–proton deep inelastic scattering probes the proton's gluon content.

means that low-momentum fraction partons are virtually all gluons, which is in line with less direct means of estimating gluons in the proton. It also makes intuitive sense: if the gluons carry around 50% of the nucleon momentum and if, as other experiments have shown, the number of gluons having a large fraction of the proton momentum is modest, then there must be huge numbers of soft gluons in the proton, Figure 8.8.

The soft sea

Understanding the behavior of the proton's gluons, sea quarks, and anti-quarks at low proton momentum fractions is "one of the most diffi-cult and important problems of QCD," according to Russian theorist Eugene Levin, a pioneer in the field. "At least, it should be solved before QCD will be a real theory of the strong interactions at high energy," he said. As if to underscore Levin's comments, experimenters at HERA have discovered a surprising and massive rise in the proton's soft quark inhabitants.

CERN experimenters, probing down to momentum fractions of 0.01 using muon scattering on stationary hydrogen and deuterium tar-gets, had already seen hints of what was to come. In 1992, the same year that the NMC group published their results, HERA experimenters presented preliminary deep inelastic scattering data at the San Miniato conference in Italy. These pushed the quark momentum fraction way

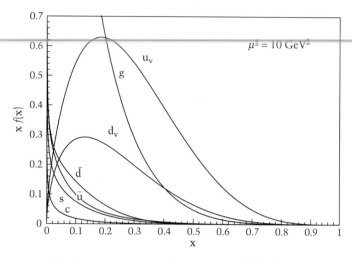

FIGURE 8.8 Measured proton quark and antiquark distributions, multiplied by momentum fraction x, plotted against x. The "v" subscript denotes valence quarks, and "f" denotes gluon, or particular quark or antiquark flavor distributions. This plot shows the small but measurable strange and charm quark content of the proton. It also shows that the up antiquark content is not the same as the down antiquark. Combining the quark and antiquark distributions in this plot yields the structure function F_2. Credit: Particle Data Group.

down to 0.0001, a hundredth of the NMC's figure, the previous best. A new chapter in the study of the proton had been opened.

To knock out quarks having such low momentum fractions, experimenters need photons having a wavelength long enough to probe these longer wavelength quarks, yet able to transfer a lot of energy to the struck quark. One option might be to build an electron beam of stupendous energy that strikes a stationary hydrogen target in the normal way. But the beam energy would need to be too large to be realistic, besides which the scattered electron would be at such a shallow angle it would be hard, if not impossible, to see it. A better option would be to build a collider that would not only give collision products easily accessible to detectors surrounding the collision point, but could do the job by colliding beams of more reasonable energies. And this is the route taken at HERA, where the 27 GeV electron beam colliding with a 920 GeV proton beam is equivalent to a 55 000 GeV electron beam hitting a stationary target. The Zeus and H1 detectors, enveloping the collision points can, in addition to measuring the scattered lepton, track the strongly interacting collision products, a significant break with the tradition of measuring just the scattered lepton.

While it might seem unnecessarily pedantic to go chasing the first one percent or so of the structure function plot, there are plenty of people who would like to see this part of the graph filled in. After all, this is where most of the sea is concentrated. For the pragmatists, good quality calculations of a wide range of processes, such as proton–proton scattering into Z and W bosons, bottom and top quarks, and the putative Higgs particle depend on knowing the population of low momentum fraction quarks and gluons. And tests of sum rules, embracing contributions from *all* momentum fractions, demand this information too. Theorists have produced widely varying estimations of what the plot should look like. Their efforts, hampered by a dearth of data, focus on the gluons, since these in turn drive the sea quark and antiquark population.

The HERA "low x" results create difficulties for a pre-QCD approach called Regge theory that, for all its faults, offers a workmanlike description of some important cross-sections, for example the total cross-sections for proton–proton or proton–antiproton scattering. Regge theory, named after the Italian theorist Tullio Regge, describes hadron scattering in terms of the exchange of families of mesons, and is more comfortable with a roughly constant quark population. However, a rapidly rising quark population was anticipated in the work of four Russian theorists, Yanko Balitsky and Lev Lipatov at the Leningrad Institute for Nuclear Physics, and Victor Fadin and Eduard Kuraev at the Budker Institute for Nuclear Physics, Novosibirsk, as long ago as 1975. They developed a prescription, encapsulated in the so-called BFKL equation, for including contributions from layer upon layer of ever-softer gluons, giving the gluon population evolution in terms of their momentum fraction. The alternative DGLAP evolution equation approach (the acronym derives from the names of Yuri Dokshitzer, Vladimir Gribov, Lev Lipatov, Guido Altarelli, and Giorgio Parisi) allows for inter-conversions of quarks and gluons and, applied to the gluon population, describes how it changes as the probe wavelength varies.

Neither approach can predict the gluon population from thin air, and both need to be fed a measured distribution to start with. Both can, however, fit the new data, given a judicious choice of starting distribution or tweaking of the basic model. One difficulty is that the predicted distribution is for gluons, whereas measurements typically yield the total quark content, which is one step removed. Continually improving direct measurements of the gluon distribution itself, for example at HERA through charmed quark–antiquark pair production or J/ψ production, will help test theorists' models by pinning down the gluon distribution more exactly. Meanwhile theorists are trying to understand the link between the two approaches, which at some stage each must meld smoothly into the other. They have also had to face the BFKL model's prediction that the gluon population could explode, threatening unitarity. To beat this, theorists postulate the idea of saturation, which says that, as the population density of gluons in the proton goes up and up, at some point they begin to interact and recombine with one another. Thus gluon recombination limits the gluon content of the proton. Crude back-of-the-envelope estimates suggest that saturation may only come into play just beyond the observational capabilities of HERA. However, some argue that the low momentum fraction gluons might crowd round valence quarks, forming "hot spots" that HERA might be able to see.

QUARKS AND THE NUCLEUS

Nature has added a further twist to the internal machinery of nucleons: the distribution of quarks in a nucleon that is part of a nucleus is *different* from the quark distribution inside a free nucleon. In other words, quarks inside a nucleon are sensitive to the nucleon's environment, presumably the quarks and gluons of neighboring nucleons being palpable in some way. This is the so-called EMC effect, named after CERN's European Muon Collaboration, who first observed the difference when they measured nucleon structure functions using deep inelastic muon scattering from both iron and deuterium targets. Their initial shocking results, reported in 1983, have been refined and extended by them and several other groups since, showing that the disparity depends on the momentum fraction of the struck quark in a not-so-simple way, Figure 8.9. In addition, for elements ranging from helium to lead, the pattern of nucleon structure function distortions is similar for the heavier nuclei, with differences linked to differing overall densities of the nuclei.

Initially, theorists harbored hopes of identifying some simple mechanism that would explain the new results. Instead, confronted once again with the specter of non-perturbative QCD, they have

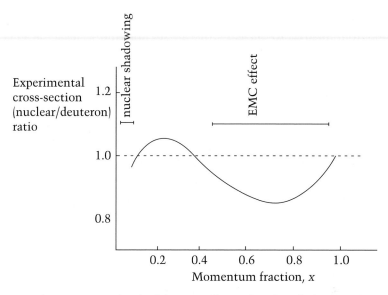

FIGURE 8.9 Sketch of the EMC effect and nuclear shadowing. The vertical axis is the ratio of the nuclear cross-section to that of deuteron. If a nucleon inside a nucleus responded to the probe photon in the same way as a nucleon in a deuteron, the plot would follow the horizontal dotted line.

sculpted various QCD-inspired models that attempt to account for the characteristic dip in the higher momentum end of the quark distribution. Some proffer a view of the nucleus as a collective of multi-nucleon clusters, with each cluster a multiple of three quarks. Others suggest that, for heavier nuclei, the region of confinement, the size of the cage trapping quarks and gluons, extends beyond the nucleon, even out to the boundary of the nucleus itself. Then again, mesons, which in the conventional view bind nucleons together to build nuclei, might tie up some of the quark momentum that would normally be found within nucleons. Something more is needed to explain the downturn in the heavy nuclei cross-section at low momenta, the low x dip in Figure 8.9. This so-called nuclear shadowing is attributed to low-momentum long-wavelength gluons from neighboring nucleons spilling out and interacting with one other. It's a version of the gluon recombination that might make for gluon saturation in lone nucleons, mentioned in the previous section.

The challenge presented by nuclear shadowing and the EMC effect is to find a comprehensive QCD description of both nucleon and nuclear structure functions. The reward would be insight into the transition from nucleon to nuclear structure. In principle at least, QCD is the basic theory with which to explain all nuclear structure. In practice, however, theorists understand nuclei in terms of a variety of nuclear structure models well removed from the intricacies of QCD. "Ironically QCD has not yet given much insight into the strong interaction problems like the nuclear forces which at the start motivated the research," wrote Harald Fritzsch. According to Wilczek, the relationship between QCD and nuclear physics parallels that between QED and chemistry. There, the full-blown quantum field theory of electromagnetic interactions is the fundamental theory of chemical interactions, but in reality most chemists work with bonds, orbitals, and the like.

Chemistry, in fact, provides a graphic model of how QCD fits in with nuclear physics. Neon, the stuff of illuminated signs, is a gas at room temperature. It's a really model element, its atoms having just the right number of electrons to make them incredibly unreactive. Yet at $-246\,°C$, neon becomes a liquid, and a couple of degrees colder it becomes a crystal. This means that, although each atom appears to be a near-perfect non-interacting, chargeless sphere, there is in fact some kind of residual force between them. The trick to understanding the origin of this force is to remember that each atom is not actually a solid structureless ball, but is composed of a nucleus and electrons. In fact, when two atoms are close to one another, they can and do feel one another's charges. The electron cloud on each is shifted ever so slightly relative to the nucleus: suddenly, each atom's positive charged core no longer coincides with the center of its electron cloud. As a consequence, each atom becomes a little electric dipole, resulting in an attractive force between closely spaced atoms. This is the van der Waals interaction at work, named after the Dutch physicist Johannes van der Waals.

To cross to nuclear physics, replace charge-zero neon atoms with color-charge-zero nucleons. There's no direct color-based force between them, just as there's no direct electric charge-based force between the neon atoms. But neighboring nucleons can distort one another's color charge clouds, in a way analogous to the neon atoms' ability to distort each other. In parallel with the van der Waals interaction responsible for the force between neon atoms, there is an analogous force arising from the deformed color-charge clouds. It is this van der Waals-like force that holds nucleons together in the atomic nucleus.

Nuclear forces emerge as the result of a kind of off-centering in the strong inter-quark forces holding individual nucleons together. Quantum chromodynamics might ultimately provide the underpinning of our understanding of the nucleus but, to coin Wilczek's phrase, the role of QCD in practical problems of nuclear structure is essentially that of "holy water."

A LESSON REVISED

Studying the proton's structure through deep inelastic scattering not only helps particle physicists understand how it's put together, it provides them with a test bed for QCD. One of the best tests of all involves looking at the way the distribution of quarks inside nucleons is actually at variance with the simplest picture.

In deep inelastic scattering, a probing photon sees point-like quarks, and so far as the simple or, in the jargon, the naive parton model is concerned, that's the end of the story. But as the wavelength of the probe photon falls, it sees a quark plus an emitted gluon, or at a finer level a quark and quark–antiquark pair, and so on. The smaller the probe wavelength, the more particles the probe sees. As the number of particles increases, the parent nucleon's momentum has to be spread more ways, and the probability of finding a quark having a particular fraction of the proton momentum will change. The QCD-improved parton model suggests large numbers of low-momentum constituents, and structure functions that in fact *do* depend on the probe wavelength, but in a characteristic way.

Theorists analyze this in terms of the radiation of a gluon by a soon-to-be-struck quark. The quark's momentum, originally a fraction of the parent proton's momentum, is reduced by the amount of momentum carried off by the gluon. So the chance of the probe photon encountering a quark with the expected fraction of the proton momentum is modified by the probability that the quark radiates a gluon. The key point is that the probability of a quark radiating a gluon increases the smaller the probe wavelength. This means that, according to QCD, the measured distribution of quarks having a certain fraction of the proton momentum will, in fact, depend on the wavelength of the probe photon. The original lesson was scaling, that structure functions depend on some dimensionless quantity reflecting the fact that the scatterer itself is point-like. This is the lesson that needs revising: scaling is violated, but in a particular way described by QCD. Scaling lends support to point-like scatterers. Scaling violations lend support to QCD.

Gluon emission is just one of the sources of scaling violation in the proton's quark distribution. Contributions also arise from gluons spawning more gluons, or gluons giving rise to more quarks and anti-quarks that in turn become the struck parton. The structure functions are said to evolve, in other words change slowly, with probe wavelength. This evolution is calculable within perturbative QCD, based on the various probabilities of quarks emitting gluons, gluons emitting quarks, and gluons emitting gluons. The resulting DGLAP evolution equations have already made their debut in the discussion of the proton's soft gluon population.

It's perhaps surprising that the first suggestions of scaling violations are as old as QCD itself: for example, in their great paper announcing asymptotic freedom in gauge theories, published in 1973, Gross and Wilczek showed that what we now know as QCD predicts scaling violations. Soon after came the publication of the first experimental sign of scaling violations, in deep inelastic scattering of muons from an iron target at Fermilab. Five years later CERN experimenters used neutrinos' intrinsic handedness to measure scaling violations in the third, parity-violating structure function, providing additional support for QCD in a subtly different context. Measuring scaling violations, which also have a key role in the fragmentation of quarks into strongly interacting particles, is now big business, with vast amounts of data gathered from a variety of experiments. Scaling violations don't just offer a test of QCD. The gentle leaning of the structure function towards lower values of the momentum fraction as the probing photon wavelength drops is directly related to the distribution of gluons within the nucleon. In fact, working backwards, measurements of scaling violations provide a way of extracting the gluon distribution.

Theorists cannot calculate the form of the structure functions from scratch using perturbative QCD, but they *can* calculate how they evolve, in contrast to the probe wavelength-independence expected from perfect scaling. The degree of this scaling violation depends on the strength of the coupling, and hence on the scale parameter lambda, the one number of QCD. All of which helps make the scaling violations witnessed in deep inelastic scattering one the best available tests of QCD, some would say *the* best.

A standard way to compare experiment and theory is to begin with a measured distribution of quarks at some specified probe wavelength. Using the DGLAP evolution equations, refined to include more complex diagrams, this experimental starting distribution forms the basis of a predicted quark distribution at different probe wavelengths, and these can be compared with experiment. Either a value for the scale parameter can be imported from elsewhere, or the whole fitting procedure juggled to give a best-fit set of values for various parameters, including the scale parameter.

Of the inevitable complications, one of the most subtle, interesting and pervasive is that due to additional gluon exchanges involving the final state of the "basic" process. For example, in Figure 8.10(a), showing deep inelastic scattering, an additional gluon links the proton fragment with the scattered quark. In Figure 8.10(b), showing a Drell–Yan process for proton–antiproton scattering, the proton fragment is linked to an internal antiquark from the antiproton. For reasons stemming from the arcane mathematics involved, these types of processes are dubbed "higher-twist" effects.

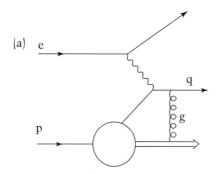

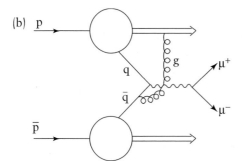

FIGURE 8.10   Higher-twist effects involving additional interactions with spectators in (a) deep inelastic scattering and (b) proton–antiproton collisions.

It's a question of timing. When, for example, a photon probes a proton at short wavelengths, everything happens very fast. This is because a virtual photon having a short wavelength will also have a short lifetime, dictated by the uncertainty principle. A quark struck by a short-wavelength photon is dislodged cleanly, according to this so-called impulse approximation. But when the interaction is less than a bolt from the blue, when both photon wavelength and time scale are longer, there's time for additional interactions to come into play. These mean that the original participants become linked, or correlated, in some way with what would normally be merely spectators to the original scattering. Bank robbers run: to walk would be to invite (higher-twist) interactions with spectators.

Higher-twist effects cannot be calculated using perturbation theory, since they are intrinsically more important at the long-wavelength low-energy end of the scale where perturbation theory's reliability becomes doubtful. It's this, and the fact that so many processes can involve higher-twist effects, that makes them so challenging and interesting. Theorists can only estimate the magnitude of higher-twist contributions to a particular process, or model them based on features in structure function data, or at least identify experiments where the impact of higher-twist contributions is likely to be minimal.

Complications such as higher-twist effects aside, scaling violations remain an important test bed for QCD: "The observed scaling violations are in such excellent agreement with the predictions from QCD that we can speak of a spectacular success of the theory," wrote Fritzsch in 1992.

HITTING PROTONS WITH LIGHT
Slice the deep from deep inelastic scattering, and some surprising things happen. When firing an electron at a proton to probe its interior, the parameters of the experiment can be tuned so that the photon mediating the collision is real, or at least very nearly real, so-called "quasi-real," as opposed to virtual. For example, real or quasi-real photons result when the scattered electron leaves the collision scene virtually undeflected. Scattering with a real or quasi-real photon is called photoproduction.

And thus is Pandora's box opened. In photoproduction, nonperturbative effects are ubiquitous, higher-twist approaches to understanding them largely irrelevant, Bjorken scaling non-sensical, and the proton's quark content almost invisible. Yet, somehow, photoproduction has to dovetail seamlessly with deep inelastic scattering. All of which makes photoproduction very intriguing.

Something odd happens to the photon, too, which in a strange way also helps to understand the proton's structure. A photon is accompanied by a cloud of virtual particles, electrons and positrons, quarks and antiquarks even, born of the vacuum and returning to it according to the rules of the quantum world. For a real or nearly real photon, these quantum excursions last in excess of around 10^{-25} s, and so the photon's virtual quarks and antiquarks, and their gluonic progeny, have time to engage in strong interactions. In short, the photon, the quintessential elementary particle, behaves as though it is *not* a simple elementary particle. It can appear to contain quarks and antiquarks, and thus looks like a composite particle along the lines of a proton. Photons can therefore suffer strong interactions. In the quantum world, nothing is ever as it seems: look harder, and the view always changes.

Interactions in which photons behave as whole, indivisible photons are called direct photon interactions, Figure 8.11(a). In contrast, photons interacting via the quarks and gluons associated with them are termed resolved photons, Figure 8.11(b). In a sense, a resolved photon is a kind of fragmented photon.

If a resolved photon interacts via quarks and gluons, this then implies some sort of a distribution of quarks and gluons "in" the photon. In other words, there should be a photon structure function, in parallel with the nucleon's structure function. How then to measure such a thing? The answer is to follow the example of the nucleon: take a photon and look at it under a microscope. In practice, this means deep inelastic scattering, using electron beams to create a virtual photon probe to examine a real or quasi-real photon. This may sound implausible, but the implementation is simple enough. Just collide an electron beam with a positron beam, and rig the experiment so that what's observed corresponds to a virtual probe photon from (say) the electron beam, in the normal way of deep inelastic scattering, and a photon from the other beam, the positron, which is real or nearly real, Figure 8.12. Then the probe virtual photon sees a target photon resolved into quarks and antiquarks.

This type of process was first seen in 1981, by the PLUTO group using the DORIS electron–positron collider, later at PEP, and has since been explored at LEP and at other electron–positron colliders. The photon does indeed have a structure function: in other words, the *quark*

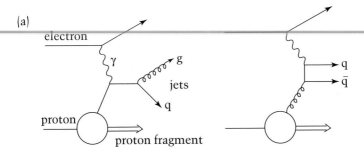

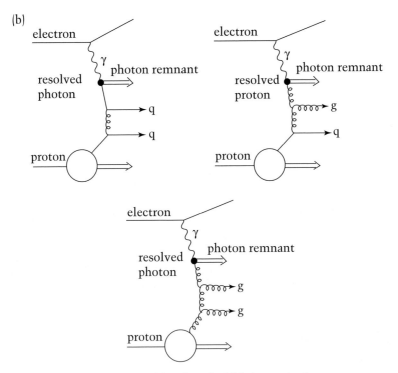

FIGURE 8.11 Direct (a) and resolved (b) photons in photon–proton scattering. The shaded blobs in diagrams (b) indicate resolved photons, which give rise to a photon remnant.

content of the *photon* is real and measurable. Results from LEP confirm a gently bulging quark distribution, with most quarks having middling fractions of the parent photon's momentum. This is to be expected, because the whole process *has* to revolve around the spawning of a

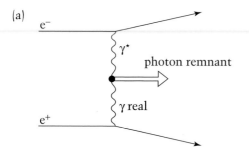

(a)

e⁻

γ*

photon remnant

γ real

e⁺

FIGURE 8.12 Deep inelastic scattering measures the photon structure function. An incoming electron creates a virtual photon probe, γ*, which interacts with the resolved photon from the positron (a). A simple contribution to the resolved photon blob is depicted in (b).

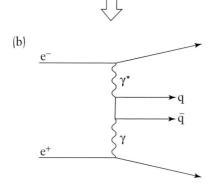

(b)

e⁻

γ*

q

q̄

γ

e⁺

quark and an antiquark from a single photon, whose momentum is simply split between them. Though the visible quark content of the photon rises slowly as the probe wavelength falls, in line with QCD predictions, it's less clear what happens at low momentum fractions. Theorists speculate that the quark distribution will show some sort of a big rise at low momentum fractions, mimicking that seen in protons. But so far, down to momenta of just one quarter of one percent, there's no sign of any such rise.

One of the attractions of studying the photon's structure function is its perceived simplicity. In 1977 Princeton University theorist Ed Witten showed that by using QCD it's possible, in principle at least, to predict the actual distribution of quarks within the photon. This is over and above showing how the quark content of the photon evolves with changing probe wavelength, and is something new: there's no corresponding way of deriving the quark content of nucleons from "pure thought." The difference is that, for the photon, the whole story has a simple starting point, the photon–quark–antiquark intersection, in

Feynman-diagram-speak. There's no such simple starting point for the proton.

Particle physicists once hoped that the photon structure function might provide an especially clean way of measuring the coupling strength parameter of QCD, which features in a simple and appealing way in Witten's prediction of the photon's quark distribution. Their hopes were dashed, washed up on the rocks of assumptions that had no place in a world of real experiments, and it turns out that predicting the measurable quark content of the photon from pure theory doesn't pan out. However, the attraction remains that the photon is a simpler object than the nucleon, and understanding the hadronic structure of the former should, the reasoning goes, shed light on the latter.

The photon's quark content, revealed by electron scattering, makes for a photonic schizophrenia when real or quasi-real photons interact with protons. Photons can interact directly, or they can interact as though they are a source of quarks and gluons, and both must be taken into account. Compared to a direct photon interaction of similar complexity, a resolved photon interaction spawns an additional jet, directed approximately along the direction of the incoming photon, Figure 8.13. For photoproduction at HERA, the proton remnant jet follows the proton's original course, while the photon remnant from a

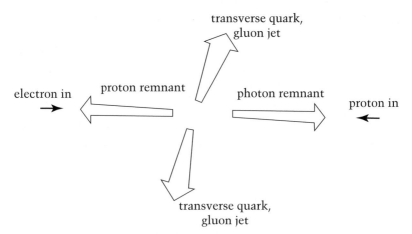

FIGURE 8.13 Protons colliding with resolved photons give four-jet events, with remnant jets from both the proton and the photon, and transverse jets from quarks and gluons. At HERA, the pattern of jets is biased in the direction of the proton, due to the proton's large momentum.

resolved photon interaction follows the scattered electron's path. Both the simplest direct and resolved photon processes should result in two jets produced at large angles relative to the colliding beams, so-called "high transverse momentum jets," but what distinguishes between direct and resolved is the additional photon remnant jet of the latter. Both types of event have been seen at HERA where, under the right conditions, resolved photon scattering is more probable than direct. The photon remnant is clearly visible at HERA, but hidden in earlier fixed-target experiments, due to its proximity to the beam.

Deep inelastic scattering from a real photon, and real photon collisions with a proton, both involve just a single real photon. It should, however, be possible to collide a real photon with a second real photon, and create matter. This is expected in QED, even though the laboratory creation of matter from light–light scattering was only seen for the first time in 1997, when a team working at SLAC created electron–positron pairs from undeniably real photons. This they achieved by colliding an immensely powerful laser beam head-on with SLAC's high-energy electron beam. The oncoming electrons knocked light photons back the way they had come, increasing their energy in the process and turning them into gamma photons. These then collided with incoming laser beam photons, creating electron–positron pairs.

Nobody doubted it would happen, but it's reassuring to see confirmation that the vacuum can thus be made to "spark," as Adrian Melissinos, one of the team members, put it. Five years earlier the AMY collaboration at TRISTAN, the Japanese electron–positron collider, did a little vacuum sparking of their own. Their photons were of the almost-real kind, in this way a little different from the SLAC real photon experiment, but the AMY group was able to use light–light scattering to pull hadrons from the vacuum. They collided electrons with positrons, both with very shallow scattering angles, so that the virtual photons mediating the interaction were nearly real, Figure 8.14. They found that strongly interacting particles emanated from the collision point in numbers that only made sense if one or even both photons were acting as a source of both quarks *and* gluons: theirs was the first indisputable sighting of resolved photons in something other than electron – photon deep inelastic scattering, and the first signal that photons contain gluons. In due course they, the TOPAZ group, and various LEP collaborations have been able to distinguish jets of hadrons coming from collisions of quasi-real photons: in all,

(a) direct

(b) double resolved

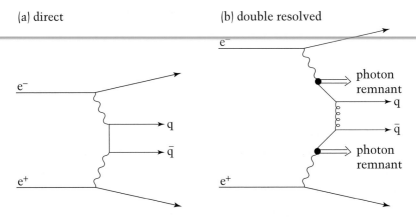

FIGURE 8.14 "Light–light" scattering can proceed via (a) two direct photons, (b) two resolved photons or one direct and one resolved photon (not shown). The three cases give two, four, or three jet events, respectively.

there can be up to four hadron jets, one each from the two "basic" quarks, gluons or antiquarks, and one or two arising from photon remnants.

Collide two quasi-real photons together and it's even possible to spark proton–antiproton pairs. This is no idle notion – it has been seen at several laboratories, for example at the CESR electron–positron collider. Thus photoproduction throws up another marvel of the quantum universe, which can build a particle as complex as a proton from the rubble of colliding "simple" photons. We live in a Universe that can effortlessly conjure a complete and perfect hydrogen nucleus, in all its splendor and intricacies, from empty space.

SPIN: A CRISIS

Occasionally, nature hands out a free gift. Ingenious experimenters at HERA have put one such present to work in an attempt to solve a great riddle of particle physics – the origin of the spin of the nucleon.

Electrons or positrons circulating in a storage ring throw off synchrotron radiation. As a by-product, the electrons in the ring become transversely polarized. In other words, their intrinsic spins become oriented across their direction of flight, a feature first detailed by two Russian physicists, Arsenii Sokolov and Igor Ternov, both of Moscow State University, back in 1963. The spin alignment is a consequence of photon emission being easier for those electrons that flip their spins so that their spins finish up aligned antiparallel to the direction of the bending magnetic field. It's this spin alignment that is nature's freebie, but only recently have accelerator physicists put it to use. The catch is that to be really useful, the spins of the electrons – or positrons – need to point *along* the flight path, so-called longitudinal polarization, not across it.

An experiment at HERA, dubbed HERMES, (*HERA Me*asurement of *S*pin, or, for aficionados of Greek mythology, the son of Zeus and ambassador of the Gods) achieves just this. HERMES was built to study spin interactions. Whereas the H1 and Zeus detectors sit astride proton–positron collision points, only the positron beam passes through HERMES, which watches for collisions between positrons and its own custom-built stationary target. A special combination of magnets upstream of the detector twists the spin orientation of the incoming positrons, so that they are longitudinally polarized by the time they reach the target. A second magnet combination, or spin rotator, twists the spin direction back again downstream of the target, so that those positrons continuing their journey round the ring appear unchanged. This is the first time longitudinal polarization has ever been produced in a high-energy ring accelerator. Achieving it was no small feat: the near-perfect alignment demanded of the focusing magnets round all 6.4 km of the HERA ring, for instance, left some people skeptical that the technique would ever work. The first data, published in 1997, proved that it does.

The data also vindicated a second novel technology exploited by HERMES. The target is a cloud of gas sitting in the way of the beam. There is no container for the gas, so there's no chance of obscuring the real signal with collisions between the beam and some container material. Instead, gas is piped to the target region, and is removed by powerful pumps so that it does not fill the beam pipe and destroy its vacuum.

The target gas atoms in the earliest runs were polarized helium-3 atoms. Making such a gas of atoms having polarized nuclei, which is what the experiment needs, is a subtle business. At HERMES, experimenters used a weak radio-frequency field to excite helium-3 atoms into a temporarily stable higher energy state, in which the atoms' electrons are no longer paired in the lowest energy configuration. These atoms were then polarized by shining polarized laser light on them.

target atoms

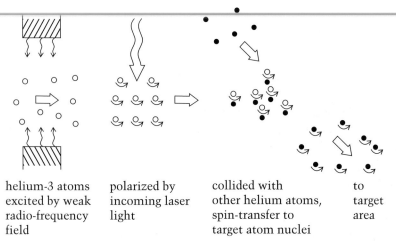

| helium-3 atoms excited by weak radio-frequency field | polarized by incoming laser light | collided with other helium atoms, spin-transfer to target atom nuclei | to target area |

FIGURE 8.15 Creating a polarized helium gas target at HERA. Those atoms destined for the target are shaded for clarity.

Next, the excited atoms were made to collide with other helium atoms that became the *target* atoms. In the collisions, the spin tied up in the electron motion of the excited atoms transferred to the *nuclei* of the target atoms. It's these atoms, having polarized nuclei, which were used as the target, Figure 8.15.

One big advantage of using gas targets such as helium-3, deuterium, or hydrogen is that the resulting signal is uncluttered, relative to those taken with more traditional solid targets such as solid butanol or ammonia. In ammonia, for example, only 3 of the 17 nucleons can be polarized, so most of the events spotted in the detector arise from collisions with unpolarized parts of the target, and this dilutes the interesting signal. For hydrogen and deuterium, all the nucleons are polarized. For helium-3, the two protons of the nucleus are paired in a zero-spin configuration, the lone neutron being responsible for the polarization. With no simple neutron target available, a polarized helium-3 nucleus is the best available substitute, even though only one in three nucleons can be polarized. HERMES and two SLAC experiments, E142 and E154, have all used a polarized helium-3 gas target.

To do a typical polarization experiment, the spins of the target and probing beam are lined up first pointing in the same direction, and then pointing in opposite directions. For HERMES, this is achieved by

(a)

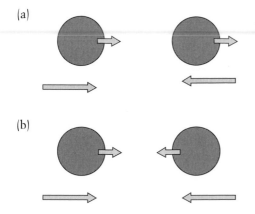

FIGURE 8.16 Colliding particles with spins parallel (a) compared to spins antiparallel (b) reveals the spin content of the target nucleon.

(b)

reversing the orientation of the target spin, by flipping the polarization of the laser light used to create the target. The difference in the cross-sections measured in the two cases, spins parallel and spins antiparallel, reflects the spin content of the target nucleon, Figure 8.16.

So, what did HERMES find? According to HERMES, up valence quarks spin in the same sense as the parent proton, whereas down quark spin is oriented in opposition. But there is a shortfall: the HERMES team confirmed that just 25% to 30% of the spin of the proton and neutron is due to the spins of the quarks inside. From theory, the figure should be more like 60%, or even 100%. This mismatch is not satisfactory. In fact, as originally revealed, it was so unsatisfactory it gave rise to a new phrase in particle physics: the spin crisis.

The name spin crisis is now a little outdated, since the crisis has been around long enough to abrogate the immediacy implied by the word crisis: the patient is chronic, rather than acute. But the core problem, how to explain the nucleon spin, remains, even though a better name these days is perhaps "proton spin problem." As it loomed in 1988, however, when the EMC at CERN reported experiments in which they scattered polarized muons from polarized protons, the spin crisis was stark. EMC's finding – "that the total quark spin constitutes a rather small fraction of the spin of the nucleon" – came as a major shock. In fact, their results even allowed for the prospect that none of the nucleon's spin came from the quarks. On top of that, the EMC's conclusion was at odds with a 1983 SLAC experiment that, more limited in scope, gave no warning of the surprise to come.

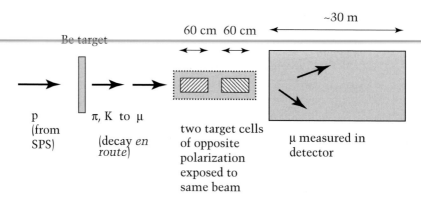

FIGURE 8.17 Schematic of the SMC, polarized muon-polarized proton experiment. Pions and kaons, created by protons striking beryllium, decay to give automatically polarized muons. Two oppositely polarized target cells, cooled to 50 mK, are exposed to the muon beam simultaneously.

An expectation of something big, say 60% or 100%, when experiment returns something small and actually statistically consistent with zero, is not a situation that any scientist accepts laying down. Following in the footsteps of the EMC was CERN's Spin Muon Collaboration, or SMC, who were charged with the task of getting to the bottom of the spin issue. Like the EMC, they exploited another of nature's free lunches: pions, created by bombarding a beryllium target with protons from CERN's SPS, decay via the weak force to give muons which are congenitally polarized, a natural longitudinally polarized probe, Figure 8.17. Meanwhile, at SLAC, several groups of experimenters used a polarized electron probe to explore spin structure. The SLAC data are the more precise; the CERN data in turn provide greater cover. In particular, SMC could reach smaller x, right down to x values of 0.003, where the struck quark carries just 0.3% of the parent nucleon's momentum.

The SLAC, SMC, and HERMES experiments are all deep inelastic scattering experiments: a lepton ricochets off the target into a detector, where its energy and trajectory are measured. This time, however, the target and the probe particle – electron, positron, or muon – are both polarized. And the objective of the exercise is to extract not just structure functions, the momentum distributions of the scattering centers inside the nucleon, but also *polarized* structure functions.

There are two "common" polarized structure functions which, like their unpolarized counterparts, both depend on the host nucleon momentum fraction. The most easily accessible to experimenters is the longitudinal structure function. This manifests itself as the difference in scattering probabilities between a probe and nucleon with spins aligned either parallel or antiparallel along the probe particles' direction of motion. The difference in the two cross-sections is proportional to the longitudinal structure function. Alternatively, using a transversely polarized target, in which the target spin is oriented across the beam direction, experimenters can extract the other player, the transverse spin distribution of the nucleon.

The longitudinal structure function has a very neat and tidy quark model interpretation as the total longitudinal spin due to all quarks and antiquarks in the nucleon. All momentum fractions are included and, in the usual way, the up quark contribution carries four times the importance of the down quarks, due to the difference in electrical charges of the two quarks. The structure function describes the distribution of the difference between quarks whose spins lie parallel to the host proton and quarks whose spins lie antiparallel to that of the proton. Including u, d, and s quark flavors and all three corresponding antiquarks covers all bases, even if the simplest view of the nucleon, comprising three quarks, suggests using just the u and d quark spin contributions.

Experiments show that the simplest quark model approach, which gives such respectable agreement with measured magnetic moments, seems to run aground when it comes to predicting the nucleon spin. According to this approach, the difference between the u quark spins parallel to the host proton's spin, and that opposite to the proton's spin, is $4/3$. The corresponding figures for the d quark and s quark are $-1/3$ and 0, respectively. The total is therefore one unit of spin. In other words, 100% of the nucleon spin is due to the three valence quarks, at variance with what experimenters see. In short, the simple quark model is too simple to explain the spin of the nucleon.

Sum rules were made for breaking
Predicting the form of the spin structure functions from pure thought is something theorists would love to be able to do, but cannot. What they *have* done is write down a couple of sum rules, expressing properties of the spin structure function aggregated over all possible momentum fractions.

One of these rules, due to Bjorken, stands out like a beacon above the sea of unknowns that is spin structure physics. The Bjorken sum rule expresses the difference between the longitudinal spin structure functions of the neutron and proton in terms of weak interaction parameters. That the neutron and proton spin structures are related in some way is not a huge surprise perhaps, given that protons and neutrons are linked by isospin symmetry. And recalling that a neutron converts into a proton via a weak decay process motivates the appearance of weak interaction parameters.

Bjorken put together his rule in 1966, years before QCD, using quite general arguments rather than some detailed quark model or QCD calculation. He wasn't thrilled with the results of his labor, branding the result "worthless," assuming that no experiment could ever check it out. In this he underestimated the sagacity of experimentalists, who have recently been able to confirm that Bjorken's rule is correct. That checking required neutron spin data, which first became available in 1993, with really accurate data only appearing in 1997.

With the emergence of QCD, Bjorken's sum rule is revealed as one of its predictions, the modern version including corrections that depend on the strong coupling parameter. The fact that the suitably QCD-corrected Bjorken sum rule agrees with modern proton and neutron data to within a few percent is seen as a success for QCD: a discrepancy with the Bjorken rule would be a serious problem.

In 1974, John Ellis at Caltech and Robert Jaffe at MIT, pointed out what was for them a difficulty with the Bjorken rule. Checking out Bjorken's rule requires both proton *and* neutron data, but, at that time, there was no chance of any neutron data for years to come. So Ellis and Jaffe set out to find a new sum rule that they could apply to either protons or neutrons alone. Their Ellis–Jaffe sum rule, along similar lines to Bjorken's, ignored QCD and virtually all quark model ideas, but to make progress they had to insert one additional assumption: that strange quarks and antiquarks contribute nothing to the nucleon spin. They did not even assume that there are no strange quarks in the nucleon, only that if they are there then their spins cancel out, so their net spin contribution is zero. Given that there shouldn't be much in the way of strange quarks in the nucleon anyway, this does not seem too harsh.

Subject to QCD corrections added in more recent times, the Ellis–Jaffe rule is a plausible hypothesis, though not a prediction of QCD.

Which is perhaps a good thing, since the rule is violated. It gives a value for the longitudinal spin structure function that differs from the value quoted by experimentalists. It was in terms of such a discrepancy that the EMC originally unleashed the spin crisis. In the language of quarks, the Ellis–Jaffe rule implies that around 60% of the nucleon spin should be due to the spins of the quarks themselves – a value that is double what the experimenters find.

Where has all the spin gone?

Though it falters in the face of experiment, the Ellis–Jaffe sum rule is not weak enough to simply be dismissed. Yet it is not strong enough to force some total theoretical re-vamp. The fact that it disagrees with experiment is best viewed as an opportunity to learn something.

For one thing, translating experimental determinations of the longitudinal structure function into statements about the amount of spin carried by quarks is a ticklish business. In fact, the subtleties are such that, for several years, SLAC experimenters and the SMC disagreed in their conclusions concerning the total fraction of nucleon spin carried by quarks. Only in 1997 was the misunderstanding finally laid to rest, an issue of analysis and extrapolation rather than actual data.

One of the big questions, which formed part of that debate and which continues to tantalize, centers on how to incorporate contributions from the full range of possible quark momenta. The difficulty is that it is quarks having a tiny fraction of the nucleon momentum which dominate the quark contribution to nucleon spin, yet these are the contributions experimenters find hardest to measure. In fact, getting an answer to compare to theory means extrapolating to both ends of the momentum range, where there are no measurements, an exercise entailing some acts of faith at the low end of the momentum range in particular. If there is something going on here that experimenters have not seen, and theorists have not dreamed of, then the numbers might be all wrong. This could explain the failure of the Ellis–Jaffe rule.

Then again, maybe the extrapolations are fine and there really is a difference. Then it could be that the assumption of no strange quark contribution to the spin is wrong. The Ellis–Jaffe prediction and experiment can be shoehorned into line if strange quarks contribute around 10% of the proton spin, oriented *opposite* to the proton's own spin.

But perturbative QCD calculations for the generation of sea quarks make no such prediction for strangeness polarization. On top of that, unpolarized deep inelastic scattering experiments suggest that strange quarks carry a modest 3% of nucleon momentum, which raises the question of why the strange quark spin contribution might be so much larger.

Or, perhaps the strange quark contribution to the nucleon spin is small and spin from other sources holds the key. These might be the relative motion of quarks (in other words quark angular momentum) or gluon spin, or relative motion between gluons (gluon angular momentum), none of which feature in the simplest quark model. Recently, theorists including Anatoly Radyushkin, of the Jefferson Lab in Newport News, Virginia, have introduced so-called generalized parton distributions that attempt to embrace the correlations between partons in the nucleon. On offer is some understanding of the contribution of relative parton motion to the overall nucleon spin, and even the distribution of quarks and gluons across the nucleon, a kind of nucleon tomography. But the price is better experiments that can identify the collision debris.

HERMES belongs to a new generation of spin structure experiments designed to do just that. To date, spin experiments have focused on measuring just the scattered probe particle. But HERMES is specifically designed to be able to detect and identify individual fragments emanating from the target nucleon as a result of the probing. These particles, for example pions and kaons, will contain the quark that was struck, Figure 8.18. So, for example, the number of kaons that contain strange quarks will indicate the number of collisions involving strange quarks in the sea of a target nucleon. In this way, experimenters can pin down the strange quark contribution to the nucleon's spin. Indeed, HERMES experimenters have shown that the contribution of the sea quarks is small, and they have discounted the notion that strange quarks contribute a spin in opposition to that of the parent nucleon. They also saw little evidence of cancelations between the spin contributions of sea quarks and those of the valence quarks, a route that some researchers proposed as a way of understanding the spin crisis.

Experimenters are also exploring ways of extracting information about the gluons' potentially significant contribution to nucleon spin. Theorists can calculate how spin structure's dependence on the probe

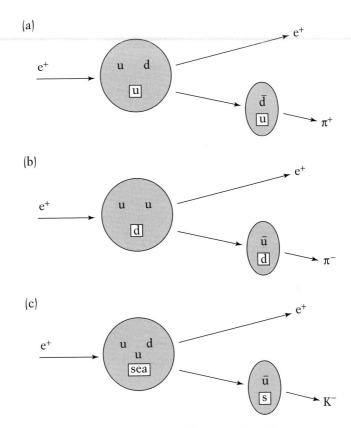

FIGURE 8.18 Identifying pions and kaons produced in muon–proton or positron–proton collisions allows experimenters to measure contributions from different quark flavors. The extra quarks and antiquarks appearing in the pions and kaons come from the vacuum, discussed in Chapter 6.

wavelength is in turn influenced by the spin distribution of the gluons. Experimenters are already trying to exploit this relationship to extract gluon information, while other experiments also at an early stage are seeking to measure the "poor-relation" transverse spin structure function. According to the simple quark model, this also has a simple and tidy interpretation: it's zero, assuming quarks are longitudinally polarized. The first, and rather tentative, measurement of transverse spin structure came from SMC in 1994, a second from the E143 experiment at SLAC in 1996. A result different from zero would show spin-dependent correlations between quarks and gluons inside the nucleon, an example of a higher-twist effect. And HERMES experimenters have confirmed that gluons really do contribute to nucleon spin, spinning in the same sense as their host particle.

Simplistic quark model ideas, which seemed to work well enough for baryon magnetic moments, suggest that 100% of the nucleon spin is due to that which sits on the three valence quarks. Assuming no strange quark polarization in the nucleus, the Ellis–Jaffe sum rule instead uses general principles to suggest that quark spins are responsible for 60% of the total. Experiments give a number somewhere between 25% and 30%. Fortunately, the Bjorken sum rule, a QCD prediction for the difference between the spin structure of the proton and neutron, fares rather better. But, after all that, the big question still remains – exactly where is the spin of the nucleon? The money is on a helping of spin from the quarks themselves, with a generous portion from gluons, served up with the relative motion of quarks.

POMERONS: THE RESURRECTION

The fact is it's not necessary to look to the latest spin structure or sea quark distribution data for challenges to throw at QCD. Even before QCD's birth, experimentalists had started gathering information on basic proton–antiproton scattering – information that remains something of a dilemma for QCD, since the theory has yet to fully accommodate it. The strong interaction data, on total cross-sections and elastic cross-sections, for example, found a *description*, at least, in terms of Regge theory and in particular in terms of an entity called the Pomeron. Nothing succeeds like success, and success has kept the Pomeron afloat for over three decades, for what it does it does well. But the traditional Pomeron and QCD are uncomfortable bedfellows. Though QCD might be the fundamental theory, the embarrassing fact is that Pomerons outperform it for most high-energy proton–proton and proton–antiproton collisions.

So what is a Pomeron? Results from HERA, in particular, have shed new light on this mysterious object that most physicists are wary of calling a particle. It's a kind of a particle, maybe, or at least a sort of recognizable state within protons and other strongly interacting particles. It's something to do with gluons. Whatever it is, the Pomeron, or some equivalent object, plays an important role in strong interactions.

A family myth

Total cross-sections measure the overall strength of an interaction. At low energies, total cross-sections for strongly interacting particles show peaks due to resonances, and in many instances an overall drop as the collision energy rises. At larger energies, however, a new feature comes into play, first noticed by experimenters at the Serpukhov proton synchrotron, 90 km south of Moscow, back in 1971. There's an across-the-board rise – in proton–proton, pion–nucleon, kaon–nucleon collisions, and more besides – in cross-section with increasing collision energy. It's as though particles expand and become more opaque at high energies.

A theory of the strong force should be able to explain this kind of basic feature. In fact, there is a simple way of at least representing it, if not explaining it, and that's Regge theory – the description of hadronic scattering in terms of exchanged mesons. In 1992, using Regge theory, two British physicists, Sandy Donnachie from the University of Manchester and Peter Landshoff from the University of Cambridge, published a simple power-law equation describing a whole gamut of strong interaction total cross-sections over a huge energy range. So Regge theory can't be all bad. Besides, though it may not be a fundamental blue-ribbon theory, and it doesn't sit too easily alongside QCD, it is none the less a product of a whole clutch of eminently reasonable physical principles, so can never be dismissed out of hand.

Part of the Donnachie–Landshoff result corresponds to the conventional Regge meson family approach. One such family, for example, includes the ρ, a_2, ρ_3, and a_4 mesons, another comprises the ω, f_2, ω_3, and f_4 mesons. This exchange mechanism accounts for the falling total cross-section typical of the lower energy end of the scale. The rise in cross-section at higher energies is described by the exchange of a second, related entity, introduced to perform exactly this task in 1961 by Geoffrey Chew and Steven Frautschi, and named after the Russian physicist Issak Pomeranchuk. It's called the Pomeron.

Pomeron exchange is exchange of a *mythical* family of mesons. That is, the Pomeron fits in with the Regge theory scenario as exchange of a set of particles whose properties do not tally with any known mesons. At the very highest energies, Pomeron exchange should dominate, and numerous experiments testify that it does. As an explanation of an underlying mechanism, Pomerons are less than convincing. But as a description, however, Pomerons work well, better than QCD in their particular domain.

The pomeron revealed

That domain is the so-called soft interactions, in which little momentum is passed between colliding participants. The bulk of the total cross-section is soft, likewise elastic scattering, in which particles simply bounce off one another. Soft interactions spell trouble for QCD, because the energies involved are too small to guarantee perturbation theory's validity. But one class of soft interactions is providing insight into the existence and nature of the Pomeron, and hinting, at least, at common ground between it and QCD.

In some proton–antiproton collisions, one particle might be hit rather harder than in a purely elastic collision, causing it to ring like a bell, then decay or fragment. Such a collision, in which one initial particle emerges unscathed though with slightly reduced momentum and the other breaks up, is termed diffraction scattering, or sometimes diffractive dissociation: diffractive scattering makes up about 15% of proton–(anti)proton collisions. In a typical proton–antiproton diffractive scattering event, one initial particle, say the antiproton, continues on with something like 95% of its original momentum. In the opposite direction is a shower of particles corresponding to the fragmentation of the other particle, in this instance the proton. Significantly, the fragments will be directed into one hemisphere of the reaction region about the beam direction, with large parts of the detector devoid of particles.

In 1985, Peter Schlein at the University of California Los Angeles and Gunnar Ingelman at CERN reasoned that Pomerons, since they mediate a strong interaction, should themselves be composed of gluons and perhaps even quarks. According to their scheme, in the proton–antiproton diffractive collision outlined above, the antiproton throws off a Pomeron and continues on its way. The proton then smashes into the Pomeron, spawning a shower of particles that, since the Pomeron carries only a fraction of its parent's momentum, are projected in the direction of the proton beam. The proton–Pomeron collision itself is to be viewed as a full-on smash between a gluon in the Pomeron, and a quark or gluon in the proton, Figure 8.19. So in general at least some diffractive scattering events, in which either the proton or antiproton remains intact, should show distinctive patterns of jets with pairs of jets arising from the Pomeron–proton or Pomeron–antiproton "subcollision." This contrasts with jet production in which *both* proton *and* antiproton are smashed.

(a)

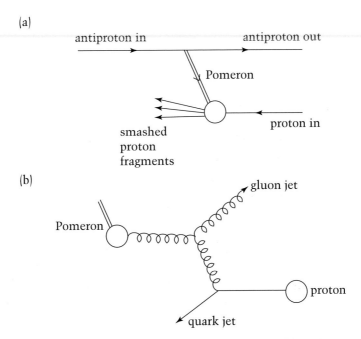

(b)

FIGURE 8.19 (a) Diffractive proton–antiproton scattering via Pomeron exchange. (b) Hard scattering "sub–collision" between a gluon from the Pomeron and a quark from the incoming proton.

The UA8 collaboration at CERN looked for, and found, exactly this kind of diffractive scattering with jets, their first results from a test run appearing in 1988. With more data, the UA8 team, including Schlein, showed that pairs of jets are produced back to back and with characteristics typical of QCD jets. But what they could not do was demonstrate the validity of dividing the entire process into two parts, a Pomeron emission plus a scattering from the Pomeron's constituents. Nor could they measure the implied structure function of objects inside the Pomeron, or test the idea that the Pomeron itself is somehow distributed inside the proton.

HERA and the Tevatron have since entered the fray, with the CDF and D0 collaborations spotting diffractive events in proton–antiproton collisions at the Tevatron, and showing that the Pomeron is indeed mostly gluons, but with a measurable quark content. The Fermi-lab experimenters have, in addition, identified collisions in which a Pomeron launched from a proton collides with a second Pomeron from the antiproton. At HERA, diffractive scattering means that the proton

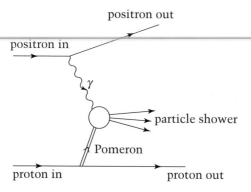

FIGURE 8.20 Diffractive deep inelastic scattering at HERA.

continues on its way unscathed, and in the Pomeron picture emits a
Pomeron that collides with the incoming positron, Figure 8.20. This
gives rise to deep inelastic scattering events in which the proton does
not break up – these so-called diffractive deep inelastic scattering events
were first identified by the H1 collaboration in 1995. Their results support the division of the overall scattering into a Pomeron emission
part, followed by hard scattering from point-like constituents within
the Pomeron. In addition they, and the ZEUS collaboration, were able
to measure the distribution of Pomerons within the proton and the
distribution of the point-like scatterers within the Pomeron itself.

   After three decades as a mythological device for describing scattering data, the Pomeron has assumed some sort of cloak of reality,
though it remains rather enigmatic. Some believe that it is a state, or
set of states, within the proton, and others that it's a bound state of gluons. Maybe it's a glueball – certainly an attractive option in terms of
economy of physical ideas. Other researchers think that the Pomeron
is just a useful piece of terminology for processes whose natural explanation lies within QCD, a piece of terminology that one day will be
consigned to the scrap heap. For some, from the more traditionalist
camp, the Pomeron remains firmly part of Regge theory, and is intrinsically non-perturbative in nature. This is the Donnachie–Landshoff
or non-perturbative Pomeron, or soft Pomeron, and is sometimes pictured as a pair of (non-perturbative) gluons. Pomerons that engage in
hard collisions are attractive to practitioners of perturbative QCD, however, leading to widespread support for the alternative hard Pomeron,
alias the perturbative Pomeron, or BFKL Pomeron. This child of QCD

is tackled using the BFKL equation to incorporate contributions from a whole ladder of gluons.

Whatever the true nature of the Pomeron, one thing is certain. The processes it successfully describes, elastic and diffractive scattering, are part of strong interaction physics. The strong force is supposed to be described by QCD. Therefore, eventually, the Pomeron and QCD must somehow unite.

9   Much ado about nothing

"Nature abhors a vacuum," wrote the irreverent French physician and satirist, François Rabelais. If "vacuum" has its conventional meaning, the insides of a box with the air pumped out and the lights switched off, then Rabelais turns out to have been ahead of his time: nature does abhor a vacuum, so much so that it has filled the vacuum up. The ether, the mythical substance that nineteenth-century scientists believed filled the void, is a reality, according to quantum field theory.

Field theorists do not abhor the vacuum, they redefine it. In quantum field theories such as QED and QCD, the vacuum of everyday speech becomes instead the condition of lowest energy, that which remains when everything is switched off, a level below which the system cannot go. It is like the residual hiss of the music center amplifier that endures even when the volume dial reads zero, or the dregs left in the bottom of the "empty" coffee cup, or perhaps the last, tiniest ripples on the "flat" pond surface.

There was a time when the vacuum state, where all that remains is a residual energy that quantum mechanics term "zero-point energy," was presumed to be of little physical consequence. It was simply about where one set the zero of some energy plot, and it was changes relative to this lowest state that showed up in experiments. In due course, however, the vacuum has revealed itself in some surprisingly direct ways. On the grandest of all scales, the vacuum may be a source of the background energy density of the cosmos, expressed through something called the cosmological constant of general relativity, and may therefore have a bearing on the overall structure of the Universe. In QCD, physicists now realize that the vacuum is a crucial aspect of the theory, a key player in its global symmetry, with roles in confinement and understanding the creation of a new state of matter, a "gas" of free quarks and gluons called the quark–gluon plasma. But the best place to really see the vacuum strut its stuff, however, is in electrodynamics.

THE PUSH OF EMPTINESS

Sit two conducting plates parallel to one another in a closed box. Remove all the air, any source of heat, vibration, and so on, and see what happens to the plates. By rights, nothing should happen: the plates are in a vacuum and there is nothing there to cause anything to happen. In practice, the plates are forced together, and the source of that force is the residual energy of the empty space around them.

This is not some theoretical fancy. Experimentalists can see this attractive force between plates, called the Casimir effect after Dutchman Hendrik Casimir. Though Casimir's explanation dates from 1948, the first really convincing demonstration of the effect was not until 1997, in an experiment by Steve Lamoreaux, then at the University of Washington in Seattle. Lamoreaux's experiment, which started out as a student project, was cobbled together from bits and pieces of equipment lying around his laboratory. In essence, it was a kind of torsion balance. One plate, curved to circumvent the problem of keeping two flat plates parallel, faced a second flat plate secured to one end of what was, in effect, a freely rotating rod suspended by a fine wire about its mid-point. The other end of the rod formed part of an electrical capacitor, so the balance arm could be held steady by applying small voltages across the capacitor. To do the experiment, which was so sensitive it could sense his presence in the room by the tilting of the floor, Lamoreaux inched the curved plate towards the flat plate attached to the rod, and measured the voltage that he needed to apply to the capacitor to keep the rod steady. The force between the two plates clearly showed the plate-spacing-dependence characteristic of the Casimir force.

The force pushing the plates together is tiny. For two centimeter-square flat plates a micron apart, the Casimir force roughly equals the gravitational attraction between two mugs of coffee separated by a finger's width, or the electrical attraction holding an electron in a hydrogen atom. The origin of the Casimir force lies in the quantum zero-point energy, residual energy that can never be removed or switched off. In quantum mechanics, zero-point energy can be viewed as a consequence of Heisenberg's uncertainty principle: if, in its lowest energy state, a collection of quantum objects was truly at rest, there would be no uncertainty in the momentum (it is zero for all) and the uncertainty principle would be violated. In the language of quantum fields, the field, for example the electromagnetic field, can never quite be switched off. Instead, it fluctuates without ever vanishing to zero, manifesting as virtual photons spanning a range of energies that, in a fleeting moment, pop into and then out of existence. This, then, is nothing, the state of lowest energy, the vacuum: a seething cauldron of virtual photons.

This lively ether, which is responsible for "noise" in lasers, also explains the force between the plates in Lamoreaux's experiment. Between the plates, the possible range of photons is limited since, viewed as a wave, a photon must fit in the space between the plates. So photons having a wavelength more than half the inter-plate spacing are not allowed. Outside the plates, however, there is no such restriction, and even the longer wavelength photons get a go at popping into and out of existence. The relative excess of photons outside the plates results in a net overall "radiation pressure" that pushes the plates together. In other words, squeezing the quantum system into a confined space modifies the zero of energy, and it is this change that's visible as the inter-plate force. Put yet another way, the vacuum is not simply empty, but undergoes fluctuations: introducing objects, even objects that are large on the scale of quantum effects, can modify those fluctuations in a measurable way.

One consequence of the Casimir effect is that photons traveling between the plates move at a velocity *greater* than the conventional speed of light. It turns out that quantum uncertainties in emission and reception times of light pulses prevent superluminal information transmission, so the Casimir effect offers slim pickings for science fiction aficionados. At a more mundane level, as a result of the Casmir effect, the solvent pentane spreads on water, whereas some other similar chemicals form globules. Another consequence is the propensity of superfluid helium-4 to creep up the walls of its container. The Casimir effect also contributes to the properties of colloids, systems such as inks and clay slurries, in which tiny particles are dispersed in a fluid. In the 1940s Jan Overbeek, at the Philips Laboratory in Eindhoven, the Netherlands, was studying just such a system, a dispersion of powdered quartz, when he discovered a gap in the prevailing theory of colloids. The modification Overbeek introduced intrigued a colleague, one Hendrik Casimir. Casimir realized Overbeek's modification was due to the vacuum fluctuation effect that now bears his name.

The role of the vacuum is most easily witnessed via quantum electrodynamics, but it is a general feature of quantum field theories.

Impose some boundary on the vacuum, for example conducting plates or particles of ink, and the fluctuating quantum fields are modified. The trick is to know how to spot the consequences.

Nobody has done a QCD equivalent to Steve Lamoreaux's two-plate Casimir test, but the QCD Casimir effect is a reality. One place where it rears its head is in the MIT bag model, discussed in Chapter 7 as a model of nucleons. Here, the Casimir effect makes up a few percent of the total energy of the system. But perhaps the real moral of the Casimir tale is that the vacuum has significant and perhaps unexpected physical consequences, and that these must be taken seriously. For QCD, the vacuum is everywhere, spoiling symmetry, spawning exotic new particles, and implicating itself in confinement.

SYMMETRY, QCD, AND EMPTY SPACE

A humble grain of salt sat dormant on the kitchen table is a lesson in how misleading commonsense can be. The apparently quiescent grain is in fact buzzing with vibrations caused by heat-induced agitation of its atoms. Much as light waves are quantized as photons, these vibrations of a crystal lattice come in elemental packets called phonons. A single grain of salt at room temperature is buzzing with around 10^{18} of these phonons.

The presence of phonons can be traced to symmetry, or rather a lack of it: the crystal lattice, an ordered regular array, lacks the smooth translational and rotational symmetry of the interactions binding atoms together in the crystal. Magnetism – particle physicists' favorite example – shows a similar effect.

As already discussed in Chapter 5, when iron is cooled from above 770 °C to below that temperature, it spontaneously magnetizes. Rank upon rank of individual atomic magnets in a single domain align with one another, the mutual aligning force between neighbors exceeding temperature's drive to randomize their relative orientations. For a chosen domain, there is one obvious observation that turns out to have deep and far-reaching consequences: there is no way to predict which direction the little atomic magnets will democratically select as the one along which they will all align. To put this in perspective, there is nothing in the force between neighboring atoms, or in the master equation describing the interaction between neighboring atoms, which says anything about any particular direction. The magnetic system selects one direction from infinitely many possibilities *spontaneously*. This is an example of what physicists term spontaneous symmetry breaking: the symmetry of the lowest energy state is less than that of the interaction in question.

Some people say that the expression "spontaneous symmetry breaking" is a little misleading. The lowest energy state, or ground state, of the magnet, that in which all atomic magnets are aligned, does not share the rotational symmetry of the basic interaction between those atomic magnets. So the symmetry of the interaction is just hidden, but hasn't gone away. Hidden symmetry, they argue, is maybe a better expression than spontaneous broken symmetry. But it's the latter that has taken hold.

Spontaneous symmetry breaking features throughout particle physics, and a little more magnetism shows why. In the ground state, the directions of the atomic magnets in a ferromagnet all point the same way. If, for instance, all the spins are pointing up, a horizontal slice through the magnet is a little like a wheat field, each stem of wheat representing the arrow of an atomic magnet, the head of wheat being the arrow head. Mimicking the way a gentle summer breeze will send ripples through the wheat field, warmth applied to the iron sends ripples through the orientation of the atomic magnets, so that tips of the direction arrows of a plane of atoms move in a way that mimics the wave-like motion of the heads of wheat.

In much the same way that light waves are quantized, so too are these magnetic "arrow orientation waves," giving rise to massless particles called magnons. Magnons are analogous to photon "particles of light," and reveal themselves in neutron scattering experiments.

Magnons are a consequence of spontaneous symmetry breaking. Quite generally, a quantum system in which symmetry is hidden, a system in which some continuous global symmetry visible in the master equation of the theory is absent from the ground state, invariably contains massless, spinless particles such as phonons or magnons. Particles such as magnons and phonons that owe their origin to spontaneous symmetry breaking are collectively known as Goldstone bosons, after the Cambridge physicist Jeffrey Goldstone, who deduced their existence back in the early 1960s.

In the magnetization example, each of the magnetization directions is equally available and has equally low energy, so in principle a rotation from one to another would cost no energy. But the atom magnets have selected just one. Then the rotations that might have otherwise been available to turn one ground state into any other find themselves reduced to simply causing the wobbles, the "arrow orientation waves," about the chosen direction, and these (quantized) waves are the magnons. So the "hidden" rotational symmetry of the interaction resurfaces as the source of magnons.

Spontaneous symmetry breakdown with attendant massless particles is also a feature of QCD. Here, the lowest energy state is the vacuum, which in a field theory plays the role of the "ground state" of a system of atomic magnets. And the "massless" particles turn out to be old, familiar friends: the pions.

Mesons from nothing

It works like this. Compared to the energy typical of strong interactions, the masses of the three lightest quarks – the up, down, and strange – are approximately the same. If the quark masses *are* the same then, so far as QCD is concerned, the three quark flavors are freely interchangeable since QCD cares little for quark flavor. This represents a global symmetry of QCD, a symmetry under interchange of three quark flavors, and means that the theory automatically respects the key isospin and strangeness conservation rules.

In fact in strong interactions typical energies are so high that it raises few eyebrows to set these same three quark masses to zero. So take the equality argument one step further and insist that the up, down, and strange quarks not only have equal mass, but that they are *all* of mass zero. The result is another hike in total symmetry.

Then something quite new happens. Since these quarks are now massless, they travel at the speed of light. Depending on whether their spins are oriented in the direction of travel, or in opposition to it, the quarks are *either* right- or left-handed, respectively. But there is no simple handedness-changing capacity in QCD for massless quarks. A left-handed quark can emit a gluon but must remain a left-handed quark, whereas a right-handed quark remains a right-handed quark. In fact, left-handed and right-handed massless quarks lose the power to talk to one another, and remain in their separate worlds. Yet there are still three different flavors, so now the flavor mixing, which mathematically means flavor rotations, can happen for left and right quarks independently. So going massless means a further twofold increase in symmetry, over and above the global symmetry of QCD with three equal-mass quarks.

Transformations that act on something left-handed to give something that is still only left-handed, or on something right-handed to yield something right-handed, are termed chiral, from the Greek *kheir*, meaning hand. Driving a car to town is a chiral transformation, since what started out as a left (or right)-handed journey finishes up as a left (or right)-handed journey, despite all kinds of "interactions" along the way. Particle theorists describe symmetry operations that stir left-handed quarks into more left-handed quarks, and corresponding transformations for the right-handed world, as chiral symmetry transformations.

Global chiral symmetry appears not to make a big splash in real-world strong interactions, however. If it did, then every hadron would have a partner of the same mass, and various other properties, but differing in intrinsic symmetry with respect to mirror reflection. But physicists don't see these "parity partners." Instead, the chiral symmetry of QCD is hidden, or in other words spontaneously broken.

The source of that symmetry breaking is the vacuum. The vacuum fails to respect the chiral symmetry of the theory, and thus the vacuum's symmetry is less. In the ferromagnetism example above, the aligned multitudinous atomic magnets gave a lowest energy configuration having a reduced symmetry relative to the interaction between those magnets, yielding Goldstone bosons. The chiral QCD case runs exactly parallel, the vacuum providing the analog of the lowest energy state, and its disregard for chiral symmetry mimicking the aligned atomic magnets' disdain for the rotation-independence of the interaction.

In the ferromagnetic case, symmetry breaking results in magnons. In QCD, the result of breaking chiral symmetry is a fistful of mesons: three pions, π^0, π^+, and π^-, four kaons, K^0, $\bar{K}^0$, K^+, and K^-, and last but not least the η. When the symmetry that is broken is exact, then the Goldstone bosons are massless. If the symmetry is not exact, but only approximate, then the result is, instead, low-mass spinless particles called pseudo-Goldstone bosons, instead of massless Goldstone bosons. Real up, down, and strange quark masses, though small on the energy scale of strong interactions, are not exactly zero, so real QCD does not have an *exact* chiral symmetry only an *approximate* chiral symmetry. So chiral symmetry breaking in QCD gives particles having a small mass, the eight mesons listed above. Through this mechanism, meson masses are linked to the (current) quark masses.

Indeed, chiral symmetry breaking is the route to estimating the up quark mass as 4 MeV and the down quark mass of 7 MeV, the values given in Table 4.6 in Chapter 4. The effective masses of quarks bound into observable particles such as pions, protons, and neutrons, in other words the constituent masses of around 300 MeV, are due to the interaction of quarks with the surrounding chiral-symmetry-breaking vacuum. In this way, spontaneous symmetry breaking is responsible for effective quark masses. The chiral symmetry aspects of QCD are also the launch pad for another calculational tool that works over longer distance scales where conventional QCD perturbation theory breaks down. This is chiral perturbation theory, which takes as its starting point the perfect chiral symmetry of a hypothetical QCD with massless quarks, then allows actual real masses to enter as perturbations. The resulting effective field theory is a fruitful source of low-energy results, including the quark mass estimates above.

So what does the vacuum of QCD look like, and how can this "empty space" have symmetry-breaking properties? Magnetism once again provides a clue. There, the Goldstone particles, the magnons, are ripples in the very magnetization whose alignment along some arbitrary direction breaks the symmetry. In QCD, mesons arising from symmetry breaking imply that the vacuum must be some kind of quark–antiquark backdrop, which physicists have dubbed the quark condensate. It's ripples in this quark condensate vacuum that are manifest as real, observable mesons.

The quark condensate is an ocean of quark–antiquark pairs, the spins of a pair opposed so as to guarantee Lorentz invariance and minimum energy. So the condensate builds in a correlation between the quarks and antiquarks that pop into and out of existence courtesy of the zero-point energy. Left-handed quarks now know about right-handed antiquarks across a finite-sized portion of space determined to be roughly the size of a nucleon, and it is this that spoils chiral symmetry, since left and right communicate. (Figure 9.1.)

Actually, the quark condensate turns out to be just part of the story. The vacuum is also home to a gluon condensate and several other components besides. It's through these condensates that the background quark and gluon presence enters the QCD sum rule method so useful for calculating hadron properties, as discussed in Chapter 7. However, the structure of the vacuum of QCD remains one of its great unresolved mysteries. Its riddles notwithstanding, theorists believe it's

FIGURE 9.1 Yoichiro Nambu, pictured here in 1994, was one of the first to appreciate the role of the vacuum and spontaneous symmetry breaking. Another of Nambu's many contributions to particle physics was the suggestion of the color quantum number, along with Han and, independently, Greenberg. Credit: The University of Chicago.

pretty certain that the vacuum plays a key role in confinement. In the language of the vacuum, the dual-superconductor model of confinement, also discussed in Chapter 7, envisages a pion as a quark–antiquark pair linked by a tube of chromo-electric field, and living in a vacuum that is a condensate of chromo-magnetic monopoles and antimonopoles, a chromo-magnetic superconductor. When theorists really do understand confinement, they will presumably then hold the key that unlocks the vacuum's secrets, at least in part. Or is it that they must understand the vacuum first, and confinement will follow? Whichever it is, no one is betting on a quick solution. Theorists have been pondering quark condensates and symmetry breaking since even before the dawn of QCD, stemming from groundbreaking work in the 1950s on pion decay by Marvin Goldberger and Sam Treiman. The 1960s was an era when symmetry reigned: there was Goldstone's theorem, and later the application of spontaneous symmetry breaking ideas to gauge symmetries by Peter Higgs and others, ultimately vital to the creation of the electroweak theory, and of course Gell-Mann and Ne'eman's Eightfold Way, quarks, and later Greenberger's color. But so far as the strong interaction was concerned, the progress of the 1960s was achieved in the absence of an actual theory of the strong force. By the time QCD came along, theorists already knew that a strong force theory should have an approximate chiral symmetry, and that that symmetry should be broken spontaneously courtesy of a quark–antiquark condensate linked to pion masses and decay rates. The fact that QCD could naturally accommodate the required symmetry patterns was a factor in its rapid assent.

Symmetry accounting

For all its apparent successes, chiral symmetry and its breaking actually harbors a problem. A little symmetry bookkeeping reveals what's wrong. Equal-mass up, down, and strange quarks endow the master equation of QCD with an additional symmetry, as described above. This symmetry is the group $U(3)$ which, ignoring technicalities, is just the product of two old friends, $SU(3)$ and $U(1)$. Making the quarks massless introduces a second copy of $U(3)$, in other words a *second* copy of the product $SU(3) \times U(1)$. One $SU(3)$ portion is that of the Eightfold Way, grouping strongly interacting particles into families. The pions and other such mesons familiar from these family groupings are linked to the spontaneous breaking of the second $SU(3)$ portion. And one $U(1)$ portion guarantees baryon number conservation. So all looks well, except for a single little $U(1)$ factor that seems to play no role in the cast of observed particles. Perhaps it's another victim of symmetry breaking, yes? Apparently not – there's no corresponding lightweight broken symmetry type of meson either. Tidying up what might seem a little technical detail concerns particle theorists so much they have given it a name, "the $U(1)$ problem." It was Weinberg who first appreciated the significance of the $U(1)$ problem, and who insured it stayed on the theoretical play list. The resolution of the $U(1)$ problem points to another, and perhaps surprising, layer of vacuum structure.

WINDING THROUGH THE VACUUM

One might have thought that, this far into the story of QCD, every last little piece of symmetry – or symmetry breaking – would have been tracked down and brought to account. Not so. This one last $U(1)$ symmetry, currently running wild, would again imply a swathe of particles absent from the real world, partners to all the common species such as protons identical in every way but having opposite intrinsic symmetry under mirror reflection. The absence of such parity partners suggests that the $U(1)$ symmetry is in fact broken in some way, but the question then is how, given that there's no suitable meson flagging the presence of spontaneous symmetry breaking. The answer reveals an unexpected richness in the vacuum of QCD. A little gardening will show how.

Picture a peaceful, sunny, suburban weekend, Chardonnay by the pool, bees buzzing round the flowerbeds, birds singing in the trees. Invariably somebody breaks the spell by starting up a lawn mower. Motor mowers are the worst, simply because they are so noisy. Electric mowers are quieter but, for their owners, have a built-in problem: the power lead.

Mowing a lawn where there's a couple of trees in the way, and maybe a greenhouse and a few other obstructions, is something of a chore with an electric mower because the lead trails and tangles around them. Ultimately this limits movement: there's no way of pulling the lead through the trunk of a tree, and no way of getting the lead over the top. The only way of unwinding the lead is to retrace your steps. Of course, making a wide birth around the tree makes little difference, the lead still gets tangled, so the results of a short path and a long path are, in this sense, equivalent. Going round the same tree twice in the same direction only makes things worse, and clearly results in a different final problem than going round just once.

All of which is totally obvious to any gardener, who therefore has an intuitive grasp of homotopy, one of the big ideas of the branch of mathematics called topology. Topology is about classifying spaces without regard to actual distances between points. Topologists have expended a great deal of time and effort devising what they call "topological invariants," labels they use to tell if two spaces are equivalent or distinct, in a topological sense. Their sense may not map exactly to commonsense: to a topologist a soccer ball is the same as an American football ball, and a teacup is the same as a (annular) doughnut. A topologically inclined gardener might go once round the tree, and assign this single winding a "winding number" of one. Twice round the tree and the winding number becomes two, and so on. From a topological perspective, a wide path or a short path round a tree is the same since both yield a winding number of one. Slightly more formally, two paths are equivalent if they can be continuously deformed into one another without cutting. A double loop round a tree is distinct from a single loop round a tree, and both are distinct from another path that loops round two trees, since the different paths cannot be deformed one into another due to obstacles, in this case trees.

Drive the mower round a tree on a circular path back to where you started, and it might look as though nothing has changed, and yet the starting and finishing states are subtly different: they differ by a wind around an obstacle.

Split an electron such that it passes partly to one side of a current-carrying coil, and partly to the other side, then recombine the two parts of the electron beyond to complete the circuit. The result is a closed electron path around an obstacle. When an electron wave is split and then recombined, the result is an interference pattern. When it is split and passed either side of a solenoid before recombination, the interference pattern changes: this is just the Aharonov–Bohm effect encountered in Chapter 3. It's exactly equivalent to simply transporting an electron right round a solenoid. In that case, the starting and finishing points might be physically indistinguishable, but the electron has been modified by an amount that depends only on the magnetic field enclosed in the solenoid, though not on the details of the path or the distance from the coil. The mower is sensitive to the non-trivial topology of the lawn, in the sense that it is a flat plane punctured by trees and other obstacles. The Aharonov–Bohm effect signals the non-trivial topology of the magnetic field-free vacuum: it's a plane with a hole in it, the puncture provided by the solenoid.

The Aharonov–Bohm effect shows how non-trivial topology of the vacuum can have physical ramifications. For QCD, too, there is a non-trivial vacuum topology that turns out to have physical effects.

Using technology that's second nature to mathematicians, it's a simple matter to add a whole new portion to the master equation of QCD that exposes the topological nature of QCD's vacuum. This so-called topological term is certainly gauge and Lorentz invariant, and thus satisfies the most basic symmetries of the theory, yet it is something new and unusual, since it makes no contribution to any Feynman diagram, and is not associated with a flow of particles from the established particle zoo.

In 1975 Alexander Polyakov and his colleagues at the Landau Institute for Theoretical Physics in Moscow became the first to show how this extra topological portion broadens the picture. For QCD, their work reveals a gluonic entity that starts out far away and long ago, and finishes up far away and in the future, but with a gauge transformation "twist" acquired *en route*. Though the initial and final field strength vanishes, in between there is a localized region where the field has some amount of positive energy. The animal that Polyakov and his colleagues unearthed is now called an instanton, an "energy lump" made of gluons, a ripple in the gluon fields. It is a very special kind of ripple, however.

The story goes that, in 1834, engineer John Scott Russell was riding beside the Edinburgh–Glasgow canal watching a boat being drawn by two horses. The boat suddenly stopped, forcing a large rounded water wave forward from its bow. Intrigued, Russell was able to follow the wave on horseback for a mile or two until he lost sight of it. In subsequently describing this peculiar solitary wave to the British Association, Russell was setting out the first account of what is now called a soliton.

Solitons are, as their name suggests, solitary waves. Unlike the ordinary waves resulting from a stone thrown into a pond, which spread and die away, solitons maintain their shape and size. And when two solitons collide, they pass through each other and emerge beyond with their original shape. In the 1960s, solitons came to the attention both of physicists, for whom they provide a natural candidate for a model elementary particle, and to mathematicians, who found them an Aladdin's cave of mathematical riches.

One example of a soliton has already debuted. The magnetic monopole that 't Hooft and Polyakov introduced in 1974 is a soliton, and is a cousin of the instanton. The instanton is another example of a soliton. As solitons, instantons can swill around without melting away, passing through each other with impunity. Their remarkable stability is linked to their topological character. They carry a number, a "topological charge," which cannot simply be destroyed but is conserved. It's this topological nature that guarantees their safety. It's a bit like a flat strip of paper taped to a table at both ends, but with a single 180° twist introduced before the ends are fastened. The region of the twist will tend to be localized, and though it can be moved back and forth along the strip, the twist will remain a twist until the ends are released or the strip is cut. The twist, the topological feature in this toy example, fulfils the role of a topological charge. Instantons are gluonic solitary waves whose energy is localized in both space and time – hence the "instant" of instanton – and which are endowed with some kind

FIGURE 9.2 Alexander Polyakov giving his first-ever talk on instantons in the United States, 1978. Polyakov was one of the first to appreciate the relevance of topological ideas in Yang–Mills theory. Credit: Courtesy Alexander Polyakov.

of kink, or topological feature, measured by a conserved topological charge. (Figure 9.2.)

Usually, a conserved charge is related to a continuous symmetry, expressed by the invariance of the master equation under some symmetry group. Conventionally, charges that change in time give rise to currents. For example, in Maxwell's electromagnetism the drop in electrical charge contained in some closed region equates to the total electrical current crossing the boundary of that same region. In classical theory, Maxwell's equations derive from the master equation by exposing it to path variations to find the minimum energy route, an example of the principle of least action at work.

But changes in paths are, in the main, exactly the kind of thing that makes no difference to a topological property. All paths that differ from one another by "a little bit" are equivalent so far as a topological property is concerned, just as a lawn mower lead tightly encircling a tree is equivalent to one that's loosely coiled: what counts is the number of windings.

The instanton-spawning extra term added to the master equation of QCD exhibits this kind of path insensitivity, in addition to lacking the usual manifestation of a continuous symmetry as a charge plus current. Yet there *is* a conserved quantity, whose genesis lies in the way two spaces can be overlaid, a situation for which the lawn mower lead is actually a prototype. In the case of the lawn mower lead, the straight lead can overlay a circle time and again, each additional looping representing something new if the circular template embraces a tree. For instantons, the looping of one space onto the other, where one space is that of the color symmetry group and the other the universe in which

QCD functions, turns into infinite possible loopings that can be encoded in the gluon wave. This "looping number," a mathematically precisely defined winding number, is the topological charge.

The suggestion is that this new type of contribution, with its new breed of conserved quantity, caters for new configurations in quantum theory. This is because the Feynman path integral seeks to embrace *all* contributions from *all* possible configurations. There is every reason to expect the path integral to be sensitive to the topological term. And it is.

In quantum mechanics, a system might start in its lowest energy state, the ground state, then undergo some transition or process before finishing up, ultimately, back in its ground state. In quantum field theory, that starting and finishing point is the vacuum, and it's in terms of the vacuum that Feynman's path integral finds its proper expression. The path integral was introduced in Chapter 3 as a measure of the probability of making transitions from one configuration of field quanta to another via all intervening paths. It's better to say that the path integral is the vacuum-to-vacuum transition amplitude, expressed as a sum over all paths connecting initial and final states. In other words, according to quantum field theory, the starting point for a process is the vacuum state, from which field quanta are created before going about their business prior to their ultimate demise: earth to earth, ashes to ashes, dust to dust, so to speak. The path integral calculates the amplitude of this happening, embracing all possible paths weighted according to their contribution to the master equation, and that amplitude squared gives the probability.

What's new is the prospect of paths containing kinks, paths that are topologically distinct in the sense that they cannot be continuously deformed into one another. An instanton then corresponds to a path linking initial and final vacuum states that differ in winding number. That is, instantons provide a route for the system to finish up "twisted" in some way relative to its starting point. Therefore, since instantons link vacua having different winding numbers, and those winding numbers range right up to infinity, "the" vacuum in fact comprises an infinite number of apparently identical but topologically distinct vacuum states, each labeled by a different winding number.

The notion of many topologically distinct states finds a parallel in an infinitely long mower lead wrapped round and round a tree time and again. Each particular tangle is distinct both from all those with fewer windings, and from all those with more windings. Getting from one tangle to another is possible with a little bit of cheating – by looping the lead over the top of the tree. For the vacuum, there is a similar barrier: each of the topologically distinct vacuum states is separated from all the others by an energy barrier. Now, in quantum mechanics, energy barriers are not the obstacles that they can be in the classical world. A classical frog in a sufficiently deep classical bucket cannot escape. But a quantum frog in a bucket can ooze through the side of the bucket. This quantum tunneling, as this leakage effect is called, features, for example, in radioactive alpha-decay and tunneling electron microscopes. Instantons provide the leakage, the quantum tunneling, by linking one vacuum state to another topologically distinct vacuum state. The true vacuum of QCD is in fact a blend of all possible equal-energy vacuum states distinguished by different winding numbers, the relative contributions of each state to the true vacuum controlled via a parameter θ. The vacuum of QCD turns out to be this infinite, twisted morass called the theta-vacuum.

This parameter, theta, is *another* parameter of QCD. It determines the size of the topological term that enters the master equation of QCD. Its value is small, experimenters know that much. And theorists know, for reasons that will become apparent, that if this were not so they would have another disaster on their hands. There must be some pretty powerful reason why particle physicists have invited such a potentially rough player into their game as this topological term. And there is.

Energy lumps rescue QCD
Sometimes, symmetry is a quantum victim. That is, sometimes symmetries apparent in the "classical" version of some theory do not survive quantization. It turns out that this is exactly the fate of the left-over $U(1)$ symmetry in quantum chromodynamics. The trick is to track down the source of the symmetry spoilage.

The point is this. If this $U(1)$ symmetry is destroyed by the usual symmetry breaking linked to pions and kaons, then there should be a corresponding meson. To fit the mold, this particle should not only possess the properties of the real-world η meson, but it should also have a mass comparable to that of the pions, since they would share a common parentage. However, the η meson is already earmarked as one of the Goldstone particles and the next candidate with the right properties, the η', is much too heavy. Nature appears not to have supplied a particle matching that expected from the spontaneous breaking of this apparently spoiled symmetry. One way out of the conundrum is to invoke a *new* symmetry-breaking scenario, something that can at once spoil the symmetry, as required, but which does not demand an attendant meson that is a sibling of the pions. The $U(1)$ problem is thus resolved.

Instantons provide just the right symmetry spoilage. A massless gluonic whorl, an instanton, flips a right-handed quark into a left-handed one. This flipping of quark handedness is the breaking of constant-handedness, or chiral, symmetry. So instantons break this symmetry, but without demanding a lightweight meson. The $U(1)$ problem evaporates, and QCD lives on to fight another day.

The magnitude of the topological contribution to the master equation of QCD, that portion which includes the instantons, is controlled by the theta parameter. So it's no great surprise that theorists sometimes describe the symmetry spoilage in terms of theta. Theta selects one particular blend of vacua to give the vacuum state. In a sense, theta picks out a direction in the vacuum, and one direction, selected from the infinite possibilities on offer, is a symmetry breaking. In much the same way, an elephant atop a (very strong but flexible) flag pole will make the pole bow out in one particular direction, and in so doing break the original 360° rotational symmetry. Instantons emerge as fluctuations in theta, similar to the fluctuations in the direction of atomic magnet orientations that become magnons in magnetic systems. The difference is that the fluctuations in theta embrace a succession of different possible vacua, which in fact are not identical in all possible ways. This is because the generator of these fluctuations is not mindful of the fundamental color symmetry of QCD. The net outcome is that instantons are unphysical. Instantons are not true particles in the sense of electrons and pions.

The instanton's role in the $U(1)$ problem was unearthed by Gerard 't Hooft, with further details of the vacuum's structure flowing from the efforts of Roman Jackiw, Claudio Rebbi, Curtis Callan, Roger Dashen, and David Gross. The result was a flurry of intense interest that, however, was short-lived as theorists realized that instantons were not going to divulge their secrets easily. "Despite initial hopes, the discovery of instantons has not led to much improvement in our ability to do quantitative calculations in quantum chromodynamics," wrote Weinberg, two decades after 't Hooft's ground-breaking work. "On the other hand…it has produced spectacular qualitative changes in our understanding of quantum chromodynamics and other gauge theories."

The fact that instantons are not real particles does not necessarily mean that they are invisible. There is a chance that experimenters might spot their imprint in deep inelastic scattering experiments, by exploiting the fact that instantons flip quark handedness. An electron and proton might then collide to give a very unusual mix of quarks: a photon from the electron might scatter off a gluon in the proton to give a selection of gluons and, say, right-handed quarks and antiquarks, Figure 9.3. Encouraged by computer simulations of instanton effects, particle physicists at HERA have championed this route, though they still lack a usable, unambiguous instanton test. An instanton is likely to create a miniature fireball, they believe, giving rise to around ten or so quarks and gluons, in addition to a single quark jet. The fireball remnants are likely to give more sideways-directed

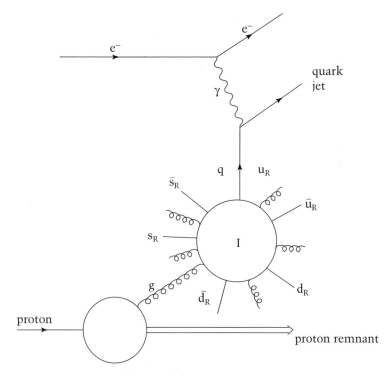

FIGURE 9.3 Schematic diagram of the putative main instanton (I) contribution to electron–proton deep inelastic scattering at HERA.

energy and more charged particles than is normal in an "ordinary" scattering event: the trick is to learn how to manipulate the data to reveal instanton events hidden amongst the multitudes. HERA is also a good place to study collisions between polarized protons and polarized leptons. Physicists argue that, because of the handedness-flip of a quark interacting with an instanton, the instanton contribution to spin-dependent cross-sections should be distinct from the better-known quark–gluon collisions. Whatever the route, discovering instantons would be a dream come true – an entirely new breed of object, a topological object, and a pointer to the real and intricate structure of the QCD vacuum.

Experimentalists have something else to contribute besides the hunt for instantons: they have a handle on the theta parameter of QCD. Theta is not only important in the structure of the vacuum; it is a secret door through which a terrifying specter might enter. For the theta parameter controls the overall magnitude of the topologically based part of the theory, and that portion has the unpleasant property of violating both mirror reflection symmetry and time reversal symmetry. The strong interaction is supposed to respect both these symmetries. So theorists face a quandary: admitting the prospect of exciting topology of the vacuum, and thus fixing the $U(1)$ problem via instantons, means endangering sacrosanct symmetries.

The way experimenters can get a value of theta is via a quantity called the electric dipole moment of the neutron, the tendency of a neutron to twist when exposed to an electric field. The important point is that measurements of this quantity imply a limit on the value of theta, and that limit is about 10^{-9}. In other words, theta is small, possibly even zero, but all experimenters can say for sure is that it is less than one billionth. A theorist's first guess for a value of theta might be more like one: understanding the reason for such a tiny value is a whole new problem, known in the industry as the "strong CP problem."

Perhaps theta is zero. Setting it to zero by hand is cheating, and leads to other complications. Roberto Peccei and Helen Quinn at Stanford University suggested another round of symmetry breaking as a route to explaining why theta is so small. Their idea was taken up by Weinberg and independently by Wilczek, leading to the introduction of yet another particle, a physical, spinless particle called the axion, christened thus by Wilczek after a household detergent. When theorists' original vision of the axion turned out to be at odds with experiment, they decided that axions should have a very small mass, perhaps just a few micro electronvolts, suggesting that experimenters had been looking in the wrong place. And they estimated that the number of axions around us could be huge, bathing the Universe in perhaps a million million axions per cubic centimeter. This is a scenario of instant appeal to cosmologists, who have seized on axions as a candidate for what they term dark matter, the Universe's missing mass. The movement of galaxies within clusters of galaxies implies the presence of more mass than astronomers can actually see, and is just one piece of evidence suggesting that around 90% of matter in the Universe is invisible, simply because it does not radiate light. That dark matter is real is beyond question, according to cosmologists, but nobody knows what it actually is. It may be something as mundane as lots of small dark objects such as planets or dead stars. Perhaps it's a generous population of massive neutrinos. Or it might be axions.

In 1983 Pierre Sikivie of the University of Florida pointed out that it should be possible for axions to convert into photons in a strong magnetic field. This opened the door to a new generation of axion-hunting experiments. One such experiment, based at the Lawrence Livermore National Laboratory in California, and led by Leslie Rosenberg and Karl van Bibber, was essentially a giant supercooled radio receiver tuned to look for radio-frequency photons arising from axions of specified masses, Figure 9.4. Another, led by Seishi Matsuki at the University

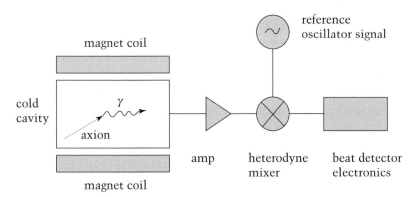

FIGURE 9.4 The conventional way to catch an axion. Axions convert to photons in the resonant cavity when subjected to a strong magnetic field. The amplified high-frequency signal is mixed with a reference signal to produce beats, picked up by the detector.

of Kyoto, likewise relied on axions converting into photons, but used an atomic beam technique to detect those photons. A new variant is CAST, the CERN Solar Axion Telescope, which will point a 10 m long magnet (a decommissioned LHC test magnet) end-on at the Sun: solar axions should then convert inside the magnet into X-rays that will be detected at either end. So far, though, despite numerous searches, nobody has found an axion. And, according to some researchers, they never will. Which of course is all the more reason for looking.

A VACUUM EPILOGUE

There is a simple two-part take-home message to distil down from all of this. Firstly, the vacuum is not some inanimate emptiness but has a structure, and contributes observable effects. Secondly, the vacuum of QCD may be twisted and kinked in a way that is invisible to the usual perturbation approach, and which represents a whole essential layer of the theory relevant to confinement, instantons and a great deal more besides. The vacuum is more than it's cracked up to be.

10 Checkerboard QCD

There's a mountain in quantum chromodynamics, an obstacle that, sooner or later, is found to be blocking almost every path. It's simply this: how to deal with QCD at low energy?

This is the problem reflected in the sliding interaction strength of QCD. As discussed in Chapter 5, at high energies and short distances, the coupling parameter is relatively small and the conventional mathematical technique of perturbation theory suffices. But at low energies, or correspondingly larger distances – about a fermi – the QCD coupling parameter is no longer small and perturbation theory breaks down. To beat the mountain, theorists need some ways of doing QCD calculations without resorting to perturbation theory. Such tools might one day open the path to proving confinement, for example, and to exploring the nature of the QCD vacuum. Theorists would be able to calculate the properties of protons and other composite particles from their basic quark building blocks, and could then work out exactly how quarks and gluons fragment into the jets of particles seen by experimenters in their mighty detectors. The list of possibilities is long indeed.

The bad news is that, in general, there is no known chunk of mathematical technology that will do the trick. But the good news is that by using some of the largest, most powerful computers ever built it's possible to get some way, perhaps a long way, towards the goal of non-perturbative QCD, of calculating with QCD but dispensing with conventional perturbation theory.

Quantum chromodynamics simulation on a computer, called lattice QCD, offers enormous potential when it comes to quantifying the low-energy long-distance properties of the strong force. But computer calculations are surprisingly expensive and time consuming. And never was the small print so important.

QCD ON A COMPUTER
The basic idea of lattice QCD, introduced by Kenneth Wilson in 1974, appears simple enough. Though space and time are continuous, Wilson's vision was to divide them up into chunks, and represent

them using points on a lattice. Some clocks divide up time, the second hand clicking from one time point to the next around the clock face. Stepping-stones across a river achieve the same effect for space. Setting out both space and time as points on a lattice, then allowing the spacing between the lattice points to shrink away to nothing while simultaneously ballooning the total volume of the lattice to infinity, returns the true, continuous world.

Pinning QCD to the points of a lattice automatically helps to control the infinities that plague the theory and necessitate the renormalization program in the perturbation theory of the "conventional" route. This is because, for a lattice to provide an adequate description of quarks living there, the quark's wavelength (recalling, for a moment, that a quark is a matter wave) must be at least double the lattice spacing. So the finite lattice space tallies with a lower length scale limit, and this is helpful since it's short length scales that are the source of (ultraviolet) infinities. The basic strategy, then, is to first write down QCD for some finite lattice spacing and finite total lattice volume. The next step is to do some interesting non-perturbative calculation, using numerical simulations on a big computer, and the final step is to recover the real world by letting the lattice spacing shrink to nothing and the lattice volume run off to infinity. Sounds easy.

The lattice to be inhabited by quarks and gluons has four dimensions, one time and three space, and is not a real physical lattice, but instead is represented by sets of numbers inside a computer. The lattice can nevertheless be pictured as a grid of intersecting lines, shown in two dimensions in Figure 10.1. Where two lines cross is a lattice point, and at each such point is a number representing the probability that a quark

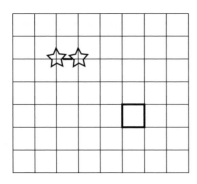

FIGURE 10.1 On the lattice, quarks live where two lattice lines cross, a lattice point. Two quarks, the stars, are shown in the figure. Gluons live on the links between lattice points. In the figure, a gluon joins the two quarks, and in addition a pure gluon state comprising a closed gluon loop is depicted.

resides there. Between lattice points are lines linking those points, and it's on these links that gluons live. Figure 10.1 shows two particularly simple configurations. One shows a single link connecting two quarks. The other is a closed loop of links without any quark content, and this represents a gluons-only system of gluon self-interaction.

The result is a computing problem of truly awesome proportions. The typical number of lattice points, governed by computer size, might be 48 in each of the time and three space directions, making $48^4 = 5\,308\,416$ lattice points in all. Since there are thirty-two varying quantities to keep track of in the problem, the total summation is equivalent to calculating the volume of an irregular region within a box having almost two hundred million sides. The problem is reduced to something more manageable by telling the computer to "roll a dice" in what's called a Monte Carlo technique. Then averaging over a limited set of quark, antiquark, and gluon configurations produces an approximate value of the summation. The different configurations are selected using "informed dice throwing" to home in on the most important contributions. These tricks make the problem realistic, but at the price of introducing statistical errors.

Lattice QCD uses a Monte Carlo sampling estimation method to attempt a brute-force approximation of the path integral for QCD. By contrast, perturbation theory seeks to approximate the path integral as a series of terms via Feynman diagrams of ever-increasing complexity.

In practice, to calculate the mass of a particle, say a proton, lattice simulators feed in a number of lattice points, say 32 for each of the three spatial directions, and often more for the time direction, perhaps 56, the quark masses and the interaction strength. They then introduce an initial configuration, which might be a "hot start" of randomly assigned values, or a "cold start" of uniform values. A Monte Carlo program then works on this initial configuration to create several hundred "independent" configurations. The likelihood that, for example, two up quarks and a down quark (i.e., a proton) can emerge and migrate forward along the time axis of the lattice can then be computed from the average values of contributions from each configuration of quark, antiquark, and gluon fields. Since the distance a particle such as a hadron will propagate in the lattice falls away both in time and with its mass, with heavier particles traveling shorter distances, the decay in the likelihood of the "lattice proton" making its journey can be used to extract a value of its mass.

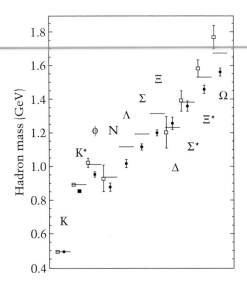

FIGURE 10.2 The light hadron spectrum using the quenched approximation. The short horizontal lines are experimental values. The open squares are Don Weingarten's GF11 results. The black dots are the CP-PACS results. Credit: CP-PACS.

The results come out looking pretty convincing. Figure 10.2 shows a table of lattice-calculated particle masses, the light hadrons, compared to their measured values shown as short horizontal lines. In fact, the figure shows two sets of different lattice estimates for each particle. The first set, shown as open squares, are from Donald Weingarten and his colleagues at IBM's Thomas J. Watson Research Center in Yorktown Heights, New Jersey. The other set (shown as black dots in the figure) is more recent, from the CP-PACS collaboration in Japan. Both forced their computer-generated value of the K meson mass to agree with the measured value, and then extracted a strange quark mass for calculating the remaining particle masses. In a nearly ideal world, both computer-generated values would agree with each other, and in a truly perfect world both would agree with the measured value as well. As it is, lattice QCD comes out no more than a few percent wide of the right answer, so perhaps the world is not too far from ideal. So has lattice QCD proved that QCD is true, and can it calculate all that we want? Certainly lots of people are very enthusiastic about it. But there is always that small print to watch out for.

The trials of lattice life

"There is no such thing as a problem, just challenges and opportunities." A bit of corporate wisdom there, from a salesperson who, incidentally, hails from one of the world's giant computer companies with a stake in the world of lattice QCD. It's a world that offers plenty of opportunities, for example the demonstration of confinement, or the calculation of glueball masses from scratch, to name but two. And there is no shortage of challenges to share round.

Some of those challenges are inherent in the lattice approach. Two sources of error in lattice studies that simply will not go away are the finite size of both the lattice spacing and, to a lesser extent, the total volume of space-time represented by the lattice. The capacity and speed of computers feature in both, especially the lattice spacing which needs to be as small as possible.

The lattice spacing is critical, since any physically meaningful result from a lattice calculation, for example an estimation of particle masses, has to be independent of the spacing. Otherwise, the estimation is a feature more of the lattice than the real world. However, that point will only be reached if the physics being simulated has no important size features comparable to or shorter than the lattice spacing. So a small lattice spacing is vital if answers are to be accurate and meaningful. Lattice practitioners typically use a spacing of around one tenth of a fermi.

Given that the size of the available computer fixes the total number of lattice points, the lattice spacing is linked to the total volume of the simulation. This is because the total volume of the little box represented by the lattice is simply the product of the number of lattice points, and the spacing between them. However, the larger the box the better, with a box something like two and a half fermis across being about right to make sure that whole particles such as protons are fully contained. With a small box, the box size itself will begin to distort the answer. But a large box implies a large lattice spacing, running counter to the need for the smallest possible lattice spacing: lattice QCD is full of compromises, and the lattice community's thirst for increasingly larger and faster computers knows no bounds.

Independent of lattice size effects, the mere use of a lattice can cause some headaches. Solid-state physicists know about doing calculations on a lattice. They have a nature-supplied lattice to work with, the crystal lattice, in which atoms are set out in regular three-dimensional arrays containing millions upon millions of atoms. The electrical and mechanical properties of crystalline materials, for example, depend profoundly on this neat atomic ordering.

Given that a lattice arrangement can be responsible for the properties of collections of atoms, it comes as no great surprise to discover that forcing quarks and gluons to inhabit a lattice introduces new features. In fact it causes a serious problem. The mere presence of the lattice means that extra quarks, replicas of the "real" ones, accidentally enter the play. These are an issue, as Wilson himself saw. He attacked this problem by introducing an extra factor, now dubbed the Wilson term, into the master equation of QCD as implemented on the lattice. This extra portion allows the additional, unwanted set of quarks to be conveniently removed. In massaging the equations in this way, Wilson was exploiting the silver lining of the cloud of problems inherent in the lattice program: when the master equation of QCD is rewritten for the lattice, entire families of new pieces can be added; these should not affect the true physics, since they disappear when the continuous real world is recovered. The Wilson term is one such addition.

The Wilson term resolves one problem, the replicated quarks, but at the expense of introducing another. The root of the new problem is the chiral symmetry applicable to massless quarks, which, as already encountered in Chapter 9, leaves left-handed quarks as left-handed quarks and, likewise, right-handed quarks as right-handed quarks. Even with massless quarks, the Wilson term breaks this symmetry, causing a number of technical headaches. In particular, quark mass can begin to appear as if by magic in the lattice calculation, and can only be made to go away by the undesirable and artificial practice of carefully fine-tuning parameters in the simulation.

Wilson had fallen victim to what was eventually seen to be an underlying ground rule. That rule says that it's impossible to build a single-quark lattice QCD that preserves the all-important color symmetry without simultaneously wrecking chiral symmetry. Some simulators use the Wilson approach, but there is another popular option. John Kogut, at the University of Illinois, Urbana, and Leonard Susskind at Stanford University, devised a way of spreading QCD across a lattice in such a way that chiral symmetry is preserved, but at the expense of having four species of quark for every actual quark flavor. In essence, a single quark is smeared across four neighboring lattice sites. There's no problem with quark mass having to be artificially spirited away, and near-massless quarks are a natural feature of their so-called "staggered fermion" approach. There are, though, some technical downsides, the most obvious being the relationship between each group of four staggered lattice quarks and actual up, down, strange, and charmed quarks. It might be a little messy, but the staggered fermion approach has come to dominate large-scale simulations of QCD. However, developing and improving the way in which the master equation of QCD is spread on the lattice is a subject of consuming interest in lattice circles.

The lattice spacing and limited volume of the box containing the quarks and gluons are the two main systematic errors inherent in the lattice approach. The "systematic" means that, once the spacing and box size are given, no amount of computer time is going to yield a better answer. Another source of error (helped by more computer run-time) is the statistical error riding on the back of the Monte Carlo technique. All three of these errors are present in all lattice calculations. And as if that wasn't bad enough, there is a fourth error, present in essentially all lattice calculations to date, which lattice theorists actually introduced deliberately. It's called the quenched approximation.

Permitting a lattice computation to unfold to include all possible quark contributions turns out to be hugely expensive in terms of computer time. This is partly because so many quark and antiquark pairs can emerge during the calculation, but is mainly due to the immense amount of work involved enforcing the Pauli exclusion principle, which quarks and antiquarks must obey. In 1981 Weingarten and Giorgio Parisi (at the University of Rome) simplified the problem by striking Solomon-like through the troublesome, complicated portion of the calculation and simply replacing it with the number one. This is the quenched approximation. Though light on physical justification, it is a simple expedient that seemed to work. Physically, the quenched approximation, sometimes called the valence approximation, amounts to neglecting all the virtual quark–antiquark pairs popped out of the vacuum from gluons. In other words, sea quarks are neglected – it's as though their mass is infinite – and only valence quarks feature. (Figure 10.3.)

The good thing is that, using the quenched approximation, lattice practitioners can get answers in a reasonable time. Virtually all early lattice results were generated using the quenched approximation, including those depicted in Figure 10.2. The trouble is, these results are not quite right. For example, the 1998 results from the Japanese CP-PACS collaboration, shown in Figure 10.2, are consistently about 10% wide of the known particle masses. Lattice researchers know that in the main this is the fault of the quenched approximation, which they

FIGURE 10.3 Giorgio Parisi talking at a conference in Paris, 2001. Parisi's interests extend beyond quantum chromodynamics to embrace such diverse topics as aggregation, glassy, and other disordered systems, and non-equilibrium physics. Credit: Courtesy John Iliopoulos.

now understand to be wrong or, at least, too much of an approximation. The obvious response is to do so-called "full QCD" calculations, incorporating dynamical quarks, in other words allowing for the full quark population complete with pair creation and annihilation. The catch is that, relative to quenched calculations, full QCD calculations require hundreds of times more computing power. Lattice researchers at most major centers have now moved beyond the quenched approximation to realistic, full QCD calculations. They have already generated improved results for the light hadron masses, accurate to within a few percent, and promise much more besides. All they need are smart algorithms and very, very fast computers.

The need for speed

"There are three things you need in research: money, money, and money," so the saying goes. Good ideas help, of course, but in lattice QCD ideas alone are not enough. You have to have the cash to build big enough computers to implement and test those clever algorithms, and to generate some meaningful answers in something like a reasonable amount of time.

Lattice QCD demands rather more than a home personal computer or a researcher's desktop workstation. It hungers after the biggest and fastest computers available. The machines that lattice researchers use are often custom-built and dedicated to the job, typically sharing their load amongst a nest of separate processors harnessed together. They are parallel supercomputers.

One of the first of this breed of QCD computers was created by Weingarten and his colleagues at IBM. They called their machine GF11. It had 566 processors working in parallel on the one single problem, essentially 566 computers all pulling together, and could perform 11 billion arithmetic operations every second. And this explains the name GF11: the unit of calculational speed for a computer, a crucial parameter in any discussion of high-performance computational science, is the number of floating point operations per second, crudely, the number of times the machine can multiply two big numbers together in a single second. Since computers are so fast, a more useful unit is megaflops, or Mflops, which stands for "*m*illions of *f*loating-point *o*perations *p*er *s*econd." With machines like Weingarten's come billions of flops, called gigaflops, or more succinctly Gflops. Hence GF11 for an eleven Gflops machine. (Figure 10.4.)

Begun in 1983, GF11 took eight years to design and build. It produced a value for the mass and decay rate of a glueball after *two years* of continuous calculating on 80% of the machine. According to GF11, the spinless glueball should have a mass of 1740 MeV, a figure very much in line with the mass of glueball candidate $f_0(1710)$. And GF11's 1994 spectrum of the light hadrons, shown in Figure 10.2, became something of a standard against which other collaborations pitted their own results.

One of those other groups, whose data also appears in Figure 10.2, is the CP-PACS (Computational Physics by Parallel Array Computer System) collaboration based at the University of Tsukuba, Japan. Though not dedicated entirely to lattice QCD – the CP-PACS computer spends time on astrophysics and condensed matter physics too – come the end of the 1990s this machine was the most powerful computer doing QCD. Its two-thousand-plus independent processing nodes are basically souped-up commercial Hewlett Packard computers, connected in a three-dimensional 17 by 16 by 8 cubic grid, with a distinctive

FIGURE 10.4 Man and machine. Don Weingarten pictured with GF11 in
1993. The two rows of "filing cabinets" behind Weingarten each house
half of GF11's processors. In the left foreground one of the processor
racks has its cover removed. Credit: IBM.

pattern of communication links between the different nodes. Theoretically capable of delivering 614 Gflops, CP-PACS can manage 368.2
Gflops on test problems, which hints at some of the subtleties of these
great computers – their speed on real problems is sometimes a far cry
from their listed, or theoretical, capabilities. Another subtle point is
ease of use, which for some very large computers has posed a problem due to the specialist computer languages they demand. The CP-
PACS computer, on the other hand, can work efficiently with programs
written in standard high-level languages, Fortran, C, and C++, with
only the innermost loops demanding specialist "assembler" coding for
top speed and efficiency. Built at a cost of 22 million US dollars, the
machine looks like a room full of prosaic but very expensive filing cabinets. CP-PACS is typical of its generation: a custom design that is the
result of a large collaboration of physicists and computer scientists.
Never before have physicists had to be so *au fait* with the niceties of
machine architecture.

There is a scattering of these giant QCD computers around the
globe. High up in the QCD league table is the 180 Gflops, 12 288
node QCDSP machine at the RIKEN Brookhaven Research Center, at

Brookhaven National Laboratory. Also in New York is the 120 Gflops, 8 192 node QCDSP machine at Columbia University. And setting a new standard for cheap megaflops is the APEmille squadron of machines, with two 250 Gflops and two 128 Gflops units at INFN, Italy's National Institute of Nuclear Physics, in Rome, and a 250 Gflops unit at DESY.

While both the QCDSP and APEmille computers are homegrown affairs, based on armies of identical, custom-built processing nodes connected in exotic arrays to enhance the machines' lattice abilities, other significant machines include large-scale commercial computers. Other groups pursue another tack, the workstation farm. The idea here is simple, to lash together large numbers of ordinary workstations or PCs, linked by ordinary computer networks, and spread the calculation amongst them. Farms cannot yet match the performance of dedicated machines, but they are cheap and flexible, and can be easily modified to take advantage of the rapid progress in computer technology.

As the world's clocks ticked over and a new millennium dawned, just three computers had crossed the psychological teraflops (Tflops) barrier, with the most powerful computer on Earth being Intel Corporation's ASCI Red, at Sandia National Laboratories in Albuquerque, New Mexico. Yet at 2.379 Tflops, even this machine, dedicated, in essence, to explosion-free nuclear weapons testing via simulation, would not be enough to satisfy the hunger of the lattice QCD community for computer power. In 2004 the fastest QCD computer in the world was installed in Edinburgh, the UKQCD collaboration's QCDOC, which stands for QCD On a Chip. That's a reference to the custom-built chips, tailored to do QCD calculations, which give this machine, and its sister at Brookhaven, a speed of 5 Tflops. It doesn't end there – larger machines are already on the horizon: who will be the first to propose a Petaflops QCD computer?

FRUITS OF THE LATTICE

Using all that computer grunt to calculate the masses of a bunch of particles listed in data tables for all to see may not seem very adventurous. Yet reproducing the spectrum of the light hadron masses will always have a special place in the hearts of particle physicists. The reason is simple: to be able to calculate the mass of the proton and other common particles from pure theory is one of the great goals of particle physics. The day that physicists can do it is the day they know they are really getting to grips with understanding how the world works at the most fundamental level. And for QCD, with its particular challenges at low energies, calculating the mass of a composite particle *ab initio* epitomizes much of what is difficult and unknown about the theory. There is also a more pragmatic reason – the masses are known to good accuracy and present themselves as an obvious first place to try out lattice QCD. Ever since the birth of lattice QCD, the hadron mass spectrum has been the place to take new ideas and new algorithms for a test drive.

Lattice jockeys have evaluated many other quantities, however. It's possible to extract a value of the strong coupling parameter, for example, along with masses for the four lightest quarks, and a very tightly constrained estimate of the bottom quark mass. Lattice calculations can also help shed light on the "$\Delta I = 1/2$ rule," encountered in Chapter 7, in which strong interaction effects encroach on weak decays. A badge in the lapel comes from studying the decay of the D_s meson, which is made of a charmed quark and strange antiquark (D_s^+) or charmed antiquark and strange quark (D_s^-). The D_s meson's decay constant, a measure of its tendency to decay via the weak interaction, was *predicted* by lattice calculations to be 220 ± 30 MeV, in comfortable agreement with the experimental value of 241 ± 32 MeV, though this apparent success carries a quenched approximation caveat. Lattice researchers are particularly interested in calculating properties of the B meson system, in parallel with experimental facilities that will tie down B meson physics to great accuracy.

Perhaps some of the most glamorous lattice calculations are those targeting glueballs and hybrid mesons. According to quenched lattice calculations, there should be an entire spectrum of glueballs, kicking off with a spin 0 glueball of mass approximately 1.7 GeV, then a spin 2 version of mass around 2.4 GeV, with a plethora of heavier glueballs with various quantum numbers. But, as discussed in Chapter 6, spotting glueballs is tough. One route would be to hunt for a glueball having "exotic" quantum numbers, in other words quantum numbers corresponding to an object outside the framework of the conventional quark model. According to lattice predictions, the lightest such state weighs in at over 4 GeV, which is very heavy even by glueball standards, and may be the reason it has not been found by experimenters. Hybrid mesons – also introduced in Chapter 6 – are entities such as the $\pi_1(1400)$, a $q\bar{q}g$ combination that lies outside the conventional quark model. The computer simulations referred to in Chapter 6 are indeed lattice QCD calculations, and these suggest that the lightest exotic meson having the quantum numbers of the $\pi_1(1400)$ should have a mass between 1.8 and 2 GeV, which is too high. For both glueballs and hybrid mesons, matching lattice output with the real world remains a captivating challenge.

Computer confinement...
The possibilities of lattice QCD are enormous, even if some topics, such as calculating nucleon structure functions and the fragmentation of quarks and gluons into jets, are in their infancy. The lattice, for example, is pretty well the only place to learn about instantons. But what of the ultimate prize: confinement?

Back in the early days, things looked very hopeful. At the start of the 1980s, lattice results showed an interaction energy between a quark and an antiquark doubling with a doubling separation, in the requisite manner. So, confinement is proved. Well, not quite. The catch looks like a technicality, but is important enough to ensure that the celebratory champagne remains firmly corked. The conclusion that QCD on a lattice is confining stems from work on a relatively coarse lattice. The spacing between lattice quarks is therefore large, so their interaction strength is relatively strong. To make the transition to reality, that lattice spacing has to be shrunk down to nothing, and this is where the problems start. As the lattice spacing shrinks, so too does the quark–antiquark spacing, which means that the strength of the interaction between the quarks becomes smaller and smaller, asymptotic freedom at work again. What's missing is proof that during this shrinkage, the confinement property holds good – and because of the sliding interaction strength this cannot be taken for granted. Though computer simulations *suggest* that confinement really does persist as the lattice spacing shrinks, nobody has been able to *prove* it. In case all this seems academic, it's worth remembering that QED *also* shows confinement on a coarse lattice. If confinement persisted in the zero lattice spacing limit of QED it would be a disaster, proving that electrons should be just as confined as quarks and, as a consequence, that there should be no such thing as television as we know it. In the case of QED, however, researchers have been able to show that there's a jump, a phase transition, from the confining coarse-lattice QED, to a non-confining version on the smooth world that we inhabit. The challenge, yet to be met, is to prove beyond doubt that for QCD there is *no* such transition.

There are other ways in which lattice simulations might help resolve the confinement conundrum, however. Some researchers claim that lattice results support the dual-superconductor model of confinement, introduced in Chapter 7. Here, the vision is for the chromo-electric force linking the quark and antiquark of a meson to be squeezed into a tube by a surrounding sea of monopoles. So the monopole condensate expels the chromo-electric lines of force from the vacuum, and is therefore responsible for confining force tubes. Confinement is then about transitions from a normal vacuum to a superconducting vacuum. According to this picture, surrounding a quark is a small region, or bubble, of normal vacuum – an idea that has a certain resonance with bag models of baryons. When a meson's quark and antiquark are close together, their bubbles of normal vacuum overlap. The force between them is at its most modest, and is slave to antiscreening effects. If the quark and antiquark are pulled apart, the bubbles of normal vacuum become a dumbbell of normal vacuum surrounded by superconducting vacuum, the neck of the dumbbell being the color force tube linking the pair.

...and deconfinement

There is another way of thinking about the whole confinement issue. So far as seawater, meteorites, niobium spheres, and the everyday world is concerned, quarks (and gluons) are confined. But must they always be confined? Are there special conditions under which quarks and gluons can be free? And, if so, is it possible to initiate the jump from the confined world to a deconfined one?

Particle physicists believe that free quarks and antiquarks *are* possible, and that just such a jump is a reality, but only under very special conditions. The idea is that, when ordinary particles such as protons and neutrons are vigorously squeezed together, they lose their individual identities, like so many grapes in a wine press. The inner constituents of each become exposed to those of the others as the composite particles overlap. Where once the quarks of a proton, say, sought and found mutual color neutralization, now other quarks from other participants become equally available. The notion of individual hadrons becomes redundant under these conditions, and is instead replaced with that of a gas, or "plasma," of free, unbound quarks and gluons.

Though asymptotic freedom contributes to the emancipation of quarks in this quark–gluon plasma, theorists believe a new mechanism also comes into play due to gluons screening the color force between quarks. To borrow again from the world of solid-state physics, in atomic matter, the electric field due to a positively charged ion is modified by the electrons around it. Due to their negative charge, these surrounding electrons partially screen the positive ions. If the screening is strong enough to suppress the influence of the positive ions over a distance that's short compared to the spacing between them, some electrons can become free to roam the material, forming an "electron sea" that results in electrical conduction. However, if the screening is longer range, electrons remain pinned to ions, and the material is an insulator. But if the range of the screening in an insulator can be modified relative to the distance between ions, it should be possible to loosen some pinned electrons, and turn an insulator into a conducting metal. Such insulator–metal transitions, first suggested by Nevill Mott in 1968, are now well-known and can be brought about by a range of effects, in particular by pressure. Pressure reduces the spacing between atoms, raising the electron concentration. This means more screening. At first sight, it might look at though nothing will happen, since the reduced spacing between ions should surely mean that a pinned electron will be even more likely to stay pinned. But the increasing electron concentration makes for a screening effect that increases more strongly than the diminishing spacing. The screening wins out, so squeezing can turn an insulator into a metal: at a pressure of 1.4 million atmospheres and a temperature of 2300 °C, for example, hydrogen becomes metallic and conducts electricity.

The transition from insulator to conductor via electrical charge screening offers up a model of the transformation from hadrons to quark–gluon plasma via color screening. As particles such as protons and neutrons are squeezed together, any one quark finds itself practically rubbing shoulders with many others, and not just with those from its home nucleon. But the squeezing creates a rising gluon density, in analogy with the rising electron density of the Mott transition, that screens any one quark from the influence of those close by. Quarks (and antiquarks) that started out as part of some composite hadron thus find themselves cut loose and released into the mêlée, where they barely feel one another's presence. And much as electrical charge flows in a conductor as a consequence of loose electrons, this quark–gluon plasma will support the flow of color charge. Then again, deconfinement might actually be altogether simpler. Polyakov and Susskind have both suggested that the strings binding quarks together simply melt at high temperatures.

Lattice calculations suggest that there will be a jump, a phase transition, from a state of identifiable particles, such as protons and neutrons, to this new state of matter, the quark–gluon plasma, and that the jump will be abrupt. Since this phase transition is from a regime in which quarks and gluons are confined to one in which they are not, it's called the deconfinement transition. Evidence for a jump out of a confined phase is, of course, another strand of evidence for the existence of a confined phase to jump out of. According to lattice researchers, the deconfinement transition should occur at temperatures of around 10^{12} °C, roughly 10^5 times the temperature of the Sun's core, and densities approximately 10^{20} that of normal matter or about ten times the density of normal nuclei. All the combinations of temperature and density marking the jump to the quark–gluon plasma from the hadronic phase (which includes regular nuclear material) delineate a "phase boundary" on a temperature–density plot. Reaching the plasma state requires extreme temperatures at relatively modest densities, or gigantic densities at more modest temperatures.

And merely reaching the plasma may not be the end of the story. There is another layer of structure to consider, over and above the deconfining transition. In Chapter 9, the ferromagnet debuted as a prototype of spontaneous symmetry breaking, the alignment of myriad atomic magnets providing a lowest energy state lacking the rotational symmetry of the basic interaction between neighboring atomic magnets. Crucially, this happens below a certain temperature – raise the temperature of the magnet and the alignment vanishes. The magnet then jumps out of its ferromagnetic state into a state in which it looks like any ordinary metal.

Imagine doing the corresponding thing with the vacuum of QCD. Dump enough energy into the vacuum, and the correlation between quarks and antiquarks in the quark condensate will be destroyed as the condensate "melts." With the quark condensate gone, all signs of chiral symmetry breaking vanish, and chiral symmetry should be restored. At high temperatures, then, there should then be *another* phase transition, in addition to the deconfinement transition, marking the onset of what theorists call chiral restoration, the recovery of the chirally symmetric regime where left-handed and right-handed quarks remain aloof from one another. In the broken chiral symmetry phase of ordinary particles, such as protons and neutrons, quarks have an effective, or constituent, mass of around 300 MeV as a consequence of chiral symmetry breaking. Above the chiral restoration temperature, in the chirally symmetric phase, quarks are nigh on massless: their mass is merely the so-called current mass values, 4 MeV and 7 MeV for the up and down quarks, respectively. Theorists expect that, as a result, relative to their tabulated values some particles created via a quark–gluon plasma will have different masses, lifetimes, and a modified spectrum of decay products as a consequence of chiral restoration. What remains uncertain is whether the deconfining and the chiral restoration transitions are related, and perhaps even occur under the exact same conditions.

In another twist, recently the idea of the quark condensate, central to the explanation of chiral symmetry breaking, has metamorphosed into something yet more potent. In the conventional picture of the vacuum above, left-handed quarks condense with right-handed antiquarks of the same flavor, and vice versa. But in 1999 Frank Wilczek, Mark Alford, Krishnas Rajagopal, and Thomas Schäfer came up with something totally and daringly different. In very dense nuclear matter, they argued, quarks should pair up more in a way that mimics the pairing of electrons in conventional superconductivity. The result is a "color superconductivity" based on "quark Cooper pairs."

At the heart of the matter is a new breed of quark condensate that links quarks of *different flavors and colors*. Now this is fighting talk: usually in high-energy physics the perfect symmetry of color is held aloft from the not-so-perfect symmetry of flavor, but not here. The new condensate blends them together, and the result is dubbed "color–flavor locking."

The rationale is that the combined color charge of two quarks can be reduced by bringing them together, offering the basis of an attraction that fulfils the attractive role of the mediating lattice for electrons in solid-state superconductivity. Like friends passing in the street, oppositely directed quarks find themselves dallying as quark Cooper pairs. These gain an overall energy advantage by working in step, in other words condensing into a color superconducting state, just as real Cooper pairs form a superconducting state in solids.

The consequences for such a vacuum are numerous and far reaching. Right away, color and flavor in isolation are no longer good symmetries, with only certain combinations surviving – those that are locked together. Gluons acquire mass via the Meissner effect, much as photons acquire an effective mass in solid-state superconductivity. The electric charges of particles are altered, with some gluons even acquiring charge. In fact, because particle charges take on integer values and long-range color-based forces vanish through gluon masses, the theory has the key ingredients of confinement. The other big-ticket item, chiral symmetry breaking, is also taken care of, with right-hand flavor symmetry locked to color and thence to left-hand flavor symmetry, so left- and right-handed transformations are no longer independent. All up, what happens in the limit of very high density, according to this theory, has some properties that match what might be expected of a lower density world populated by real particles. Perhaps what goes on in the everyday world of quarks in protons and neutrons is some kind of limit of this superdense one. So far, color–flavor locking lacks support from experiment or simulation, but its introduction has sparked a wave of interest from theorists.

CREATING A NEW STATE OF MATTER

Particle physicists believe that nature was able to supply the requisite environment for a quark–gluon plasma in the first few microseconds following the Big Bang. One way to recreate that environment is in a "little bang" here on Earth, by hurling together bunches of quarks and gluons to produce a miniature fireball. This offers the prospect of creating a bead of quark–gluon plasma that will then expand and cool, with pions and other particles condensing out as it does so. Experimenters think they can create such a plasma, and detect the condensing particles that ensue; some claim they have already done it.

Guided and inspired to a large extent by lattice QCD calculations, physicists at New York's Brookhaven National Laboratory are attempting to re-enact in miniature the earliest moments in the life of the Universe. Here, in 1999, experimenters at the Relativistic Heavy Ion Collider (RHIC) began colliding together bunches of quarks and gluons in the form of very fast, heavy ions in the hope of creating a full-blown quark–gluon plasma.

The RHIC is a new kind of collider, bringing together high-energy beams of ions rather than single particles such as electrons and positrons. It uses the tunnel built for the canceled Isabelle proton–proton collider project, and relies on the existing AGS accelerator as an injector. Around 1600 superconducting magnets bend and focus a beam of ions, which can range up to heavy gold ions, around the 3.8 km ring.

Ions, stripped of their electrons by their passage through thin sheets of metal foil, have both a high electric charge and the same sign charge as those in the oncoming beam. The first feature means that the ions within a beam tend to repel their circulating neighbors, making for a fat beam, so an unusual feature of the RHIC is its correspondingly wide beam pipe, 8 cm in diameter. The second feature is that the like charges of the two beams means there's no avoiding two complete counter-rotating beams with their own beam pipes and magnets. The beams, with energies up to 200 GeV per nucleon, intersect and collide, the collisions watched by four detectors: PHENIX, BRAHMS, PHOBOS, and STAR. When looking for quark–gluon plasmas, these detectors are confronted with an extraordinary flood of up to ten thousand particles, mostly pions, for a single ion–ion collision.

One way experimenters hope to pick out the plasma is by looking for a *drop* in the creation of J/ψ particles in heavy ion collisions, an idea proposed back in 1986 by Helmut Satz at the University of Bielefeld, Germany, and Tetsuo Matsui at MIT. The idea is that charmed quarks and antiquarks, created as quark–antiquark pairs in the fireball, fly apart and lose each other. As the plasma cools, instead of combining to form J/ψ mesons, the charmed quarks and antiquarks will instead find themselves partnering lighter quarks in the scramble to form up into observable particles. So if this mechanism holds good, then the moment the experimental conditions become right for the creation of the quark–gluon plasma, the hit rate for J/ψ particles will suddenly fall away. Light quarks and antiquarks will lose their partners in the same way, but find replacements easily. What makes the charmed quark special is its large mass, which insures charmed quarks are relatively rare.

Early lattice-based calculations suggested that the CERN SPS should be able to create the right mix of high temperature and matter density to spark a plasma. Experiments began at CERN in 1986, initially using sulfur ions to see how many J/ψ mesons would be created in conditions theorists deemed unlikely to touch off the plasma. In 1994, the SPS was upgraded to use lead ions, these heavier projectiles giving a larger energy density in collisions and, said the theorists, enough to bring the plasma within reach. Data gathered by the NA50 team, and by others at CERN, indeed showed a drop in J/ψ production, in fact enough of a drop that by 1997 it seemed probable that the quark–gluon plasma is a proven reality. A key feature of the data is that not only is there a drop, it's a sharp drop. This is to be expected from a transition to a new state of matter, and counts against more mundane "soaking up" explanations of a falling J/ψ rate, which predict a more gradual fall in the J/ψ signal. CERN experimenters have also seen other hints of plasma formation, including an increase in the production of strange particles. In particular, Ω baryons are created at 15 times the "normal" rate, for example their production rate in proton–proton collisions. Even so, particle physicists have been leery of claiming to have found the quark–gluon plasma. It will be the RHIC, built in response to the quest for the quark–gluon plasma, that will decide if the plasma is for real.

The RHIC offers the prospect of packing enough energy into a sufficiently small volume to get well into plasma country, creating a plasma both large enough and long-lived enough for its workings to fully unfold. One of the indicators of the plasma will be its size, up to 40 fermis across and substantially larger than the few fermis of the ions creating it. Physicists expect this swelling both as a result of the initial heating, and as a counter to the drop in disorder as the plasma condenses back into observable particles, the expansion guaranteeing the overall rise in disorder demanded by the rules of thermodynamics. Experimenters plan to access the size of the plasma through the relative timings of the legions of pions emitted from its volume, a pion version of a trick astronomers use with light to determine star sizes. Those same pions may also provide another signature of the plasma, a sudden rise in their number as experimenters ramp up the collision energy and cross the threshold for plasma creation (Figure 10.5).

SPACE, THE FINAL FRONTIER

When a star between two and three times the mass of our own Sun finally runs out of nuclear fuel, it starts to cool. The nuclear inferno that once created enough internal pressure to keep the star inflated is extinguished, but the force of gravity remains. Under its influence, the particles that make up the star begin to hug one another closer and closer, until eventually the vast mass of the star collapses into a spheroid, called a neutron star, just a few kilometers across. A neutron star's density is colossal. A neutron star, as its name implies, comprises mainly neutrons: an even heavier star will form a black hole.

Neutron stars that rotate become visible through the intense beam of electromagnetic radiation they emit, which sweeps across the

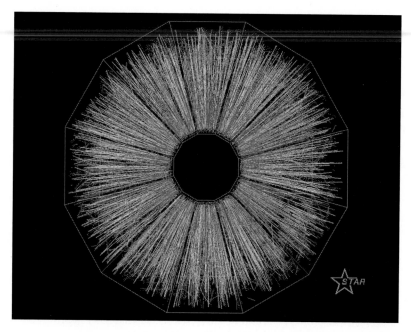

FIGURE 10.5 Analyze this. Collisions between gold beams at the RHIC, recorded here by the STAR detector, broke the record for the number of particles created in a subatomic collision. Scientists at the RHIC are searching for signs of the quark–gluon plasma. Credit: Courtesy of Brookhaven National Laboratory–STAR Collaboration.

Universe like the beam from a lighthouse sweeping the ocean. Here on Earth, this beam is witnessed as a regular pulse; a rotating neutron star is a pulsar.

Its equator extended by its own spin, as a neutron star ages it loses energy via radiation out into space, so over time it slows up and becomes less flattened by the centrifugal effect of the spin. This means that its internal density grows even larger. The extra squeezing due to the rise in density as the star slows can, for some rotating neutron stars, turn the nuclear matter within into a form of cold quark–gluon plasma called quark matter. This is the kind of thing that Wilczek and his collaborators had in the back of their minds when they explored QCD at high nuclear density. A possible signature of the creation of a "quark star," as proposed by Brookhaven National Laboratory's Norman Glendenning and his collaborators, will be a pause in the otherwise steady drop in the pulsar's pulsing frequency.

With an estimated one in a hundred pulsars passing through this phase, astrophysicists have now joined the race to track down deconfined matter. Even cosmologists are interested in the quark–gluon plasma, since if it did indeed feature in some early epoch of our Universe, its imprint may be left on the gravitational waves permeating space. Gravity, dismissed as irrelevant to the story of QCD at the start of this book, may indeed be a part of the story through the gravitational collapse of stars or the space-time ripples that make the whole Universe shudder. There is no escaping the unity of physics.

Appendix 1
A QCD chronology

1864 Maxwell unifies electricity and magnetism, and postulates that light is an electromagnetic wave.

1896 Becquerel discovers radioactivity, a consequence of the weak interaction.

1897 Thompson discovers electrons.

1900 Planck discovers quantum nature of the Universe.

1905 Einstein explains photoelectric effect in terms of light quanta.

1905 Einstein introduces special relativity, a system of mechanics applicable at speeds close to the speed of light.

1909 Geiger and Marsden observe large-angle scattering of alpha particles directed onto gold foil.

1911 Rutherford interprets Geiger and Marsden's results as evidence that atoms have nuclei, and devises his famous formula for scattering of charged particles from a point-like atomic nucleus.

1911 Millikan measures electron charge.

1912 Hess discovers cosmic rays.

1912 Wilson invents the cloud chamber.

1913 Bohr introduces his model of the atom, a nucleus with electrons in orbits around it. He explained the hydrogen spectrum and suggested radioactivity is a nuclear property.

1918 Weyl introduces the term "gauge invariance" as part of a doomed attempt to unify gravity with electromagnetism, and sees the link with charge conservation.

1918 Noether's theorem links symmetries to conserved quantities.

1919 Rutherford turns nitrogen into oxygen and uncovers the first evidence of the proton.

1920 Rutherford deduces that protons are a component of the atomic nucleus, and speculates on the existence of the neutron.

1921 Chadwick and Bieler see first evidence of strong force, in alpha particle scattering from hydrogen nuclei.

1923 de Broglie suggests wave–particle duality applies to electrons.

1925 Heisenberg introduces matrix quantum mechanics.

1925 Pauli introduces the exclusion principle.

1925 Goudsmit and Uhlenbeck propose that electrons have spin to explain the observed pairing of lines in atomic spectra.

1926 Schrödinger writes down his wave equation of quantum mechanics.

1926 Wigner introduces group theory to quantum mechanics.

1927 Heisenberg introduces the uncertainty principle.

1927 Davisson, Germer, and Thompson see wave-like behavior of electrons using diffraction in crystals.

1927 Unsöld uses Schrödinger's quantum mechanics to calculate helium's energy levels.

1927 Fock introduces the modern form of gauge transformation into QED.

1927 Dirac introduces quantum electrodynamics and quantum field theory.

1928 Gamov, Gurney, and Condon explain alpha-decay as a quantum tunneling effect.

1928 Dirac's equation of relativistic quantum theory.

1928 Wideröe builds the world's first resonance accelerator, the ancestor of cyclotrons and synchrotrons.

1928 Weyl outlines the connection between gauge invariance and charge conservation in quantum theory.

1929 Weyl works with Fock's local phase change but retains his terminology "gauge invariance."

1929 Gerthsen performs first proton–proton scattering experiments.

1930 Pauli proposes the existence of the neutrino.

1930 Mott devises a formula, extending Rutherford's, for scattering of particles with spin, for example electrons from a point-like nucleus where the electron spin is taken into account.

1931 Dirac discusses magnetic monopoles in the context of quantum theory.

1931 Lawrence and Livingston invent the cyclotron.

1932 Bethe and Fermi suggest particle exchange as the mechanism of force.

1932 Chadwick discovers the neutron.

1932 Urey, Brickwedde, and Murphy discover deuterium.

1932 Cockcroft and Walton split the atom.

1932 Heisenberg suggests that the atomic nucleus is composed of neutrons and protons, and that they are two charge states of a single particle.

1933 Carl Anderson discovers the positron.

1933 Dirac's paper on transformations of wavefunctions which was to provide the key to Feynman's path integral approach.

1933 Blackett sees electron–positron pair production.

1933 Stern measures the magnetic moment of the proton.

1933 Fermi theory of beta-decay, along with the first mention of second quantized spin 1/2 fields.

1934 Stern measures the magnetic moment of the neutron and deuteron.

1934 Fermi begins extensive experimental work in neutron scattering.

1935 Yukawa predicts the π meson as the carrier of the strong force. The first field theory of the strong force.

1935 White reports first "high-energy" accelerator proton–proton scattering experiments, and shows that the proton does not behave like a point charge.

1936 Tuve and others greatly refine proton–proton scattering using automated particle detection.

1936 Breit, Condon, and Present suggest that proton–proton and proton–neutron forces are the same.

1936 Condon and Cassen, and Wigner (1937) introduce the modern form of isospin for protons and neutrons.

1936 Breit–Wigner resonance formula published.

1937 Carl Anderson and Neddermeyer discover the muon.

1937 The klystron, a powerful microwave generator, invented at Stanford.

1938 Hahn, Strassmann, Meitner, and Frisch discover nuclear fission.

1940 Pauli proves the link between particle spin and statistics.

1942 Fermi builds the first nuclear reactor.

1944 Leprince-Ringuet and L'héritier claim the first evidence of the positive K meson, the first strange particle.

1944 Veksler and McMillan (1945) introduce phase stability for accelerators.

1946 The first synchrocyclotron (350 MeV protons) begins operation at Berkeley.

1947 Powell and Perkins discover charged pions.

1947 Lamb and Retherford measure the Lamb shift.

1947 Rochester and Butler find clear examples of so-called V particles, now called strange particles.

1947 The Shelter Island conference.

1948 Schwinger, Tomonaga, and Feynman devise renormalization, and calculate the Lamb shift and the electron's anomalous magnetic moment.

1948 Kusch and Foley measure the electron magnetic moment.

1948 Snell finds evidence of neutron decay, in which an electron is emitted.

1949 Dyson presents analysis of perturbation theory in QED.

1949 Fermi and Yang suggest that mesons are built from nucleons and antinucleons.

1950 Bjorklund, Carlson, and others discover the neutral pion.

1950 Panofsky and others find evidence for the decay of neutral pions into two photons.

1950 Cosmic ray experiments provide first evidence of the production of many strongly interacting particles in a single nucleon–nucleon interaction, along with a forward-directed jet of such secondary particles.

1950 Ward and Takahashi (1957) describe the Ward–Takahashi identities, symmetry relations between Green's functions.

1950 First evidence of jets in hadron–hadron collisions from cosmic ray studies is found.

1950 Salam and others overcome the problem of overlapping divergences in QED.

1951 First evidence of the Λ, a strange baryon made up of uds quarks, from cosmic ray experiments.

1952 Heisenberg predicts that total hadronic cross-sections rise with energy.

1952 Fermi discovers the first particle resonance, the Δ^{++}, a baryon made up of uuu quarks.

1952 The first proton synchrotron, the 3 GeV Brookhaven Cosmotron, begins operation.

1952 CERN founded.

1952 Livingston, Courant, and Snyder devise alternating gradient focusing for particle beams. Used in the CERN proton synchrotron, 1959.

1952 Glaser invents the bubble chamber.

1952 Pais introduces associated production for strange particles.

1953 Bonetti and others see the first evidence for the Σ^+ and Σ^- particles, strange baryons made up of uus and dds quarks, respectively.

1953 The Ξ^- particle, a strange baryon made up of dss quarks, is discovered in cosmic ray experiments.

1953 Gell-Mann and Nishijima introduce strangeness, leading to prediction of Σ^0 and Ξ^0.

1953 Stückelberg and Peterman discover the renormalization group.

1954 Yang and Mills extend gauge principle to non-Abelian symmetry.

1954 Gell-Mann and Low use the renormalization group to study QED.

1954 The Berkeley Bevatron, a 6 GeV proton synchrotron, begins operation.

1955 The Σ^0 particle is discovered at the Brookhaven Cosmotron.

1955 Chamberlain and Segrè discover the antiproton at the Bevatron.

1956 Chamberlain and Segrè observe the annihilation of antiprotons in matter.

1956 The antineutron is discovered at the Bevatron.

1956 Reines and Cowan detect free neutrinos and antineutrinos for the first time.

1956 Hofstadter and McAllister show that protons are not point particles.

1956 Sakata proposes a model of hadrons as composites of p, n, and Λ.

1956 Utiyama generalizes Yang–Mills theory to other groups.

1956 Lee and Yang suggest that parity may be violated in weak interactions.

1957 Wu observes parity non-conservation in weak decays, indicating that the weak interaction knows the difference between left and right.

1957 Dubna 10 GeV proton synchrotron begins operation.

1958 Goldhaber discovers that electron neutrinos are left-handed.

1959 Alvarez and others discover the Ξ^0 particle, a strange baryon made up of uss quarks.

1959 Ikeda and others introduce $SU(3)$ symmetry for hadrons.

1959 DESY founded.

1959 Regge introduces Regge theory, a framework for describing strong interaction scattering data.

1959 The 28 GeV proton synchrotron at CERN begins operation.

1960 The 33 GeV Brookhaven AGS proton synchrotron begins operation.

1960 Nambu highlights symmetry breaking via the vacuum as a source of particles.

1960 Weinberg proposes convergence theorem for Feynman diagrams.

1960 Sakurai proposes first Yang–Mills theory of strong interactions.

1961 Glashow introduces the idea of a neutral intermediate boson for weak interactions.

1961 Chew and Frautschi introduce the Pomeron to describe high-energy elastic scattering data.

1961 Alvarez and collaborators find the ω meson.

1961 Goldstone theorem links the presence of massless particles, the Goldstone bosons, with hidden continuous symmetries of quantum systems.

1961 The ρ meson is found at the Cosmotron.

1961 The η meson is found at the Bevatron.

1961 Gell-Mann and Ne'eman introduce the $SU(3)$ symmetry scheme for strongly interacting particles, the Eightfold Way.

1961 Nambu suggests that the low-mass pions are a symptom of broken symmetry.

1961 Salam and Ward propose the gauge principle.

1961 Touschek conceives and builds the first electron–positron collider, with a circumference of 4 m.

1962 Gell-Mann and Ne'eman introduce the baryon decuplet and predict the Ω^- particle.

1962 SLAC founded.

1962 Lederman and others discover the muon neutrino at Brookhaven.

1962 The word hadron introduced.

1963 Institute of High Energy Physics (IHEP) founded at Serpukhov, south of Moscow.

1963 Feynman realizes the need for ghost diagrams.

1963 Sokolov and Ternov explain the link between synchrotron radiation, and the polarization of circulating charged particles in magnetic fields.

1964 Gell-Mann and Zweig introduce quarks (called aces by Zweig) as the fundamental components of hadrons.

1964 The η' meson discovered at the Bevatron and the Brookhaven AGS.

1964 Greenberg, Han, and Nambu initiate quark color.

1964 Bjorken and Glashow introduce charm, a fourth quark flavor.

1964 Cronin and Fitch see CP symmetry violation in K mesons.

1964 Higgs and others introduce the Higgs mechanism, resulting in mass for W and Z particles.

1964 Samios and co-workers discover the Ω^- particle at Brookhaven.

1965 Vanyashin and Terentev see hints of asymptotic freedom.

1965 Han and Nambu discuss color $SU(3)$ with eight vector fields (gluons) and color singlet observable particles.

1966 The SLAC linear electron accelerator is completed (20 GeV electrons onto a stationary target).

1966 Nambu further investigates color octet of "gluons".

1966 Bjorken writes down his sum rule for nucleon spin structure.

1967 Faddeev, Popov, and deWitt formulation for ghosts in non-Abelian gauge theories.

1967 Glashow, Weinberg, and Salam set out the electroweak theory, unifying weak and electromagnetic interactions.

1967 Gottfried's sum rule for electron–nucleon scattering is proposed.

1967 Serpukhov 76 GeV proton synchrotron begins operation near Moscow.

1968 Charpak invents gas wire chamber particle detector.

1968 Veneziano introduces the dual-resonance model, building on the successes of Regge's approach and paving the way for a description of hadrons as strings.

1969 Bjorken introduces scaling hypothesis for inelastic lepton–nucleon scattering.

1969 Friedman, Kendall, and Richard Taylor find first evidence of Bjorken scaling in electron–proton scattering at SLAC, suggesting point-like scattering centers identified with quarks; first evidence that scattering centers in the proton have spin 1/2.

1969 Feynman introduces the parton model of the nucleon.

1969 Callan and Gross introduce a relation between structure functions for spin 1/2 partons.

1969 Wilson postulates the operator product expansion.

1969 Gross and Llewellyn Smith invent their sum rule for neutrino–nucleon scattering.

1969 Khriplovich distinguishes between screening and antiscreening in non-Abelian gauge theories.

1970 Glashow, Iliopoulos, and Maiani exploit a fourth quark, the charm quark.

1970 Callan–Symanzik form of the renormalization group equation is presented.

1970 Nambu suggests vibrating, rotating strings as a model of hadrons.

1970 Zimmerman proves Wilson's operator product expansion hypothesis.

1971 First evidence of rising total hadronic cross-sections seen at Serpukov.

1971 The 62 GeV ISR begins operation at CERN, colliding two proton beams, the world's first hadron collider.

1971 't Hooft and Veltman prove that non-Abelian gauge theories, including those with spontaneous symmetry breaking, are renormalizable.

1971 Wilson applies renormalization group ideas to critical phenomena.

1971 Drell and Yan suggest parton model applies to hadron–hadron collisions – the Drell–Yan process.

1972 The SPEAR electron–positron collider at SLAC is completed.

1972 The 400 GeV Fermilab (Main Ring) proton synchrotron begins operation.

1972 Bollini, 't Hooft, and others introduce dimensional regularization.

1972 't Hooft's unpublished contribution to the story of asymptotic freedom.

1972 Gribov and Lipatov discuss evolution of parton distribution functions and scaling violation.

1972 Gell-Mann and Fritzsch refer to observable particles as color singlet states, and to a color octet of vector fields satisfying Yang–Mills equations, at a conference in Chicago.

1972 van der Meer introduces stochastic cooling, crucial to the storage of antiprotons prior to acceleration in proton–antiproton colliders. First used in at the CERN SppS collider in 1981.

1972 Bouchiat, Gross, and others show that a family grouping of leptons and colored quarks is essential to preserve renormalization in electroweak theory.

1973 Bardeen, Fritzsch, and Gell-Mann publish their calculation of neutral pion decay into two photons, and the R ratio for electron–positron annihilation, both important pieces of evidence for color.

1973 Fritzsch, Gell-Mann, and Leutwyler outline the advantages of a color octet of gluons.

1973 Parisi, Callan, and Gross show scaling requires asymptotic freedom.

1973 Asymptotic freedom discovered by Gross, Wilczek, and Politzer.

1973 Coleman and Gross show that only non-Abelian gauge theories can be asymptotically free.

1973 Quark confinement discussed by Weinberg, Gross, and Wilczek.

1973 Proton–proton cross-sections seen to rise with increasing energy at the CERN ISR.

1973 Weinberg initiates the Standard Model, embracing strong, weak, and electromagnetic interactions.

1973 Gross, Wilczek, and others predict scaling violations in deep inelastic scattering.

1973 Neutrino scattering at CERN shows first evidence of nucleon momentum due to gluons and antiquarks.

1973 Kobayashi and Maskawa suggest more quarks are necessary to accommodate CP violation.

1973 Weak interactions involving an electrically neutral force carrier (so-called weak neutral currents) observed at CERN. The discovery of weak neutral currents was crucial to the acceptance of electroweak theory and of gauge theories in general.

1973 Nielsen, Olesen, and others suggest tubes of trapped magnetic field in superconductors as a model for strings linking the quark and antiquark of a meson.

1974 Gaillard and Lee predict the charm quark mass.

1974 Ting at Brookhaven National Laboratory and Richter at SLAC find the J/ψ meson, comprising a c$\bar{c}$ quark pair.

1974 The "November revolution."

1974 Georgi and Glashow introduce a grand unification scheme based on the group $SU(5)$.

1974 Ellis and Jaffe write down their sum rule for spin structure for protons and neutrons separately.

1974 The DORIS electron–positron collider at DESY begins operation.

1974 The VEPP-2M electron–positron collider starts up at Novosibirsk.

1974 Wilson introduces lattice gauge theory.

1974 First observation of scaling violations in deep inelastic scattering, seen in muon–iron scattering at Fermilab.

1974 't Hooft and Polyakov discover that non-Abelian gauge theories with spontaneous symmetry breaking suggest the existence of magnetic monopoles.

1974 MIT bag model of baryons introduced by Johnson and others.

1975 Samios and co-workers find the first charmed baryons, the Σ_c^{++} comprising uuc quarks, and Λ_c^+ comprising udc quarks.

1975 Perl and colleagues at SLAC discover the τ heavy lepton, indicating a third lepton family.

1975 First evidence of jets at SLAC's SPEAR electron–positron collider supports spin 1/2 quarks.

1975 Stochastic cooling demonstrated at the CERN ISR.

1975 BFKL equation for summing gluons introduced.

1975 Shifman and others introduce penguin diagrams (named by Ellis 1977).

1975 Polyakov and others introduce instantons.

1975 Appelquist and Politzer suggest heavy quark bound states mimicking those of positronium, a positron–electron bound state.

1975 Introduction of Cornell inter-quark potential, combining a Coulomb part and a "linear" part.

1976 SLAC experimentalists discover the first meson containing a single charmed quark, the D^0.

1976 Fermilab experimenters discover that Λ baryons are created polarized in proton–beryllium collisions.

1976 First polarized electron–polarized proton scattering reported from SLAC.

1976 SPS (450 GeV proton onto stationary target) starts up at CERN.

1976 't Hooft demonstrates vacuum tunneling due to instantons.

1976 Mandelstam and 't Hooft suggest that the vacuum is a dual superconductor, a condensate of magnetic monopoles.

1976 Ellis, Gaillard, and Ross propose three-jet events as a way of seeing gluons.

1977 Lederman and others at Fermilab discover the Υ meson, comprising a $b\bar{b}$ pair.

1977 Altarelli and Parisi develop equation for evolving quark and gluon distributions.

1977 Introduction of the gluon condensate.

1977 Wilson quarks and staggered quarks introduced into lattice QCD to eliminate quark duplication.

1977 Sterman and Weinberg pioneer jets.

1977 Peccei and Quinn introduce a symmetry-breaking scheme to retain time reversal symmetry (or, equivalently, CP symmetry) in QCD with instantons.

1978 First full-scale accelerator using superconducting accelerating cavities, the SCA (superconducting linear *a*ccelerator) at Stanford.

1978 The PETRA electron–positron collider at DESY begins operation.

1978 Wilczek and Weinberg introduce the axion.

1978 The Υ meson seen in electron-positron collisions at DORIS, DESY.

1978 Shifman and others introduce QCD sum rules, a technique for calculating hadron properties developed from the operator product expansion and embracing QCD vacuum features.

1978 Feynman and Field introduce a fragmentation recipe describing how quarks and gluons give rise to jets of hadrons.

1978 CERN experimenters use neutrino scattering to provide further support for the QCD description of scaling violations.

1979 CESR electron–positron collider begins operation at Cornell.

1979 Three-jet events, in which the third jet is due to a gluon, seen at PETRA, DESY.

1979 Weinberg lays foundation of effective quantum field theory, and reformulates chiral perturbation theory as an effective theory.

1979 Creutz and others introduce Monte Carlo methods to pure gauge theory on a lattice.

1980 The PEP electron–positron collider at SLAC starts operation.

1980 The TASSO collaboration at DESY show that the gluon has a spin of 1.

1980 The Υ (4S) meson discovered at CESR.

1981 The first baryon containing a b quark, the Λ_b, seen at CERN.

1981 Parisi, Weingarten, and others introduce the quenched approximation into lattice QCD with quarks.

1981 Nielsen and Ninomiya's "No-Go" theorem proving that, if in lattice QCD certain desirable symmetries are to be retained, then additional unwanted quarks ("species doubling") are inevitable.

1981 First sign of color coherence, in electron–positron collisions at PETRA.

1981 First evidence for B mesons from CESR.

1981 DESY experimenters see first sign of photon structure function.

1981 The Sp$\bar{\text{p}}$S begins operation at CERN, colliding protons and antiprotons at a total energy of 900 GeV. This machine is an adaptation of the original SPS that directed 450 GeV protons on to a stationary target.

1982 First unambiguous evidence of jets in hadron–hadron collisions seen at CERN's Sp$\bar{\text{p}}$S.

1983 CERN's UA1 and UA2 collaborations discover the W and Z particles.

1983 SLAC experimenters make the first measurement of the sum rule for proton spin.

1983 CERN experimenters find that structure functions of bound nucleons differ from those of free nucleons (the EMC effect).

1983 First attempts to calculate hadron masses on the lattice by Parisi and others.

1984 The Fermilab Tevatron starts operation as a fixed-target proton synchrotron, the first accelerator based on superconducting magnets.

1984 CERN two-jet results in proton–antiproton collisions provide evidence for the "running" of the strong coupling parameter at high energies, and the massless vector-like nature of gluons.

1985 Schlein and Ingelman suggest that the Pomeron has structure functions.

1986 First exploratory heavy ion experiments related to quark–gluon plasma begin at CERN SPS and Brookhaven AGS.

1987 Fermilab's Tevatron starts operation in proton–antiproton collider mode, total energy 1.8 TeV.

1987 TRISTAN 64 GeV electron–positron collider begins operation at KEK, Tokyo.

1987 ARGUS collaboration at DESY sees B meson mixing.

1987 Voloshin and Shifman, and later Isgur and Wise (1989) introduce heavy quark symmetry.

1988 UA8 see first evidence of hard scattering from Pomeron constituents.

1988 The EMC group at CERN show that the nucleon spin is much less than that expected from the simple quark model (the spin crisis).

1989 The SLAC linear collider (SLC) starts up.

1989 The LEP electron–positron collider begins operation at CERN.

1989 The BEPC electron–positron collider begins operation in Beijing.

1989 CERN and SLAC announce first experimental evidence to suggest there are just three light neutrinos, implying three families and therefore a total of six quarks.

1989 AMY collaboration at TRISTAN finds first evidence of gluon self-coupling.

1990 First silicon microstrip vertex detector at a colliding beam experiment, installed at SLAC's Mark II experiment.

1990 Georgi's formulation of heavy quark theory.

1991 Confirmation at CERN that the number of light neutrinos is three.

1991 First evidence, from CERN, that the number of down antiquarks in the proton exceeds the number of up antiquarks – there is flavor asymmetry in the sea.

1991 SLD detector installed on the SLC at SLAC.

1992 HERA electron–proton collider begins operation at DESY.

1992 First jets in deep inelastic scattering seen by the E665 group at Fermilab.

1992 TRISTAN experimenters see first evidence of gluons in the photon.

1992 First silicon microstrip vertex detector at a hadron collider begins operation at Fermilab's CDF.

1992 NA35 group at CERN makes first clear determination of the size of an interaction region using correlations of the resulting hadrons ("pion interferometry").

1993 First results on the neutron's spin structure function.

1993 HERA experimenters see first evidence that protons have a huge population of quarks with low momentum fractions.

1994 CERN experimenters make first measurements of the nucleon's other spin structure function, g_2.

1994 First evidence of the top quark from Fermilab's CDF and D0 collaborations.

1994 VEPP-4M electron–positron collider starts up in Novosibirsk.

1994 Jefferson Lab, a 4 GeV continuous electron beam accelerator, begins operation in Newport News, Virginia.

1994 Tevatron sees first color coherence in proton–antiproton collisions.

1994 Weingarten's lattice calculation of light hadron masses.

1994 CERN starts lead ion beam experiments to look for quark–gluon plasma.

1994 Müller, and later Ji and Radyushkin, introduce generalized parton distributions.

1995 Pomeron structure function measured at HERA.

1995 Table of accurate baryon magnetic moment measurements completed at Fermilab.

1995 Existence of top quark confirmed.

1995 Weingarten's lattice calculation of glueball mass and decay rate.

1995 Direct measurement of gluon distribution in the proton at HERA.

1995 SMC at CERN make the first nucleon spin structure measurements which distinguish between up and down valence quarks.

1996 LEP2 begins operation at 161 GeV, producing $W^+ W^-$ pairs.

1997 First experimental evidence of running quark mass.

1997 First evidence for quark–gluon plasma at CERN.

1997 Fermilab Main Ring shut down in favor of Main Injector.

1997 First exotic meson, the $\pi_1(1400)$, found at Brookhaven.

1998 CDF finds the B_c, the last of the "ordinary" mesons.

1998 Fermilab experimenters confirm that down antiquarks outnumber up antiquarks in the proton's quark sea.

1998 First direct evidence of the tau neutrino.

1998 CP-PACS lattice calculation of light hadron spectrum.

1998 SLC shut down.

1998 First hint of neutrino oscillations suggests small neutrino mass.

1999 Brookhaven Relativistic Heavy Ion Collider (RHIC) begins operation.

1999 SLAC and KEK B-factories start operation.

1999 Wilczek and others introduce color–flavor locking.

2000 LEP shut down.

2001 CP violation seen in B mesons.

2001 Tevatron run II begins.

2002 Doubly charmed baryons seen at Fermilab.

2003 The pentaquark Θ^+ state discovered.

Appendix 2
Greek alphabet and SI prefixes

The Greek alphabet

Name	Upper case	Lower case
alpha	A	α
beta	B	β
gamma	Γ	γ
delta	Δ	δ
epsilon	E	ε
zeta	Z	ζ
eta	H	η
theta	Θ	θ
iota	I	ι
kappa	K	κ
lambda	Λ	λ
mu	M	μ
nu	N	ν
xi	Ξ	ξ
omicron	O	o
pi	Π	π
rho	P	ρ
sigma	Σ	σ
tau	T	τ
upsilon	Υ	υ
phi	Φ	ϕ
chi	X	χ
psi	Ψ	ψ
omega	Ω	ω

SI prefixes

Prefix	Factor	Symbol
exa-	10^{18}	E
peta-	10^{15}	P
tera-	10^{12}	T
giga-	10^{9}	G
mega-	10^{6}	M
kilo-	10^{3}	k
hecto-	10^{2}	h
deca-	10	da
deci-	10^{-1}	d
centi-	10^{-2}	c
milli-	10^{-3}	m
micro-	10^{-6}	μ
nano-	10^{-9}	n
pico-	10^{-12}	p
femto-	10^{-15}	f
atto-	10^{-18}	a

Glossary

Accuracy and completeness are sacrificed to the gods of simplicity and brevity.

α The electromagnetic coupling parameter.

α_s The strong coupling parameter.

Abelian group The mathematical expression of a symmetry in which two symmetry transformations give the same outcome irrespective of the order in which they are performed. The group $U(1)$ is Abelian.

action The sum over time of the difference between kinetic energy and potential energy.

AGS Alternating Gradient Synchrotron, a proton accelerator at Brookhaven National Laboratory, USA.

ALEPH One of the four detectors at LEP, the electron–positron collider at CERN, Switzerland.

AMY One of the four detectors at TRISTAN, an electron–positron collider at KEK.

anomalous magnetic moment The departure from the ideal point-particle value of a particle's magnetic moment.

antiparticle A partner to a particle, having the same mass but opposite electric charge. Positrons are the antiparticle partners of electrons; some particles such as the photon and the π^0 are their own antiparticle.

asymptotic freedom The diminution of the color force between quarks as the distance between them is reduced.

axion A feebly interacting particle, now assumed to have a minute mass, introduced to resolve a symmetry problem in QCD. Never observed, it is also a candidate for the missing mass of the Universe.

B mesons Mesons containing a b quark.

barn A unit of area, equal to 10^{-28} m^2, used to describe scattering experiments.

baryon A heavy, strongly interacting matter particle composed of three quarks, for example the proton.

baryon number A quantity conserved in all interactions, equal to the number of baryons minus the number of antibaryons.

beam pipe In an accelerator, the innermost pipe containing the particle beam.

BFKL equation A mathematical expression allowing for the inclusion of many low-energy gluons, for example when considering the scattering of electrons off protons.

Big Bang The postulated creation of the Universe as a cataclysmic fireball followed by billions of years of cooling and expansion.

Bjorken sum rule Relates the difference in longitudinal spin content of protons and neutrons to weak interaction parameters.

boson A particle having zero or integer spin, for example a photon; many such particles can inhabit the same state at the same time.

bottom A quantum number associated with the bottom or b quark, which has a bottom value of -1; sometimes called the beauty quantum number.

bottomonium states A set of mesons comprising excited states of a bottom quark and bottom antiquark pair.

bremsstrahlung Photons emitted as a charged particle loses energy, or gluons emitted as a quark (or antiquark) loses energy.

bubble chamber A type of particle detector in which particle tracks are revealed as trails of bubbles in a superheated liquid.

C The charge conjugation operation, which swaps a particle with its antiparticle. The C operation converts an electron into a positron, but leaves a photon a photon.

calorimeter An outer segment of a particle detector that measures the energy of particles emanating from an interaction.

Casimir effect A force serving to push together two conducting plates by virtue of fluctuations in the vacuum; a consequence of zero-point energy.

CDF Collider Detector Facility; one of two detectors on the Tevatron proton–antiproton collider.

CERN The European Laboratory for Particle Physics, near Geneva, Switzerland.

CESR Accelerator colliding electrons and positrons, Cornell, USA.

channel One of several routes by which some process occurs.

charm A quantum number associated with the charm or c quark, which has a charm value of $+1$.

charmonium states A set of mesons comprising excited states of a charmed quark and charmed antiquark pair.

chiral perturbation theory Relative to conventional QCD calculations, a non-perturbative technique in which the chiral-symmetry-breaking portion of QCD is treated as a small perturbation to the chiral-symmetry-preserving part.

chiral restoration The recovery of chiral symmetry that is presumed to occur at high temperatures and densities.

chiral symmetry Symmetry that acts independently on left-handed and right-handed systems.

chromo-electric force The color force between two static color charges, the color analog of the electrostatic force between two static electric charges.

chromo-magnetic force The color force between two moving color charges, in particular spinning color charges; the color analog of magnetic forces between moving electric charges.

cloud chamber An old style of particle detector in which particle tracks emerge as "vapor trails" of droplets initiated by the passage of the particle through a vapor.

coherence A relationship between waves, in which two waves of the same frequency are said to be coherent if one always lags behind the other by a constant amount.

colliding beam experiments Experiments in which one particle beam collides with another head-on. The low probability of collisions occurring is compensated by the much larger collision energy, relative to fixed-target experiments.

color The attribute of quarks and gluons that is the source of their strong interaction, much as electric charge is the source of the electrical interaction between charged particles; there are three distinct colors, red, green, and blue.

color coherence The destructive interference between successively emitted low-energy gluons, causing angular ordering of the emitted gluons and a suppression of particles between the quark and antiquark jets in a three-jet event.

color singlet A combination of colored entities, such as the three quarks of a proton, that has zero net overall color. Observable particles are color singlets.

condensate A state formed when many identical particles lock together in some way to form a lowest energy overall configuration. Quarks, monopoles, and Cooper pairs can all form condensates.

confinement A key postulate of QCD that says free quarks and gluons cannot exist, but must live in composite particles such as protons and neutrons for which the total colors blend to give a "white" particle, in technical terms a color singlet.

conservation laws Rules that describe quantities that remain unchanged during some process. Everyday examples are energy and momentum conservation. In particle physics, there are many more conservation laws, for example baryon number conservation and, for the strong interaction, isospin conservation. Conservation laws are related both to the conserved quantities used to describe and classify interactions, and to underlying symmetries of the interaction.

constituent quark mass The quark mass as it emerges from the quark model; the constituent mass of both up and down quarks is roughly a third of the proton mass.

Cooper pairs The conductors of electricity in a superconductor, comprising loose pairings of two electrons to create delocalized spin 1 particles.

coupling "constant" A dimensionless number expressing the strength of an interaction. Perturbation theory solutions are expressed as powers of the coupling constant. Frequently, the word parameter is more appropriate than constant.

covariance The property of a relationship between quantities in which that relationship remains unchanged even following some transformation, for example a Lorentz transformation, that acts on those quantities; also called form invariance.

CP symmetry The combined symmetries of particle–antiparticle interchange symmetry, the C, and right- and left-hand coordinate interchange symmetry, the P. CP symmetry links left-handed neutrinos and right-handed antineutrinos.

CP violation The spoiling of CP symmetry, evident in K meson and B meson systems; implies a unique and absolute means of describing left and right, and perhaps explains the dominance of matter over antimatter in the Universe.

CPT symmetry The combined symmetries of particle–antiparticle interchange symmetry, the C, and right- and left-hand coordinate interchange symmetry, the P, and time-reversal symmetry, the T. All field theories obey CPT symmetry, which is necessary to preserve the exact correspondence between particles and antiparticles.

critical phenomena Phenomena characterized by the loss of distinction between two phases of matter.

critical point The conditions, for example temperature and pressure, specifying where two phases of a system merge and become indistinguishable from one another.

cross-section A characteristic area controlling the probability that two particles interact.

current quark mass The small quark mass values linked to the spoilage of symmetry of a massless strong interaction theory.

D mesons Mesons containing a charmed quark.

D0 One of two detectors on the Tevatron proton–antiproton collider.

dark matter Invisible material that is thought to contribute a large fraction of the total mass of the Universe.

decay rate The reciprocal of the average time an isolated particle lives before disintegrating.

deconfinement transition The jump from a state where quarks and gluons are confined to one where they are not confined, presumed to occur at high temperatures and densities.

deep inelastic scattering Scattering of an electron (or muon or neutrino) that probes the innermost components of the target nucleon.

DELPHI One of the four detectors at LEP, the electron–positron collider at CERN, Switzerland.

DESY German accelerator center, Hamburg.

deuterium A heavy isotope of hydrogen in which the nucleus, called the deuteron, comprises one proton and one neutron, in contrast to regular hydrogen in which the nucleus is a single proton.

DGLAP evolution equations Equations that describe the way in which structure functions depend on the wavelength of the probe used to measure them.

diffractive scattering Scattering in which one participant particle scatters elastically, retaining its identity, while the other is broken up.

dimensional regularization Isolation of the infinities due to closed loops of Feynman diagrams using a continuously varying dimension parameter.

direct photon A photon that interacts as a whole photon with a charged component of the target, without first disintegrating.

divergence An infinite result in a calculation, caused either by some quantity that increases without limit, or by dividing by some quantity that drops to zero.

DORIS Old electron–positron collider at DESY, Germany.

Drell–Yan mechanism The annihilation of a quark and antiquark pair, spawning an electron–positron pair or a pair comprising some other lepton and its antiparticle partner.

drift chamber Particle detector element for mapping particle tracks; relies on measuring ionization in a gas due to transiting charged particles.

dynamical quarks In lattice QCD, quarks that pop into existence from the vacuum, as distinct from valence quarks that are present at the start of the calculation.

effective quantum field theory A non-renormalizable quantum field theory applicable over a specific energy range.

elastic scattering Scattering in which particles "bounce" of one another, giving emerging particles identical to the colliding ones.

electron A stable, point-like, fundamental particle, the carrier of the basic unit of electric charge. Since electrons have spin $+1/2$ they are fermions; their mass is 0.511 MeV. The atomic nucleus is surrounded by orbiting electrons.

electronvolt A very small unit of energy.

electroweak theory The Glashow–Weinberg–Salam theory of weak force, according to which the weak interaction is mediated by W and Z bosons.

EMC European Muon Collaboration, a research group studying muon–nucleon scattering at CERN, Switzerland.

EMC effect The modification of the quark distributions within a free nucleon when that nucleon is instead part of a large nucleus.

exclusion principle The principle that no two electrons, or two of any other species of fermions, may occupy the exact same quantum state. Also known as the Pauli exclusion principle.

exclusive process A scattering process in which all the final-state particles are detected and measured.

Faddeev–Popov ghost *See* ghost

families The groupings of fundamental "matter" particles, or fermions, into three matching sets, the first of which includes the up and down quark, the electron and its neutrino.

fermion A particle of half-integer spin, such as an electron, proton, or quark. Fermions obey the exclusion principle.

Feynman diagram Diagrammatic representations of a portion of an approximate mathematical calculation of some particle physics process.

field A quantity or influence spread over some region, rather than being concentrated at a single point.

field theory Any theory in which the quantities of interest are represented as fields, as opposed to points. In a quantum field theory, a field is an assemblage of packets of different energies.

fixed-target experiments Those in which an incoming particle beam is directed onto a stationary target, for example a block of iron, and the resulting debris is projected into a cone following the direction of the incoming beam.

flavor The type attribute of a quark. The six flavors are up, down, strange, charm, bottom, and top.

four-momentum An extension of the usual momentum of mechanics ("three-momentum") to include an extra, fourth portion, the energy, on an essentially equally footing.

four-vector A group of four numbers specifying a quantity, for example an interval in space-time. The magnitude of a four-vector is unchanged by a Lorentz transformation.

fragmentation A process by which a quark or gluon becomes a jet of strongly interacting observable particles.

fragmentation function The probability that a quark or gluon fragments to give a meson having a momentum fraction equal to the meson momentum divided by the original quark or gluon momentum.

gauge field A field introduced in order to preserve local symmetry. In quantum theory these are spin 1 bosons that carry force; the photon, W and Z particles, and gluons are all gauge fields.

gauge symmetry Symmetry under space-time dependent changes in the phase of the field.

gauge theory A quantum theory in which the interactions are determined by local symmetry properties.

general relativity Einstein's theory of gravity, relating the force of gravity to the geometry of space-time.

GeV A unit of energy, the giga electronvolt, equal to a thousand million electronvolts.

ghost A contribution to certain Feynman diagrams; ghosts correspond to unphysical particles required to preserve probability conservation in non-Abelian gauge theories.

global transformation A transformation that is the same for every space-time point.

glueball A neutral meson composed entirely of gluons.

gluon According to QCD, gluons carry the strong force. Gluons are fundamental, massless, electrically neutral, spin 1 bosons. They carry eight different color assignments. On Feynman diagrams they are conventionally represented by corkscrew curves.

Goldstone bosons Massless particles that owe their existence to spontaneous symmetry breaking.

Green's function In field theory, an amplitude for a physical process, the summation of a set of Feynman diagrams.

ground state The lowest energy state of a quantum mechanical system.

group A mathematical expression of symmetry. Formally, in mathematics a group is a set whose elements obey four particular rules.

H1 A detector at the HERA electron–proton collider, DESY, Germany.

hadron A strongly interacting particle. Hadrons containing three quarks are called baryons. Hadrons containing quark–antiquark pairs are called mesons.

handed A distinction between left-handed and right-handed. A conventional wood screw is termed right-handed because it moves forward into the wood when turned clockwise.

hard interaction One in which the momentum transfer is large.

HERA Electron–proton collider at DESY, Germany

Higgs particle A proposed but as yet unobserved particle that plays a key role in symmetry breaking and the origin of particle masses in the electroweak theory of the weak force.

higher-twist effects Contributions to particle interactions arising from correlations between particles that become significant at low-energy regimes. They are non-perturbative effects that can only be estimated.

inclusive process One in which only a single particle of those produced is identified and measured. The remaining event products are ignored.

inelastic scattering Scattering in which energy from colliding particles is used to create new ones, so the particles leaving a collision do not match exactly those entering.

infrared divergence A divergence associated with an energy that falls to zero.

instanton A transient, massless, gluon "knot," topological in origin, capable of flipping the handedness of a quark.

integrated luminosity The accumulation over time of the brightness of the beam(s) in an accelerator.

invariant A quantity unchanged by a transformation.

invariant mass The useful energy available for particle creation in some collision, having the same value for all observers, i.e., a Lorentz invariant. Denoted $\sqrt{s}$.

ISR Intersecting Storage Rings; a proton–proton collider at CERN, Switzerland, that ran from 1971 to 1984.

jet A directed, conical shower containing strongly interacting particles.

L3 One of the four detectors at LEP, the electron–positron collider at CERN, Switzerland.

Lambda (Λ) parameter A numerical constant, determined from experiment, linked to the size of the blob of space – roughly the size of a proton – over which QCD acts. The scale parameter has a value of about 200 MeV.

LEAR Low Energy Antiproton Ring, CERN, Switzerland.

LEP Electron–positron collider at CERN, Switzerland.

lepton Matter particles, such as electrons and muons, that do not experience the strong force.

lepton number A quantity conserved in all interactions, equal to the number of leptons minus the number of antileptons. Lepton numbers for electrons, muons, and taus are conserved independently.

LHC Large Hadron Collider at CERN, Switzerland.

Lie group A symmetry group corresponding to a continuous, smooth transformation, for example a rotation.

linac A linear accelerator with electrodes arranged in a straight line (hence linear) so that accelerating particles effectively surf an electromagnetic wave the length of the machine.

local transformation A transformation that is different for every point in space and time.

luminosity Brightness of the beam(s) in an accelerator.

magnetic moment A measure of a magnet's propensity to turn when subjected to an external magnetic field.

magnon Massless quantum particle associated with magnetization in magnetic materials.

master equation The master equation of QCD is an energy equation for quarks, gluons, and the color-based interactions between them. It is fed into the Feynman path integral in order to generate Feynman diagrams. Formally, and for people who know about these things, the master equation is a Lagrangian density.

matter wave The quantum wave associated with matter "particles" such as electrons.

Maxwell's equations The equations of electricity and magnetism due to James Clerk Maxwell.

mean life The average time an isolated particle exists before disintegrating. It is the reciprocal of the decay rate, and is proportional to the half-life, the time corresponding to a 50% probability of decay. The half-life is about 70% of the mean life.

meson A strongly interacting particle comprising a quark and an antiquark, for example the pion.

Meissner effect In a superconductor, the expulsion of a magnetic field from the interior of the superconducting material.

MeV A unit of energy, equal to a million electronvolts. Particle masses are given in MeV. For example, the proton has a mass of 938.3 MeV.

momentum transfer squared The (negative of the) invariant mass of the virtual particle that mediates an interaction, denoted Q^2. For example, for an electron scattering from protons, Q^2 refers to the virtual photon mediating the scattering. High values of this quantity correspond to a short-wavelength probe able to see finer details.

monopole Literally, a single pole. A magnetic monopole is a lone north or south magnetic pole, though these do not occur in nature. A monopole in QCD, for example in the dual-superconductor model of confinement, is a lone chromo-magnetic field pole.

multiplet A collection of particles related to one another through some symmetry scheme.

muon An electrically charged elementary particle of mass 105.7 MeV, created for example in the weak decay of a pion, and usually decaying into an electron.

naive parton model The basic version of the parton model, in which point-like partons each simply carry a fraction of the parent nucleon momentum, and Bjorken scaling is satisfied.

neutrino A low-mass, weakly interacting matter particle having no electric charge, and no role in the strong force. There are three types of neutrino.

neutron A strongly interacting, neutral particle made up of udd quarks. The atomic nucleus is made up of neutrons and protons.

non-Abelian group The mathematical expression of a symmetry in which two symmetry transformations give different outcomes depending on the order in which they are performed. Both $SU(2)$ and $SU(3)$ are non-Abelian groups.

non-perturbative Describes some property or process that cannot be analyzed using the approximation technique called perturbation theory. The non-perturbative regime in QCD is the low-energy, long-distance regime encountered, for example, in the binding together of quarks to form protons.

nucleon A collective term meaning neutron and proton.

OPAL One of the four detectors at LEP, the electron–positron collider at CERN, Switzerland.

OPE *See* operator product expansion.

operator product expansion Expresses the short-distance behavior of some non-perturbative process in terms of non-perturbative background features whose relative contributions are calculable in perturbation theory.

order (of a perturbation series) The power of the coupling parameter in the expression for the amplitude; power one is first order, power two is second order, etc.

P Parity.

pair production In QED, the process in which a high-energy photon is destroyed as an electron–positron pair is created.

parity The mathematical version of creating a mirror image. Under the parity operation, a left-handed particle becomes a right-handed particle, and vice versa.

Particle Data Group An international team of particle physics researchers who co-ordinate, maintain, and update the *Review of Particle Properties*, the ultimate listing of particle properties and other key particle physics information (see http://pdg.lbl.gov/).

parton model A model of the nucleon in which it is viewed as composed of simple scattering centers called partons, identified with quarks and gluons. The "QCD-improved" parton model, as opposed to the naive parton model, caters for interactions between partons, and they no longer appear strictly point-like scatters having a simple fraction of the parent nucleon's momentum.

partons Scattering centers in the nucleon identified with quarks, antiquarks, and gluons.

path integral Amplitude for the transition between quantum states expressed as a sum over all possible paths linking those states.

PEP Electron–positron collider at SLAC, USA.

perturbation theory A mathematical technique in which an approximate answer to an otherwise intractable problem is expressed as a sequence of successively smaller terms.

PETRA Electron–positron collider at DESY, Germany.

phase 1. A measure of a wave's progress relative to some origin. 2. The physical state of some substance, for example the liquid and steam phases of water, or the magnetized and unmagnified phases of iron.

phase stability The preservation of the relative timing between relativistic, accelerating particles, and the voltage doing the accelerating.

phase transition The change from one system phase to another, for example when water boils to give steam, or steam condenses into water.

phonon A packet, or quantum, of vibrational energy in a solid.

photomultiplier tube A detection device that turns tiny flashes of light into electrical pulses.

photon A "particle" of light, in other words a quantum of electromagnetic energy.

photoproduction Scattering with real or nearly real photons.

picobarn A unit of cross-section. One picobarn is 10^{-12} barns, or 10^{-40} m^2.

pion (or π meson) A type of meson composed of up and down quarks and antiquarks, the carrier of the strong force binding protons and neutrons into the atomic nucleus.

PLUTO One of the four detectors on the PETRA electron–positron collider at DESY, Germany.

polarization The orientation of a particle's spin relative to its direction of motion. In right-hand polarized light, for example, the photon spin lies along the direction of the photon's flight. In transverse polarization, the particle's spin is oriented at right-angles to its flight path.

Pomeron An exchange (pseudo) particle mediating certain high-energy strong interactions. Composed mainly of gluons.

positron The antiparticle partner of the electron, having a positive electric charge.

positronium A short-lived "atom" made from a positron and an electron.

potential Ordinarily, the energy made available as a result of a change in the configuration of a system involving forces. The change in potential with distance is the strength of the relevant force field. Potentials become force-carrying quanta in quantum field theory.

principle of least action The rule that a mechanical system behaves in such a way that its action assumes the smallest possible value.

propagator The probability amplitude for a particle to propagate from one point to another.

proton A strongly interacting, positively charged particle made up of uud quarks. Hydrogen nuclei are protons. The atomic nucleus is made up of protons and neutrons.

pseudo-Goldstone bosons Low-mass particles, for example pions, that owe their origin to the spontaneous breaking of a symmetry that is not quite exact.

Q^2 *See* momentum transfer squared.

QCD Quantum chromodynamics, the theory of the strong force.

QCD sum rules Approximate, non-perturbative constraints on hadron parameters.

QCD vacuum The state of lowest energy of QCD.

quantum field theory The most evolved form of quantum theory, in which both matter and forces are described in terms of fields that are sums over quantum waves. Quantum field theory naturally accommodates particle creation and destruction.

quantum number Simple numbers or vectors that label and identify a quantum state and the results of measurements performed on that state. Conservation laws are expressed in terms of quantum numbers.

quark The fundamental strongly interacting particle, and carrier of color charge. In addition, quarks carry electric charge, and participate in electroweak interactions. There are six types, or flavors, of quark: up (u), down (d), strange (s), charmed (c), bottom (b), and top (t).

quark condensate A sea of quark–antiquark pairs, populating the vacuum, that forms the backdrop for strong interactions.

quark–gluon plasma A "gas" of free quarks and gluons.

quarkonium A meson comprising a heavy quark–antiquark pair. The plural is quarkonia.

radiative corrections In the calculation of the probability of some process occurring, radiative corrections are modifications arising from the emission of additional particles, for example low-energy photons in an electromagnetic process.

Regge theory The expression of strong interaction scattering in terms of the exchange of families of mesons.

regularization Mathematical procedure for isolating infinities.

renormalization A program for extracting the physical content of a quantum field theory otherwise prone to giving infinite and therefore meaningless answers.

renormalization group A set of mathematical transformations that revises the parameters of a theory whilst leaving unchanged the interesting physical features.

resolved photon A photon that interacts in terms of the quarks and gluons associated with it. Loosely, a fragmented photon.

resonance An unstable particle that decays very quickly via the strong interaction into other strongly interacting particles.

$\sqrt{s}$ *See* invariant mass.

scale invariance Invariance under a change in the length of the ruler used to measure the system.

scale parameter *see* lambda parameter.

scattering The interaction of a beam of particles or radiation with some target, typically involving a redirection of at least some of the beam.

scintillator A material, used in particle detectors, that produces tiny flashes of light in response to a transiting particle.

signature Output from a particle scattering experiment that signals a particular process of interest.

silicon microstrip detector The basic element of a silicon vertex detector, relying on stripes of one semiconductor set in another to locate passing charged particles.

silicon vertex detector The inner layer of most particle detectors, relying on silicon technology to locate forks in particle tracks occurring very close to the original interaction point.

singlet A "family" comprising just a single member, built from component states whose characteristic of interest is combined in a particular way to give a zero value for the singlet. For example, a neutral atom is a charge singlet, an isospin singlet state such as the Λ has zero isospin, and observable strongly interacting particles are color singlets, meaning they have no net color yet are composed of colored quarks.

SLAC Stanford Linear Accelerator Center in Stanford, California.

Slavnov–Taylor identities Mathematical relationships between amplitudes in non-Abelian gauge theory; generalized Ward identities.

SLC Stanford Linear Collider, Stanford, California.

SMC Spin Muon Collaboration, an experimental group at CERN, Switzerland.

soft interaction One in which the momentum transfer is small.

soliton A kind of isolated wave that does not spread or fade away.

SPEAR Electron–positron collider at SLAC, USA.

special relativity The mechanics applicable to objects traveling at speeds close to that of light.

spin An attribute of elementary particles that has parallels with the spin of everyday objects such as spinning tops. The spin of elementary particles can only assume specific values. Particles having whole-number spin (in units of $\hbar$) are called bosons, those having half-number spins are called fermions.

spontaneous symmetry breaking The symmetry reduction that occurs when the lowest energy state has less symmetry than the relevant interaction.

Sp$\bar{p}$S Proton–antiproton collider at CERN, Switzerland.

SPS Super Proton Synchrotron at CERN, Switzerland.

staggered quarks In lattice QCD, the quarks resulting from a technical trick for removing accidental, unwanted replica quarks by distributing quark information across several lattice sites.

Standard Model The electroweak theory of the weak force in combination with the QCD theory of the strong force, applied to the familiar fundamental particles.

stochastic cooling The squeezing of a charged particle beam, especially antiprotons, to reduce spreading and create a well-defined beam suitable for subsequent accelerating.

storage ring A particle accelerator in which the beam follows the same roughly circular path many times, especially when the number of particles is being increased, or the beam is being saved for later use.

strangeness An attribute of strange particles, derived from the presence of a strange quark or antiquark. A strange quark has strangeness –1.

strong CP problem How to explain the small size of the parameter controling the contribution of the instanton-generating term in QCD.

structure functions Momentum distributions of the nucleon's constituents.

sum rule A relationship between, or value ascribed to, momentum-summed quark or gluon distributions within the nucleon.

supersymmetry A symmetry linking particles having integer spin with those having half-integer spin. Though experimenters have so far not found evidence of supersymmetry, theories exploiting it have created great excitement amongst theorists.

synchrotron A particle accelerator in which the frequency of the accelerating pulses, and the strength of the bending magnetic field, are increased in a synchronized way so as to keep the accelerating particles on a circular path.

tagging The use of a some feature in collision debris as a signal of a particular process.

Tevatron Proton-antiproton collider at Fermilab, USA.

theta-vacuum The purported true vacuum of QCD, a blend of topologically distinct vacuum states.

TOPAZ One of the four detectors at TRISTAN, an electron–positron collider at KEK.

topological A property that describes gross features that are independent of geometry. American footballs and soccer balls are topologically the same (both are balls) but geometrically different (only one is a regular sphere).

trigger In a particle detector, the decision-making system for deciding whether to ignore or record an event.

twist *See* higher-twist effects.

U(1) problem The problem of reconciling an apparent symmetry of QCD (a $U(1)$ symmetry) with the absence of particles such as proton partners implied by that symmetry.

unitarity The conservation of total probability in a collision process.

unitary A term used to describe transformations of particle families that preserve the overall particle content, analogous to the way that the rotation of a clock arm preserves the length of the arm.

universality The equality, irrespective of color and flavor, of the basic coupling strength between a quark and a gluon.

vacuum In an everyday sense, the nothing that remains when, for example, the air is pumped out of a sealed box. In quantum field theory, it is the state of lowest energy.

vacuum polarization The contribution to the energy of a particle such as a photon arising from the way it disturbs the vacuum around it.

valence quarks Those quarks that between them carry the quantum numbers that label their parent particle.

W particle Alternatively called the W boson, the heavy, electrically charged carrier of the weak force.

Ward identity Relationship between amplitudes that is useful for proving renormalization in quantum field theory. Sometimes called Ward–Takahashi identities or, in the non-Abelian case, Slavnov–Taylor identities.

Ward–Takahashi identities Generalized Ward identities.

Weinberg–Salam theory An alternative name for the electroweak theory.

width A measure of a particle's propensity to decay; the inverse of its lifetime. A narrow width corresponds to a long-lived particle.

Wilson quarks In lattice gauge theory, quarks having broken chiral symmetry, whose accidental, unwanted replicas have been spirited away by addition of the extra "Wilson term" to the lattice master equation.

x The fraction of a parent nucleon's momentum carried by a quark or gluon within.

Z particle Alternatively called the Z boson, the heavy, electrically neutral carrier of the weak force.

zero-point energy The residual minimum energy of a quantum state below which the system cannot go.

ZEUS A detector at the HERA electron–proton collider, DESY, Germany.

Further reading

There is a huge number of books on particle physics, and a vast literature, so any attempt at a "further reading" list is both subjective and incomplete. The titles listed below are grouped according to level.

Easy reading

These titles require no particular prior knowledge and are devoid of mathematical equations; they are listed here in no particular order.

Gerard 't Hooft, *In Search of the Ultimate Building Blocks*, Cambridge University Press, 1997.

A beautiful, slim volume from someone who contributed a great deal to QCD. Written in part as a personal account, the contents reflect 't Hooft's own particular interests.

Vincent Icke, *The Force of Symmetry*, Cambridge University Press, 1995.

Icke discusses the role of symmetry in particle physics, relativity, and quantum theory.

John Gribbin, *In Search of Schrödinger's Cat: Quantum Physics and Reality*, Bantam Doubleday Dell, 1985.

An easy-to-read overview of the mysteries of quantum mechanics.

Tony Hey and Patrick Walters, *The New Quantum Universe*, Cambridge University Press, 2003.

Lots of nice pictures make for an attractive and simple overview of quantum theory that touches on particle physics at the end.

George Gamov and Russell Stannard, *The New World of Mr. Tompkins: George Gamov's Classic Mr. Thompkins in Paperback*, Cambridge University Press, 1999.

There's nothing else quite like this. In this updated classic, first produced in two parts in 1940 and 1945, Gamov describes bank clerk Thompkin's fantasy world where Planck's constant is large and the velocity of light is small.

Richard Feynman, *QED: The Strange Theory of Light and Matter*, Princeton University Press, 1988.

Feynman explains quantum electrodynamics in a way that only Feynman could. A classic, from one of the key figures in the development of QED.

Harald Fritzsch, *Quarks: The Stuff of Matter*, Penguin 1983.
Everything you wanted to know about quarks, no more and no less. A delightful and simple-to-read book.

Andrew Pickering, *Constructing Quarks. A Sociological History of Particle Physics*, Chicago University Press, 1999.
Billed as a sociological history of particle physics, Pickering's book is a rather different take on the subject.

Gordon Kane, *The Particle Garden: Our Universe as Understood by Particle Physics*, Addison-Wesley, 1995.
A lightweight overview of the entire field of particle physics, from strong and weak interactions to theories of everything.

George Johnson, *Strange Beauty: Murray Gell-Mann and the Revolution in Twentieth Century Physics*, Jonathan Cape, 2000.
Johnson's excellent and very readable biography of Murray Gell-Mann, one of the key figures in both the introduction of quarks, and the genesis of quantum chromodynamics.

Frank Close, Michael Marten, and Christine Sutton, *The Particle Odyssey – A Journey to the Heart of the Matter*, Oxford University Press, 2002.
Beautifully illustrated offering from the collaboration that created "*The Particle Explosion*".

There are also numerous excellent articles in publications such as *Scientific American*, *New Scientist*, *Science*, *Nature*, and other journals aimed all or in part at a broader audience.

In addition, websites at the major experimental facilities offer virtual tours and some excellent, accessible material on particle physics. To start with try:

http://www.cern.ch/
http://www.desy.de/
http://www.slac.stanford.edu/
http://www.fnal.gov/

Not-so-easy reading

These titles do assume a little more science and perhaps mathematical background. Some even contain a few equations.

Robert Adair, *The Great Design: Particles, Fields and Creation*, Oxford University Press, 1989.

> Adair covers everything, strong and weak forces to cosmology. There are a handful of simple equations in the text and some not-so-trivial ones in the notes. Plenty of graphs and diagrams with an emphasis on ideas and analogies even though the style is a little like that of a text book.

Yuval Ne'eman and Yoram Kirsh, *The Particle Hunters*, Cambridge University Press, 1996.

> Were it not for a small number of equations, this book would appear under the "Easy Reading" heading. Here is a solid account of the whole of particle physics, with some interesting asides on Ne'eman, one time Israeli military attaché in London, and several other players in particle physics.

Guy D. Coughlan and James E. Dodd, *The Ideas of Particle Physics*, Cambridge University Press, 1991.

> Meant for science students who are not planning on specializing in nuclear or particle physics, this splendid book is a step-by-step guide to the key ideas and developments in particle physics. This book assumes a readership comfortable with reasonably simple mathematics and graphs, but never gets bogged down in technical details.

Abraham Pais, *Inward Bound*, Oxford University Press, 1988.

> An excellent and scholarly history of particle physics, focusing more on the early years: quarks only make an entrance 80% of the way through.

Hard core

These titles are definitely for those with a sound mathematical background and more than a passing interest in particle physics.

There are many books on quantum field theory applied to particle physics, including:

Ta-Pei Cheng and Ling-Fong Li, *Gauge Theory of Elementary Particle Physics*, Oxford University Press, 1984.

> Getting a little dated now perhaps, but it's all here. A problems and solutions book arrived on the scene in 2000.

Lewis Ryder, *Quantum Field Theory*, Cambridge University Press, 1996.

> The title says it all. This book has a slightly more geometric flavor than some.

Steven Weinberg, *The Quantum Theory of Fields*, Cambridge University Press, 1996.

> The classic from one of the great names of particle theory. Volume one deals with foundations; volume two with applications. There is also a third volume given over to supersymmetry. Grand in scope, penetrating in depth.

There are lots of worthy titles on particle physics:

Ian Aitchison and Anthony Hey, *Gauge Theories in Particle Physics*, Adam Hilger, 1990.

> Hard to beat as a simple introduction to gauge theories. Curiously, the authors opted to avoid the complexities of field theory in the first edition. By the second, they had changed their minds, perhaps bowing to the inevitable. But the principle of accessibility to the reader runs throughout.

Donald Perkins, *Introduction to High Energy Physics*, Cambridge University Press, 2000

> A classic standard text book.

Robert Cahn and Gerson Goldhaber, *The Experimental Foundations of Particle Physics*, Cambridge University Press, 1989.

> Discusses major topics in particle physics through a commentary on the key papers, plus reprints of many of those that really shaped the subject. Big on the experimental aspects, with an historical flavor.

Particle Data Group, *Review of Particle Physics*, regularly reprinted, and also available on the internet (at http://pdg.lbl.gov).

> *The* reference work in particle physics, containing the authoritative listing of particle properties.

Vladimir Ezhela *et al.*, *Particle Physics: One Hundred Years of Discoveries: An Annotated Chronological Biography*, Springer-Verlag, 1996.

> An unusual book, possibly better suited to the library than the private bookshelf, consisting of a list of annotated abstracts of a century's worth of key papers in particle physics. It also contains an extensive chronology of discoveries.

For lattice gauge theory try:

Michael Creutz, *Quarks, Gluons and Lattices*, Cambridge University Press, 1985.

> A slim, graduate student text, complete with exercises. The basics are all here, but it's getting a little old now.

István Montvay and Gernot Münster, *Quantum Fields on a Lattice*,
Cambridge University Press, 1994.

> A meatier though less readily accessible text on lattice field
> theory.

Books with a particular QCD bias include:

Walter Greiner and Andreas Schäfer, *Quantum Chromodynamics*,
Springer-Verlag, 1995.

> One volume of a whole series in theoretical physics, this book
> covers most aspects of QCD and includes numerous gory
> example calculations. Very light on experimental coverage and
> references, but if you want to see the details of lots of
> calculations then this is a good place to look. Translated from the
> German original of 1989.

Peter Zerwas and Hans Kastrup, *QCD 20 Years Later*, World Scientific,
1993.

> The two-volume proceedings of a conference held in Aachen in
> 1992 to celebrate twenty years of QCD, containing technical
> reviews of all major aspects of the subject.

Herbert Fried and Berndt Müller, *Quantum Chromodynamics*, World
Scientific, 1999.

> This is the proceedings of the fourth workshop on QCD held in
> Paris in 1998. While the name is right on the mark, this is very
> heavy duty QCD exotica, and not for the faint of heart. It is, also,
> conference proceedings and all that that implies.

Keith Ellis, James Sterling, and Bryan Webber, *QCD and Collider Physics*,
Cambridge University Press, 1996.

> A clear and comprehensive treatment of perturbative QCD that
> does not lose sight of what all the equations actually mean.
> Excellent.

Francisco Ynduráin, *The Theory of Quark and Gluon Interactions*,
Springer-Verlag, 1999.

> Big on the maths and lighter on the physics, Ynduráin's book
> covers all the usual perturbative QCD material, but in addition
> includes such QCD-biased topics as instantons, the vacuum, and
> lattice calculations. Broad coverage, but quite intense.

Mikhail Shifman, ed., *At The Frontier of Particle Physics – Handbook of
QCD*, World Scientific, vols. I–III, 2001; vol. IV, 2002.

> Running to 2535 pages in all, this is a huge (and expensive)
> collection of 38 self-contained reviews covering all aspects of

QCD, with a strong emphasis on the mathematical side. Experts only. The historical sections in volume one are especially interesting.

Günther Dissertori, Ian Knowles, and Michael Schmelling, *Quantum Chromodynamics – High Energy Experiments and Theory*, Oxford University Press, 2003.

As its title suggests, this is about the high-energy aspects of QCD, covering similar territory to the book by Ellis *et al.* This book is especially helpful for anybody interested in the nuts and bolts of comparing theory and data.

Index

Page numbers in **bold** refer to figures.

1-MONTH